AF500359

Applied Mathematical Sciences

Volume 74

Applied Mathematical Sciences

1. *John:* Partial Differential Equations, 4th ed.
2. *Sirovich:* Techniques of Asymptotic Analysis.
3. *Hale:* Theory of Functional Differential Equations, 2nd ed.
4. *Percus:* Combinatorial Methods.
5. *von Mises/Friedrichs:* Fluid Dynamics.
6. *Freiberger/Grenander:* A Short Course in Computational Probability and Statistics.
7. *Pipkin:* Lectures on Viscoelasticity Theory.
9. *Friedrichs:* Spectral Theory of Operators in Hilbert Space.
11. *Wolovich:* Linear Multivariable Systems.
12. *Berkovitz:* Optimal Control Theory.
13. *Bluman/Cole:* Similarity Methods for Differential Equations.
14. *Yoshizawa:* Stability Theory and the Existence of Periodic Solution and Almost Periodic Solutions.
15. *Braun:* Differential Equations and Their Applications, 3rd ed.
16. *Lefschetz:* Applications of Algebraic Topology.
17. *Collatz/Wetterling:* Optimization Problems.
18. *Grenander:* Pattern Synthesis: Lectures in Pattern Theory, Vol I.
20. *Driver:* Ordinary and Delay Differential Equations.
21. *Courant/Friedrichs:* Supersonic Flow and Shock Waves.
22. *Rouche/Habets/Laloy:* Stability Theory by Liapunov's Direct Method.
23. *Lamperti:* Stochastic Processes: A Survey of the Mathematical Theory.
24. *Grenander:* Pattern Analysis: Lectures in Pattern Theory, Vol. II.
25. *Davies:* Integral Transforms and Their Applications, 2nd ed.
26. *Kushner/Clark:* Stochastic Approximation Methods for Constrained and Unconstrained Systems
27. *de Boor:* A Practical Guide to Splines.
28. *Keilson:* Markov Chain Models—Rarity and Exponentiality.
29. *de Veubeke:* A Course in Elasticity.
30. *Sniatycki:* Geometric Quantization and Quantum Mechanics.
31. *Reid:* Sturmian Theory for Ordinary Differential Equations.
32. *Meis/Markowitz:* Numerical Solution of Partial Differential Equations.
33. *Grenander:* Regular Structures: Lectures in Pattern Theory, Vol. III.
34. *Kevorkian/Cole:* Perturbation methods in Applied Mathematics.
35. *Carr:* Applications of Centre Manifold Theory.
36. *Bengtsson/Ghil/Källén:* Dynamic Meteorology: Data Assimilation Methods.
37. *Saperstone:* Semidynamical Systems in Infinite Dimensional Spaces.
38. *Lichtenberg/Lieberman:* Regular and Stochastic Motion.
39. *Piccini/Stampacchia/Vidossich:* Ordinary Differential Equations in R^n.
40. *Naylor/Sell:* Linear Operator Theory in Engineering and Science.
41. *Sparrow:* The Lorenz Equations: Bifurcations, Chaos, and Strange Attractors.
42. *Guckenheimer/Holmes:* Nonlinear Oscillations, Dynamical Systems and Bifurcations of Vector Fields.
43. *Ockendon/Tayler:* Inviscid Fluid Flows.
44. *Pazy:* Semigroups of Linear Operators and Applications to Partial Differential Equations.
45. *Glashoff/Gustafson:* Linear Optimization and Approximation: An Introduction to the Theoretical Analysis and Numerical Treatment of Semi-Infinite Programs.
46. *Wilcox:* Scattering Theory for Diffraction Gratings.
47. *Hale et al.:* An Introduction to Infinite Dimensional Dynamical Systems—Geometric Theory.
48. *Murray:* Asymptotic Analysis.
49. *Ladyzhenskaya:* The Boundary-Value Problems of Mathematical Physics.
50. *Wilcox:* Sound Propagation in Stratified Fluids.
51. *Golubitsky/Schaeffer:* Bifurcation and Groups in Bifurcation Theory, Vol. I.
52. *Chipot:* Variational Inequalities and Flow in Porous Media.
53. *Majda:* Compressible Fluid Flow and Systems of Conservation Laws in Several Space Variables.
54. *Wasow:* Linear Turning Point Theory.

Jean Mawhin Michel Willem

Critical Point Theory and Hamiltonian Systems

Springer Science+Business Media, LLC

Jean Mawhin
Institut de Mathematique Pure
et Appliquee
Chemin du Cyclotron 2
1348 Louvain-la-Neuve
Belgium

Michel Willem
Institut de Mathematique Pure
et Appliquee
Chemin du Cyclotron 2
1348 Louvain-la-Neuve
Belgium

With 1 Illustration

Mathematics Subject Classification (1980): 58F05, 58E05, 70H25, 58F22, 58E07, 49A27, 58E30, 58-02, 49-02

Library of Congress Cataloging-in-Publication Data
Mawhin, J.
Critical point theory and Hamiltonian systems / Jean Mawhin, Michel Willem.
p. cm. — (Applied mathematical sciences ; v. 74)
Bibliography: p.
1. Hamiltonian systems. 2. Critical point theory (Mathematical analysis) I. Willem, Michel. II. Title. III. Series: Applied mathematical sciences (Springer-Verlag New York Inc.) ; v. 74.
QA1.A647 vol. 74
[QA614.83]
510 s—dc19
[515.3'5] 88-39058
Printed on acid-free paper.

DOI 10.1007/978-1-4757-2061-7

Originally published by Springer-Verlag New York Inc in 1989.
MyCopy version of the original edition 1989

Camera-ready copy prepared using LaTEX.
Printed and bound by R.R. Donnelley & Sons, Harrisonburg, Virginia.

9 8 7 6 5 4 3 2 1
www.springer.com/mycopy

To Margaret, Marie, Valérie, Jean
and
Eliane, Julie, Olivier

Preface

The formulation of laws of nature in terms of minimum principles has a long history that can be traced to Hero of Alexandria (c. 125 B.C.). He proved in his Catoptrics that when a ray of light is reflected by a mirror, the path actually taken from the object to the observer's eye is shorter than any other possible path so reflected. This principle was generalized by Fermat who postulated, around 1650, that light always propagates in the shortest time from one point to another, and deduced mathematically, from this principle, the law of refraction. The same Fermat anticipated the differential calculus by stating a necessary condition for the maximum or the minimum of a polynomial that is equivalent to the vanishing of its derivative.

More ambitious was the aim of Maupertuis when he enunciated, around 1750, his principle of least action as a rational and metaphysical basis for geometrical objects and mechanics. His statement was far from precise and, in the same year, Euler expressed it as an exact theorem of dynamics in an addendum to his famous book on the calculus of variations. This book contains the famous extension of the Fermat necessary condition for an extremum of a real function to the case of functionals of the type

$$y \longrightarrow \int_a^b f(x, y(x), y'(x))\,dx,$$

called the Euler–Lagrange equation after the more analytical treatment given shortly after by Lagrange.

It will take some time, during which further necessary conditions for a maximum or a minimum will be derived by Legendre, Jacobi, Weierstrass, and others to realize with Volterra and Hadamard, at the turn of this century, that the calculus of variations is just a special chapter of a theory of extrema for real functions defined on function spaces, and to create the tools necessary to formulate, in this setting, the corresponding necessary conditions.

The question of the existence of an extremum has a more recent history, a feature shared with the more general problem of existence theorems in mathematics. Gauss, who gave four demonstrations of the fundamental theorem of algebra, admitted without proof the existence of a minimum for the functional φ given by

$$\varphi(y) = \int_{\Omega} \sum_{i=1}^{n} (D_i y(x))^2 \, dx$$

over all sufficiently regular functions y whose restriction on the boundary $\partial\Omega$ of the bounded domain $\Omega \subset \mathbf{R}^n$ is fixed. It was the origin of the so-called Dirichlet principle for the existence of a solution to the Dirichlet problem with data h on $\partial\Omega$,

$$\Delta y(x) = 0, \quad x \in \Omega$$
$$y(x) = h(x), \quad x \in \partial\Omega.$$

The long-waited justification of this principle by Arzela and Hilbert, around 1900, was the stimulus for the creation of a systematic approach for getting conditions of existence for a minimum or a maximum of a functional.

After some pioneering work of Lebesgue, it became clear with Tonelli's important contributions that the lower semi-continuity introduced by Baire in another context was the right type of continuity for a fruitful abstract formulation of the calculus of variations. The systematic development of functional analysis, and in particular the study of convex sets and reflexive Banach spaces, paved the way for a systematic development of sharp existence conditions.

The creation of a general theory of periodic solutions of Hamiltonian systems as a fundamental step in understanding the structure of their solution set was one of the major motivations of Poincaré's monumental mathematical work. Besides many other contributions, Poincaré initiated the variational treatment of those questions. In particular, he made use of Jacobi's form of the least action principle to study the closed orbits of a conservative system with two degrees of freedom. He also considered the related question of the existence of geodesics. However, despite the rigorous treatment of the closed orbits of dynamical systems with two degrees of freedom by Whittaker, and the related work of Signorini, Tonelli, and Birkhoff, and despite the fact that Birkhoff minimax theory was the impetus for Morse theory and Lusternik–Schnirelman approach to critical point theory, progress toward a global variational approach for the periodic solutions of Hamiltonian systems was very slow.

A notable exception was Seifert's use in 1948 of Jacobi's form of the least action principle and differential geometry to prove the existence of an even T-periodic solution when the Hamiltonian is the sum of a kinetic and a potential energy term. This was generalized by Weinstein in the late 70s, who proved in particular, by similar methods, that an autonomous system with Hamiltonian H such that $H^{-1}(1)$ is a manifold bounding a compact convex region always has a closed orbit in $H^{-1}(1)$.

The fundamental difficulty in applying the naive idea of finding the periodic solutions of a general Hamiltonian system through the critical points of its Hamiltonian action on a suitable space of periodic functions lies in the fact, already observed by Birkhoff, that this action is unbounded from below and from above. This makes the use of the well-developed direct method of the calculus of variations (which deals with absolute minima) unapplicable, except in some particular second order systems already considered in the 20s by Lichtenstein and Hamel.

However, in the mid 60s, extensions of the minimax approach (in particular of Lusternik–Schnirelman theory) and of the Morse theory to functions defined on Banach manifolds were given by Palais, Smale, Rothe, Clark, Ambrosetti, Rabinowitz, and others. In the late 70s, Rabinowitz initiated the use of those methods in the study of periodic solutions of Hamiltonian systems. Later, a dual least action principle was introduced by Clarke and extensively developed by Clarke, Ekeland, and others. More recently, Morse theory and an extension of it due to Conley have provided further insight into those questions.

The aim of this book is to initiate the reader to those fundamental techniques of critical point theory and apply them to periodic solutions problems for Hamiltonian systems. Those illustrations have been chosen either because of their importance in the various applications in mechanics, electronics, and economics, or because of their mathematical importance. We hope that our style of presentation will be appealing to people trained and interested in ordinary differential equations. We have the feeling that critical point theory, which has been mostly developed by specialists in differential topology, partial differential equations, or optimization, should be made more popular among people working in ordinary differential equations. Of course, the variational methods developed here are directly applicable to partial differential equations problems at the expense of a substantial complication of the technical details. They can be found in a number of the references to the literature at the end of the book.

The reader interested in other aspects of critical point theory can then consult the references given in the bibliographical notes ending each chapter as well as the following surveys and monographs: [AuE$_1$], [Berc$_1$], [Ber$_2$], [Blo$_1$], [Bot$_{1,2}$], [Bre$_2$], [Ces$_1$], [Cha$_1$], [ChH$_1$], [Cla$_3$], [Con$_1$], [Cor$_1$], [Dei$_1$], [Des$_1$], [Eel$_1$], [Eke$_5$], [EkT$_1$], [EkTu$_1$], [Fen$_1$], [Fun$_1$], [Kli$_1$], [Koz$_1$], [Kra$_2$], [Lju$_1$], [Maw$_{2,3,5,11}$], [Mil$_2$], [Mor$_3$], [Moy$_1$], [Mrs$_2$], [Nir$_2$], [Rab$_{2,6,12,13,14,19}$], [Roc$_1$], [Rot$_5$], [Ryb$_1$], [Sch$_1$], [Smo$_1$], [Str$_{4,5}$] [Szu$_3$], [Ton$_1$], [Vai$_{1,2}$], [Vol$_{1,2}$], [VoP$_1$], [Wil$_{3,5}$], [You$_2$], [Zeh$_1$], [Zei$_2$].

Acknowledgments

We wish to thank the people who have contributed to the realization of this work.

The manuscript benefited from suggestions and criticisms by our colleagues C. Fabry and P. Habets, and our students A. Fonda, C. Gorez, W. Omana, and S. Tshinanga, who have read parts of the manuscript.

We are indebted to the editorial board of Springer-Verlag for carefully reviewing the manuscript.

We are very grateful to Béatrice Huberty for her accurate and superb typing of the manuscript.

Finally we wish to thank our families for their patience during the elaboration of this book.

Louvain-la-Neuve

Jean Mawhin
Michel Willem

Contents

1

The Direct Method of the Calculus of Variations

Introduction

A real function φ of a real variable which is bounded below on the real line needs not to have a minimum, as it is clear from the example of the exponential function. If we call *minimizing sequence* for φ any sequence (a_k) such that

$$\varphi(a_k) \to \inf \varphi$$

as $k \to \infty$, a necessary condition for the real number a to be such that

$$\varphi(a) = \inf \varphi$$

is that φ has a minimizing sequence which converges to a (take $a_k = a$ for all integers k). Without suitable continuity assumptions on φ this condition will not be sufficient, as shown by the example of the function φ defined by $\varphi(x) = |x|$ for $x \neq 0$ and $\varphi(0) = 1$, which does not achieve its infimum 0 although all its minimizing sequences converge to zero. In order that the limit a of a convergent minimizing sequence be such that $\varphi(a) = \inf \varphi$, we have to impose that

$$\lim_{k \to \infty} \varphi(a_k) \geq \varphi(a).$$

This will be the case if φ is *lower semi-continuous* on $\mathbf{R}$, i.e. if

$$\underline{\lim}_{k \to \infty} \varphi(u_k) \geq \varphi(u)$$

whenever $u_k \to u$.

Now, in $\mathbf{R}$, the existence of a convergent minimizing sequence is equivalent to that of a bounded minimizing sequence, a feature which is lost when we replace $\mathbf{R}$ by an infinite dimensional Banach space with its norm topology. In Section 1.1, we show that the existence of a bounded minimizing sequence still guarantees (and is indeed equivalent to) the existence of a minimum for $\varphi : X \to \mathbf{R}$, when X is a reflexive Banach space (in particular a Hilbert space) and when the lower semi-continuity property for φ holds for the weakly convergent sequences (u_k) in X. Section 1.2 shows that this weak lower semi-continuity is equivalent to the lower semi-continuity in norm when φ is convex.

If a differentiable function $\varphi : \mathbf{R} \to \mathbf{R}$ has a local maximum or a local minimum at a, then

$$\varphi'(a) = 0.$$

This is the most elementary version of the basic necessary condition in the theory of extremums. Its extension to the case of a differentiable real function on a normed space is given in Section 1.3 and a detailed study of its meaning in the case of the classical functional of the calculus of variations with periodic boundary conditions can be found in Section 1.4. In particular, for a continuous mapping $F : [0,T]\times\mathbf{R}^N \to \mathbf{R}$, $(t,u) \to F(t,u)$, such that $\nabla F(t,u) = D_u F(t,u)$ is continuous, the solutions of the problem

$$\ddot{u}(t) = \nabla F(t, u(t))$$

$$u(0) - u(T) = \dot{u}(0) - \dot{u}(T) = 0$$

are the *critical points* u (i.e., the points with $\varphi'(u) = 0$) of the *action functional*

$$\varphi_F : u \to \int_0^T [(1/2)|\dot{u}(t)|^2 + F(t, u(t))]\, dt$$

on a suitable space of T-periodic functions.

We describe in Sections 1.5 to 1.7 various conditions upon F (and possibly ∇F) which insure that φ_F has a bounded minimizing sequence. In the simple case of the scalar linear problem

$$\ddot{u}(t) = -h(t)$$

$$u(0) - u(T) = \dot{u}(0) - \dot{u}(T) = 0,$$

the necessary and sufficient condition of solvability is well known and given by

$$\int_0^T h(t)\, dt = 0,$$

i.e., h must have mean value zero. Our results are, in various directions, nonlinear extensions of this condition. For example, Theorem 1.5 implies that for a continuous $g : \mathbf{R} \to \mathbf{R}$ such that

$$\lim_{u\to-\infty} g(u) = g_- < g_+ = \lim_{u\to+\infty} g(u),$$

the problem

$$\ddot{u}(t) = g(u) - h(t)$$

$$u(0) - u(T) = \dot{u}(0) - \dot{u}(T) = 0$$

is solvable if

$$g_- < \overline{h} < g_+,$$

where $\overline{h}$ denotes the mean value of h, i.e., $T^{-1} \int_0^T h(t)\, dt$.

On the other hand, Theorem 1.6 will imply that for $g(u) = a \sin u$ (the forced pendulum equation), the above periodic problem is always solvable when $\overline{h} = 0$. Finally, a consequence of Theorems 1.7 to 1.9 is that for g non-decreasing, the above periodic problem is solvable if and only if $\overline{h}$ belongs to the range of g. This completely extends the solvability results of the linear case

$$\ddot{u}(t) = \lambda u - h(t) \quad (\lambda \geq 0)$$
$$u(0) - u(T) = \dot{u}(0) - \dot{u}(T) = 0$$

to the more general situation where the linear restoring force λu is replaced by an arbitrary continuous non-decreasing function of u.

1.1 Lower Semi-Continuous Functions

Let X be a normed space.

A *minimizing sequence* for a function $\varphi : X \to]-\infty, +\infty]$ is a sequence (u_k) such that

$$\varphi(u_k) \to \inf \varphi$$

whenever $k \to \infty$. A function $\varphi : X \to]-\infty, +\infty]$ is *lower semi-continuous* (resp. *weakly lower semi-continuous*) if

$$u_k \to u \Rightarrow \underline{\lim}\, \varphi(u_k) \geq \varphi(u)$$

$$(\text{resp. } u_k \rightharpoonup u \Rightarrow \underline{\lim}\, \varphi(u_k) \geq \varphi(u)).$$

The following properties are easy consequences of the definition:

i) The sum of two l.s.c. (resp. w.l.s.c.) functions is l.s.c. (resp. w.l.s.c.).

ii) The product of a l.s.c. (resp. w.l.s.c.) function by a positive constant is l.s.c. (resp. w.l.s.c.).

iii) If $(\varphi_\lambda)_{\lambda \in \Lambda}$ is a family of l.s.c. (resp. w.l.s.c.) functions, the function $\sup_{\lambda \in \Lambda} \varphi_\lambda$ defined by

$$\left(\sup_{\lambda \in \Lambda} \varphi_\lambda\right)(u) = \sup_{\lambda \in \Lambda} \varphi_\lambda(u)$$

is lower semi-continuous (resp. w.l.s.c.).

Theorem 1.1. *If φ is w.l.s.c. on a reflexive Banach space X and has a bounded minimizing sequence, then φ has a minimum on X.*

Proof. Let (u_k) be a bounded minimizing sequence. Going if necessary to a subsequence, we can assume, by the reflexivity of X, that (u_k) converges weakly to some $u \in X$. Thus,

$$\varphi(u) \leq \underline{\lim}\, \varphi(u_k) = \lim \varphi(u_k) = \inf_X \varphi,$$

so that $\varphi(u) = \inf_X \varphi$. □

The existence of a bounded minimizing sequence will be in particular insured when φ is *coercive*, i.e., such that

$$\varphi(u) \to +\infty \quad \text{if} \quad \|u\| \to \infty.$$

1.2 Convex Functions

A function $\varphi : X \to]-\infty, +\infty]$ is *convex* if

$$\varphi((1-\lambda)u + \lambda v) \leq (1-\lambda)\varphi(u) + \lambda\varphi(v)$$

for all $\lambda \in]0,1[$, $u, v \in X$.

The following properties are easy consequences of the definition:

i) The sum of two convex functions is a convex function.

ii) The product of a convex function by a positive constant is a convex function.

iii) If $(\varphi_\lambda)_{\lambda\in\Lambda}$ is a family of convex functions then $\sup_{\lambda\in\Lambda} \varphi_\lambda$ is a convex function.

In view of Theorem 1.1, it is important to obtain sufficient conditions for weak lower semi-continuity. We shall obtain such a condition from the following result.

Mazur Theorem. *If (u_k) is a sequence in a normed space X such that $u_k \rightharpoonup u$, there exists a sequence of convex combinations*

$$v_k = \sum_{j=1}^{k} \alpha_{k_j} u_j, \quad \sum_{j=1}^{k} \alpha_{k_j} = 1, \quad \alpha_{k_j} \geq 0 \quad (k \in \mathbf{N}^*)$$

such that $v_k \to u$ in X.

Theorem 1.2. *If X is a normed space and $\varphi : X \to]-\infty, +\infty]$ is l.s.c. and convex, then φ is w.l.s.c.*

Proof. Assume that $u_i \rightharpoonup u$ and let $c > \underline{\lim}\, \varphi(u_i)$. Going if necessary to a subsequence, we can assume that $c > \varphi(u_i)$ for all $i \in \mathbf{N}^*$. By Mazur's theorem, there exists a sequence (v_k) with

$$v_k = \sum_{j=1}^{k} \alpha_{k_j} u_j, \quad \sum_{j=1}^{k} \alpha_{k_j} = 1, \quad \alpha_{k_j} \geq 0$$

such that $v_k \to u$. Since φ is l.s.c. and convex, we obtain

$$\varphi(u) \leq \underline{\lim}\, \varphi(v_k) \leq \underline{\lim} \left(\sum_{j=1}^{k} \alpha_{k_j} \varphi(u_j) \right) \leq \left(\sum_{j=1}^{k} \alpha_{k_j} \right) c = c.$$

Since $c > \underline{\lim}\, \varphi(u_i)$ is arbitrary, we have $\varphi(u) \leq \underline{\lim}\, \varphi(u_i)$, so that φ is w.l.s.c. □

1.3 Euler Equation

The following theorem shows that in order to solve the equation

$$\varphi'(u) = 0$$

for a differentiable function $\varphi : X \to \mathbf{R}$, it suffices to find a local minimum (or maximum) of φ.

Theorem 1.3. *If $\varphi : X \to \mathbf{R}$ is differentiable, every local minimum (resp. maximum) point satisfies the Euler equation*

$$\varphi'(u) = 0.$$

Proof. Let $u \in X$ and $r > 0$ be such that

$$\varphi(u) \leq \varphi(u + v)$$

whenever $|v| \leq r$. Then, if $v \in X \setminus \{0\}$ and $0 < \lambda < r/||v||$, we have

$$0 \leq \frac{\varphi(u + \lambda v) - \varphi(u)}{\lambda}$$

and hence $0 \leq \langle \varphi'(u), v \rangle$. Since v is arbitrary, $\varphi'(u) = 0$. The case of a local maximum is similar. □

Remark 1.1. The following simple generalization of Theorem 1.3 will be useful. If $\varphi : X \to \mathbf{R}$ is differentiable and $Y \subset X$ is a vector subspace of X, then every local minimum (resp. maximum) point of $\varphi|_Y$ satisfies the equation

$$\langle \varphi'(u), v \rangle = 0, \quad v \in Y$$

i.e., $\varphi'(u) \in Y^{\perp}$. The proof is identical to the one above, except for the very last conclusion.

1.4 The Calculus of Variations with Periodic Boundary Conditions

Let C_T^∞ be the space of indefinitely differentiable T-periodic functions from $\mathbf{R}$ into $\mathbf{R}^N$.

Fundamental Lemma. *Let $u, v \in L^1(0,T;\mathbf{R}^N)$. If for every $f \in C_T^\infty$,*

$$\int_0^T (u(t), f'(t))\,dt = -\int_0^T (v(t), f(t))\,dt, \tag{1}$$

then

$$\int_0^T v(s)\,ds = 0 \tag{2}$$

and there exists $c \in \mathbf{R}^N$ such that

$$u(t) = \int_0^t v(s)\,ds + c \tag{3}$$

a.e. on $[0,T]$.

Proof. 1) If (e_j) denotes the canonical basis of $\mathbf{R}^N$, we can choose $f = e_j$ in (1), which gives

$$\int_0^T (v(t), e_j)\,dt = 0 \quad (1 \le j \le N)$$

and (2) follows.

2) Let us define $w \in C(0,T;\mathbf{R}^N)$ by

$$w(t) = \int_0^t v(s)\,ds,$$

so that

$$\int_0^T (w(t), f'(t))\,dt = \int_0^T \left[\int_0^t (v(s), f'(t))\,ds\right] dt.$$

By the Fubini theorem and (2), we obtain

$$\begin{aligned}\int_0^T (w(t), f'(t))\,dt &= \int_0^T \left[\int_s^T (v(s), f'(t))\,dt\right] ds \\ &= \int_0^T (v(s), f(T) - f(s))\,ds = -\int_0^T (v(s), f(s))\,ds.\end{aligned}$$

Hence, by (1) we have, for every $f \in C_T^\infty$,

$$\int_0^T (u(t) - w(t), f'(t))\,dt = 0.$$

In particular, we can choose

$$f(t) = \left\{ \begin{array}{c} \sin \\ \cos \end{array} \right\} (2\pi kt/T)e_j, \quad k \in \mathbf{N} \setminus \{0\},\ 1 \le j \le N$$

and the theory of Fourier series implies that

$$u(t) - w(t) = c$$

a.e. on $[0,T]$ for some $c \in \mathbf{R}^N$, and the proof is complete. □

Remarks. 1) A function v satisfying (1) is called a *weak derivative* of u. By a Fourier series argument, the weak derivative, if it exists, is unique. The weak derivative of u will be denoted by $\dot{u}$.

2) By the fundamental lemma,

$$u(t) = \int_0^t \dot{u}(s)\,ds + c$$

a.e. on $[0,T]$. As usual, we shall identify the equivalence class u and its continuous representant

$$\hat{u}(t) = \int_0^t \dot{u}(s)\,ds + c. \tag{4}$$

In particular, by (2), $u(0) = u(T) = c$, and by soustraction in (4),

$$u(t) = u(\tau) + \int_\tau^t \dot{u}(s)\,ds$$

for $t, \tau \in [0,T]$.

3) If $\dot{u}$ is continuous on $[0,T]$, then by (4), $\dot{u}$ is the classical derivative of $u = \hat{u}$.

4) It follows from (4) and a classical result of Lebesgue theory that $\dot{u}$ is the classical derivative of u a.e. on $[0,T]$.

Let $1 < p < \infty$. The *Sobolev space* $W_T^{1,p}$ is the space of functions $u \in L^p(0,T;\mathbf{R}^N)$ having a weak derivative $\dot{u} \in L^p(0,T;\mathbf{R}^N)$. Let us recall that, if $u \in W_T^{1,p}$,

$$u(t) = \int_0^t \dot{u}(s)\,ds + c$$

and $u(0) = u(T)$. The norm over $W_T^{1,p}$ is defined by

$$\|u\|_{W_T^{1,p}} = \left(\int_0^T |u(t)|^p\,dt + \int_0^T |\dot{u}(t)|^p\,dt \right)^{1/p}.$$

It is easy to verify that $W_T^{1,p}$ is a reflexive Banach space and that $C_T^\infty \subset W_T^{1,p}$.

We shall denote by H_T^1 the Hilbert space $W_T^{1,2}$ with the inner product

$$((u,v)) = \int_0^T [(u(t),v(t)) + (\dot{u}(t),\dot{v}(t))]\,dt$$

and the corresponding norm $\|u\| = \|u\|_{W_T^{1,2}}$. Let us recall that

$$\|u\|_{L^p} = \left(\int_0^T |u(t)|^p\,dt\right)^{1/p} \quad \text{and} \quad \|u\|_\infty = \max_{t\in[0,T]} \|u(t)\|.$$

Proposition 1.1. *There exists $c > 0$ such that, if $u \in W_T^{1,p}$, then*

$$\|u\|_\infty \leq c\,\|u\|_{W_T^{1,p}}. \tag{4'}$$

Moreover, if $\int_0^T u(t)\,dt = 0$, then

$$\|u\|_\infty \leq c\,\|\dot{u}\|_{L^p}. \tag{4''}$$

Proof. Going to the components of u, we can assume that $N = 1$. If $u \in W_T^{1,p}$, it follows from the mean value theorem that

$$(1/T)\int_0^T u(s)\,ds = u(\tau)$$

for some $\tau \in]0,T[$. Hence, for $t \in [0,T]$, using Hölder inequality,

$$\begin{aligned} |u(t)| &= \left|u(\tau) + \int_\tau^t \dot{u}(s)\,ds\right| \leq |u(\tau)| + \int_0^T |\dot{u}(s)|\,ds \\ &\leq (1/T)\left|\int_0^T u(s)\,ds\right| + T^{1/q}\left(\int_0^T |\dot{u}(s)|^p\right)^{1/p} \\ &= (1/T)\left|\int_0^T u(s)\,ds\right| + T^{1/q}\|\dot{u}\|_{L^p} \quad (1/p + 1/q = 1). \end{aligned}$$

If $\int_0^T u(s)\,ds = 0$, we obtain (4″). In the general case, we get, for $t \in [0,T]$,

$$\begin{aligned} |u(t)| &\leq (1/T)\int_0^T |u(s)|\,ds + T^{1/q}\|\dot{u}\|_{L^p} \\ &\leq T^{-1/p}\|u\|_{L^p} + T^{1/q}\|\dot{u}\|_{L^p} \\ &\leq (T^{-1/p} + T^{1/q})/\|u\|_{W_T^{1,p}} \end{aligned}$$

and we obtain (4′). □

Proposition 1.2. *If the sequence (u_k) converges weakly to u in $W_T^{1,p}$, then (u_k) converges uniformly to u on $[0,T]$.*

Proof. By Proposition 1.1, the injection of $W_T^{1,p}$ into $C(0,T;\mathbf{R}^N)$, with its natural norm $\|\cdot\|_\infty$, is continuous. Since $u_k \rightharpoonup u$ in $W_T^{1,p}$, it follows that $u_k \rightharpoonup u$ in $C(0,T;\mathbf{R}^N)$. By the Banach–Steinhaus theorem, (u_k) is bounded in $W_T^{1,p}$ and, hence, in $C(0,T;\mathbf{R}^N)$. Moreover, the sequence (u_k) is equi-uniformly continuous since, for $0 \le s \le t \le T$, we have

$$|u_k(t) - u_k(s)| \le \int_s^t |\dot{u}_k(\tau)|\,d\tau \le (t-s)^{1/q}\left(\int_s^t |\dot{u}_k(\tau)|^p\,d\tau\right)^{1/p}$$

$$\le (t-s)^{1/q}\|u_k\|_{W_T^{1,p}} \le C(t-s)^{1/q}.$$

By Ascoli–Arzela theorem, (u_k) is relatively compact in $C(0,T;\mathbf{R}^N)$. By the uniqueness of the weak limit in $C(0,T;\mathbf{R}^N)$, every uniformly convergent subsequence of (u_k) converges to u. Thus, (u_k) converges uniformly on $[0,T]$ to u. □

In the case of the Hilbert space H_T^1, we obtain sharp estimates.

Proposition 1.3. *If $u \in H_T^1$ and $\int_0^T u(t)\,dt = 0$, then*

$$\int_0^T |u(t)|^2\,dt \le (T^2/4\pi^2)\int_0^T |\dot{u}(t)|^2\,dt$$

(Wirtinger's inequality) and

$$\|u\|_\infty^2 \le (T/12)\int_0^T |\dot{u}(t)|^2\,dt$$

(Sobolev inequality).

Proof. Since, by assumption, u has the Fourier expansion

$$u(t) = \sum_{\substack{k=-\infty \\ k\neq 0}}^{+\infty} u_k \exp(2i\pi kt/T),$$

the Parseval equality implies that

$$\int_0^T |\dot{u}(t)|^2\,dt = \sum_{\substack{k=-\infty \\ k\neq 0}}^{+\infty} T(4\pi^2k^2/T^2)|u_k|^2$$

$$\ge (4\pi^2/T^2)\sum_{\substack{k=-\infty \\ k\neq 0}}^{+\infty} T|u_k|^2 = (4\pi^2/T^2)\int_0^T |u(t)|^2\,dt.$$

The Cauchy–Schwarz inequality and Parseval equality then imply that for $t \in [0,T]$,

$$
\begin{aligned}
|u(t)|^2 &\leq \left(\sum_{\substack{k=-\infty \\ k\neq 0}}^{+\infty} |u_k| \right)^2 \\
&\leq \left[\sum_{\substack{k=-\infty \\ k\neq 0}}^{+\infty} (T/4\pi^2 k^2) \right] \left[\sum_{\substack{k=-\infty \\ k\neq 0}}^{+\infty} (4\pi^2 k^2/T)|u_k|^2 \right] \\
&= (T/12) \int_0^T |\dot{u}(t)|^2 \, dt
\end{aligned}
$$

since $\sum_{k=1}^{\infty}(1/k^2) = \pi^2/6$. □

Theorem 1.4. *Let $L : [0,T] \times \mathbf{R}^N \times \mathbf{R}^N \to \mathbf{R}$, $(t,x,y) \to L(t,x,y)$ be measurable in t for each $[x,y] \in \mathbf{R}^N \times \mathbf{R}^N$ and continuously differentiable in $[x,y]$ for almost every $t \in [0,T]$. If there exists $a \in C(\mathbf{R}^+, \mathbf{R}^+)$, $b \in L^1(0,T;\mathbf{R}^+)$ and $c \in L^q(0,T;\mathbf{R}^+)$, $1 < q < \infty$, such that, for a.e. $t \in [0,T]$ and every $[x,y] \in \mathbf{R}^N \times \mathbf{R}^N$, one has*

$$
\begin{aligned}
|L(t,x,y)| &\leq a(|x|)(b(t) + |y|^p) \\
|D_x L(t,x,y)| &\leq a(|x|)(b(t) + |y|^p) \qquad \text{(a)} \\
|D_y L(t,x,y)| &\leq a(|x|)(c(t) + |y|^{p-1})
\end{aligned}
$$

where $\frac{1}{p} + \frac{1}{q} = 1$, then the functional φ defined by

$$\varphi(u) = \int_0^T L(t, u(t), \dot{u}(t)) \, dt$$

is continuously differentiable on $W_T^{1,p}$ and

$$\langle \varphi'(u), v \rangle = \int_0^T [(D_x L(t,u(t),\dot{u}(t)), v(t)) + (D_y L(t,u(t),\dot{u}(t)), \dot{v}(t))] \, dt. \qquad \text{(b)}$$

Proof. It suffices to prove that φ has at every point u a directional derivative $\varphi'(u) \in (W_T^{1,p})^*$ given by (b) and that the mapping

$$\varphi' : W_T^{1,p} \to (W_T^{1,p})^*, \quad u \to \varphi'(u)$$

is continuous.

1) It follows easily from (a) that φ is everywhere finite on $W_T^{1,p}$. Let us define, for u and v fixed in $W_T^{1,p}$, $t \in [0,T]$, $\lambda \in [-1,1]$,

$$F(\lambda, t) = L(t, u(t) + \lambda v(t), \dot{u}(t) + \lambda \dot{v}(t))$$

and

$$\psi(\lambda) = \int_0^T F(\lambda,t)\,dt = \varphi(u+\lambda v).$$

We shall apply Leibniz formula of differentiation under integral sign to ψ. By assumption (a) we have

$$\begin{aligned}|D_\lambda F(\lambda,t)| &= |(D_xL(t,u+\lambda v,\dot u+\lambda\dot v),v)+(D_yL(t,u+\lambda v,\dot u+\lambda\dot v),\dot v)| \\ &\le a(|u+\lambda v|)[(b(t)+|\dot u+\lambda\dot v|^p)|v|+(c(t)+|\dot u+\lambda\dot v|^{p-1})|\dot v|] \\ &\le a_0[(b(t)+(|\dot u|+|\dot v|)^p)|v|+(c(t)+(|\dot u|+|\dot v|)^{p-1})|\dot v|]\end{aligned}$$

where

$$a_0 = \max_{(\lambda,t)\in[-1,1]\times[0,T]} a(|u(t)+\lambda v(t)|).$$

Since $b\in L^1$, $(|\dot u|+|\dot v|)^p\in L^1$, $c\in L^q$, $|\dot v|\in L^p$, and v is continuous on $[0,T]$, we have

$$|D_\lambda F(\lambda,t)|\le d(t)\in L^1(0,T;\mathbf{R}^+).$$

Thus, Leibniz formula is applicable and

$$\begin{aligned}\dot\psi(0) = \int_0^T D_\lambda F(0,t)\,dt = \int_0^T &[(D_xL(t,u(t),\dot u(t)),v(t)) \\ &+ (D_yL(t,u(t),\dot u(t)),\dot v(t))]\,dt.\end{aligned}$$

Moreover,

$$|D_xL(t,u,\dot u)|\le a(|u|)(b(t)+|\dot u|^p)\in L^1(0,T,\mathbf{R}^+),$$

and

$$|D_yL(t,u,\dot u)|\le a(|u|)(c(t)+|\dot u|^{p-1})\in L^q(0,T,\mathbf{R}^+).$$

Thus, by Proposition 1.1,

$$\begin{aligned}\int_0^T &[(D_xL(t,u(t),\dot u(t)),v(t))+(D_yL(t,u(t),\dot u(t)),\dot v(t))]\,dt \\ &\le c_1\|v\|_\infty + c_2\|\dot v\|_{L^p}\le c_3\|v\|_{W_T^{1,p}},\end{aligned}$$

and φ has, at u, a directional derivative $\varphi'(u)\in(W_T^{1,p})^*$ given by (b).

2) By a theorem of Krasnosel'skii, assumption (a) implies that the mapping from $W_T^{1,p}$ into $L^1\times L^q$ defined by

$$u\to(D_xL(.,u,\dot u),\quad D_yL(.,u,\dot u))$$

is continuous, so that φ' is continuous from $W_T^{1,p}$ into $(W_T^{1,p})^*$, and the proof is complete. □

Corollary 1.1. *Let* $L : [0,T] \times \mathbf{R}^N \times \mathbf{R}^N \to \mathbf{R}$ *be defined by*

$$L(t,x,y) = (1/2)|y|^2 + F(t,x)$$

where $F : [0,T] \times \mathbf{R}^N \to \mathbf{R}$, $(t,x) \to F(t,x)$ *is measurable in* t *for each* $x \in \mathbf{R}^N$, *continuously differentiable in* x *for almost every* $t \in [0,T]$ *and satisfy the following conditions:*

$$|F(t,x)|, \quad |\nabla F(t,x)| \le a(|x|)b(t)$$

for a.e. $t \in [0,T]$, *all* $x \in \mathbf{R}^N$, *some* $a \in C(\mathbf{R}_+, \mathbf{R}_+)$, *and some* $b \in L^1(0,T;\mathbf{R}_+)$. *If* $u \in H_T^1$ *is a solution of the corresponding Euler equation* $\varphi'(u) = 0$, *then* $\dot{u}$ *has a weak derivative* $\ddot{u}$, *and*

$$\ddot{u}(t) = \nabla F(t,u(t)) \quad \text{a.e. on } [0,T]$$

$$u(0) - u(T) = \dot{u}(0) - \dot{u}(T) = 0.$$

Proof. By Theorem 1.4, we have

$$0 = \langle \varphi'(u), v\rangle = \int_0^T ((\nabla F(t,u(t)), v(t)) + (\dot{u}(t), \dot{v}(t)))\, dt = 0$$

for all $v \in H_T^1$ and hence for all $v \in C_T^\infty$. Thus, $\dot{u}$ has a weak derivative and $\ddot{u}(t) = \nabla F(t,u(t))$ a.e. on $[0,T]$. Moreover, the existence of a weak derivative for u and $\dot{u}$ implies that $u(0) - u(T) = \dot{u}(0) - \dot{u}(T) = 0$. □

1.5 Periodic Solutions of Non-Autonomous Second Order Systems with Bounded Nonlinearity

Let us consider the problem introduced in Corollary 1.1

$$\ddot{u}(t) = \nabla F(t,u(t)) \quad \text{(a.e. on } [0,T]) \tag{5}$$

where $F : [0,T] \times \mathbf{R}^N \to \mathbf{R}$ satisfies the following assumption:

(A) $F(t,x)$ is measurable in t for each $x \in \mathbf{R}^N$, continuously differentiable in x for a.e. $t \in [0,T]$, and there exist $a \in C(\mathbf{R}^+, \mathbf{R}^+)$ and $b \in L^1(0,T;\mathbf{R}^+)$ such that

$$|F(t,x)| \le a(|x|)b(t), \quad |\nabla F(t,x)| \le a(|x|)b(t)$$

for all $x \in \mathbf{R}^N$ and a.e. $t \in [0,T]$.

The corresponding functional φ on H_T^1 given by

$$\varphi(u) = \int_0^T [|\dot{u}(t)|^2/2 + F(t,u(t))]\, dt$$

is continuously differentiable and w.l.s.c. on H_T^1 as the sum of a convex continuous function (Theorem 1.2) and of a weakly continuous one (Proposition 1.2). If φ has a bounded minimizing sequence then, by Theorem 1.1, φ has a minimum so that, by Corollary 1.1, problem (5) is solvable. It remains to find conditions under which φ has a bounded minimizing sequence. When ∇F is bounded by a L^1-function for all $u \in \mathbf{R}^N$, we shall see that it suffices to require a coercivity condition for the average of F with respect to t.

Theorem 1.5. *Assume that F satisfies condition* (A) *and that there exists $g \in L^1(0,T)$ such that*

$$|\nabla F(t,x)| \leq g(t)$$

for a.e. $t \in [0,T]$ and all $x \in \mathbf{R}^N$. If

$$\int_0^T F(t,x)\,dt \to +\infty \quad \text{as} \quad |x| \to \infty, \tag{6}$$

then problem (5) *has at least one solution which minimizes φ on H_T^1.*

Proof. For $u \in H_T^1$, we have $u = \overline{u} + \tilde{u}$ where $\overline{u} = \int_0^T u(t)\,dt$ and

$$\begin{aligned}
\varphi(u) &= \int_0^T (|\dot{u}(t)|^2/2)\,dt + \int_0^T F(t,\overline{u})\,dt + \int_0^T [F(t,u(t)) - F(t,\overline{u})]\,dt \\
&= \int_0^T (|\dot{u}(t)|^2/2)\,dt + \int_0^T F(t,\overline{u})\,dt \\
&\quad + \int_0^T \int_0^1 (\nabla F(t,\overline{u} + s\tilde{u}(t)), \tilde{u}(t))\,dsdt \\
&\geq \int_0^T (|\dot{u}(t)|^2/2)\,dt - \left(\int_0^T g(t)\,dt\right) \|\tilde{u}\|_\infty + \int_0^T F(t,\overline{u})\,dt \\
&\geq \int_0^T (|\dot{u}(t)|^2/2)\,dt - C\left(\int_0^T |\dot{u}(t)|^2 dt\right)^{1/2} + \int_0^T F(t,\overline{u})\,dt
\end{aligned}$$

by Sobolev's inequality. As $\|u\| \to \infty$ if and only if $(|\overline{u}|^2 + \int_0^T |\dot{u}(t)|^2 dt)^{1/2} \to \infty$, the above inequality and (6) imply that

$$\varphi(u) \to +\infty \quad \text{as} \quad \|u\| \to \infty$$

and hence every minimizing sequence is bounded, which completes the proof. □

As an example, let us consider the scalar problem

$$\ddot{u} = a[\sin(u - b \text{ sgn } u) + \sin(b \text{ sgn } u)] + e(t)$$

$$u(0) - u(T) = \dot{u}(0) - \dot{u}(T) = 0,$$

where $a > 0$, $0 < b < \pi$, $e \in L^1(0,T)$ and $\int_0^T e(t)\,dt = 0$. In this case,

$$F(t,u) = a[(\sin b)|u| - \cos(|u| - b) + \cos b] + e(t)u$$

and hence

$$\int_0^T F(t,x)\,dt = T a \sin b|x| - T a[\cos(|x| - b) - \cos b] \to +\infty$$

if $|x| \to \infty$. Moreover, $|F'_u(t,u)|$ is clearly bounded by $2a + |e(t)|$ and the result follows.

The coercivity condition does not hold if $b = 0$, i.e., in the case of the forced pendulum equation

$$\ddot{u} = a \sin u + e(t).$$

We shall show in the next section that the corresponding existence result can still be proved by taking advantage of the periodicity in u of the right hand member of the equation.

1.6 Periodic Solutions of Non-Autonomous Second Order Systems with Periodic Potential

We show in this section that (5) is solvable when F is periodic in each variable x_i. Let (e_i) $(1 \leq i \leq N)$ denote the canonical basis of $\mathbf{R}^N$

Theorem 1.6. *Assume that F satisfies condition* (A) *and that there exist $T_i > 0$ such that*

$$F(t, x + T_i e_i) = F(t,x) \quad (1 \leq i \leq N) \tag{7}$$

for all $x \in \mathbf{R}^N$ and a.e. $t \in [0,T]$. Then the problem (5) *has at least one solution which minimizes φ on H_T^1.*

Proof. It follows from (7) and the regularity of F that there exists $h \in L^1(0,T)$ such that

$$F(t,x) \geq h(t)$$

for all $x \in \mathbf{R}^N$ and a.e. $t \in [0,T]$. Consequently, if $C_1 = \int_0^T h(t)\,dt$,

$$\varphi(u) \geq (1/2) \int_0^T |\dot{u}(t)|^2 dt - C_1$$

for all $u \in H_T^1$. As $\inf_{H_T^1} \varphi < +\infty$, it follows from this inequality that if (u_k) is a minimizing sequence for φ, there will exist $C_2 > 0$ such that

$$\int_0^T |\dot{u}_k(t)|^2 dt \leq C_2 \tag{8}$$

for all $k \in \mathbf{N}$. Let $u_k = \overline{u}_k + \tilde{u}_k$ with $\overline{u}_k = (1/T)\int_0^T u_k(s)\,ds$, it follows from (8) and Wirtinger's inequality that

$$\|\tilde{u}_k\| \leq C_3, \quad k \in \mathbf{N} \tag{9}$$

for some $C_3 > 0$. On the other hand, it follows from (7) that

$$\varphi(u + T_i e_i) = \varphi(u), \quad 1 \leq i \leq N$$

for all $u \in H_T^1$ and hence if (u_k) is a minimizing sequence for φ, $([(\overline{u}_k, e_1) + k_1 T_1 + (\tilde{u}_k, e_1), \ldots, (\overline{u}_k, e_N) + k_N T_N + (\tilde{u}_k, e_N)])$ is also a minimizing sequence of φ and we can therefore assume that

$$0 \leq (\overline{u}_k, e_i) \leq T_i \quad (1 \leq i \leq N). \tag{10}$$

Consequently by (8), (9), and (10), φ admits a bounded minimizing sequence, and the proof is complete. □

One can obtain, as follows, a useful extension of Theorem 1.6 to some forced second order systems. It is elementary to check that, for $e \in L^1(0,T;\mathbf{R}^N)$, the linear problem

$$\begin{gathered} \ddot{v}(t) = e(t) \\ v(0) - v(T) = \dot{v}(0) - \dot{v}(T) = 0 \end{gathered} \tag{11}$$

has a solution if and only if

$$\int_0^T e(t)\,dt = 0. \tag{12}$$

This solution will be unique if we impose in addition that

$$\int_0^T v(t)\,dt = 0 \tag{13}$$

and we shall denote by E the unique solution of (11) satisfying (13). Then if we consider the problem

$$\begin{gathered} \ddot{u}(t) = \nabla F(t, u(t)) + e(t) \\ u(0) - u(T) = \dot{u}(0) - \dot{u}(T) = 0 \end{gathered} \tag{14}$$

where $e \in L^1(0,T;\mathbf{R}^N)$ satisfies (12), we obtain, letting

$$u(t) = v(t) + E(t), \tag{15}$$

the equivalent problem

$$\begin{gathered} \ddot{v}(t) = \nabla F(t, v(t) + E(t)) \\ v(0) - v(T) = \dot{v}(0) - \dot{v}(T) = 0. \end{gathered} \tag{16}$$

Now, if F satisfies the periodicity conditions of Theorem 1.6, the same is true for $F^{\#} : [0,T]\times \mathbf{R}^N \to \mathbf{R}$, $(t,x) \to F(t,x+E(t))$ and hence, Theorem 1.5 applied to (16) implies the following.

Corollary 1.2. *Under the conditions of Theorem* 1.6 *for* F, *the problem* (14) *has, for each* $e \in L^1(0,T;\mathbf{R}^N)$ *verifying* (12), *at least one solution which minimizes on* H_T^1 *the functional* φ_e *defined by*

$$\varphi_e(u) = \int_0^T [|\dot{u}(t)|^2/2 + F(t,u(t)) + (e(t),u(t))]\,dt.$$

Proof. Theorem 1.6 applied to (16) implies the existence of a solution v of (16) which minimizes on H_T^1 the function ψ defined by

$$\psi(v) = \int_0^T [|\dot{v}(t)|^2/2 + F(t,v(t)) + E(t)]\,dt.$$

Therefore, u defined by (15) solves (14) and minimizes on $\psi(.-E)$ on H_T^1. But, integrating by parts we get

$$\begin{aligned}\psi(u-E) &= \int_0^T [|\dot{u}(t) - \dot{E}(t)|^2/2 + F(t,u(t))]\,dt \\ &= \int_0^T [|\dot{u}(t)|^2/2 + (u(t),e(t)) + F(t,u(t))]\,dt + \|\dot{E}\|_2^2 \\ &= \varphi_e(u) + \|\dot{E}\|_2^2,\end{aligned}$$

hence u minimizes φ_e on H_T^1. □

As an application, let us consider the periodic boundary value problem for the forced pendulum equation

$$\begin{gathered}\ddot{u}(t) + A\sin u(t) = e(t) \quad (A \geq 0 \text{ fixed}) \\ u(0) - u(T) = u'(0) - u'(T) = 0.\end{gathered} \tag{17}$$

Notice that if (17) has a solution then, integrating the equation over $[0,T]$ and using the boundary conditions, we get

$$A\int_0^T \sin u(t)\,dt = \int_0^T e(t)\,dt$$

hence

$$-A \leq (1/T)\int_0^T e(t)\,dt \leq A.$$

One cannot, therefore, hope to solve (17) for every $e \in L^1(0,T)$ and a complete explicit description of the range of $(d^2/dt^2) + A\sin(.)$ acting on T-periodic functions is still unknown. As (17) is of the form (14) with $F(t,x) = A\cos x$, the conditions of Corollary 1.2 are satisfied if

$\int_0^T e(t)\,dt = 0$ in which case (17) has, therefore, a solution which minimizes the corresponding φ_e on H_T^1.

Another application is the pendulum with a horizontal periodic external force, whose equation is

$$\ddot{u}(t) + A \sin u(t) = \cos u(t)\, e(t)$$

and can be written

$$\ddot{u} = D_u[A \cos u + e(t) \sin u]$$

so that Theorem 1.6 is directly applicable for each $e \in L^1(0,T)$.

1.7 Periodic Solutions of Non-Autonomous Second Order Systems with Convex Potential

When F is convex in x, it is possible to eliminate the boundedness condition on ∇F in Theorem 1.5 and to deduce a necessary and sufficient condition of existence when F is strictly convex in x or when $N = 1$.

We shall need the following elementary and intuitive results on convex functions.

Proposition 1.4. *Let $G \in C^1(\mathbf{R}^N, \mathbf{R})$ be a convex function. Then, for all $x, y \in \mathbf{R}^N$ we have*

$$G(x) \geq G(y) + (\nabla G(y), x - y). \tag{18}$$

Proof. By the convexity of G we have, for each $x, y \in \mathbf{R}^N$ and each $\lambda \in]0,1[$,

$$G((1-\lambda)y + \lambda x) \leq (1-\lambda)G(y) + \lambda G(x)$$

hence

$$\frac{G(y + \lambda(x-y)) - G(y)}{\lambda} \leq G(x) - G(y).$$

Letting $\lambda \to 0$ we obtain (18). □

A function $G : \to]-\infty, +\infty]$ is *strictly convex* if

$$G((1-\lambda)x + \lambda y) < (1-\lambda)G(x) + \lambda G(y)$$

whenever $G(x) < +\infty$, $G(y) < +\infty$, $x \neq y$, and $\lambda \in]0,1[$.

Proposition 1.5. *Let $G \in C^1(\mathbf{R}^N, \mathbf{R})$ be a strictly convex function. The following properties are equivalent:*

a) *There exists $\overline{x} \in \mathbf{R}^N$ such that $\nabla G(\overline{x}) = 0$.*

b) *$G(x) \to +\infty$ when $|x| \to \infty$.*

Proof. 1. If $\nabla G(\overline{x}) = 0$, it follows from (18) with $y = \overline{x}$ that $\overline{x}$ minimizes G on $\mathbf{R}^N$. Since G is strictly convex, $\overline{x}$ is unique, hence

$$\delta = \min_{|x|=1} [G(\overline{x}+x) - G(\overline{x})] > 0.$$

The convexity of G then implies that, when $|x| \geq 1$

$$\begin{aligned}\delta &\leq G\left(\overline{x}+\frac{x}{|x|}\right) - G(\overline{x}) \leq \frac{1}{|x|}G(\overline{x}+x) + \left(1-\frac{1}{|x|}\right)G(\overline{x}) - G(\overline{x}) \\ &= \frac{1}{|x|}(G(\overline{x}+x) - G(\overline{x})).\end{aligned}$$

Hence,

$$G(x+\overline{x}) \geq \delta|x| + G(\overline{x})$$

for $|x| \geq 1$ and (b) follows easily.

2. If G satisfies (b), then G has a minimum at some point $\overline{x}$ for which (a) holds. □

Theorem 1.7. *Assume that* $F : [0,T] \times \mathbf{R}^N \to \mathbf{R}$ *satisfies assumption* (A), *that* $F(t,.)$ *is convex for a.e.* $t \in [0,T]$ *and that*

$$\int_0^T F(t,x)\,dt \to +\infty \quad \text{if} \quad |x| \to \infty. \tag{19}$$

Then problem (5) *has at least one solution which minimizes* φ *on* H_T^1.

Proof. By assumption, the real function on $\mathbf{R}^N$ defined by

$$x \to \int_0^T F(t,x)\,dt$$

has a minimum at some point $\overline{x}$ for which

$$\int_0^T \nabla F(t,\overline{x})\,dt = 0. \tag{20}$$

Let (u_k) be a minimizing sequence for φ. It follows from (18) and (20) that

$$\begin{aligned}\varphi(u_k) &\geq (1/2)\int_0^T |\dot{u}_k(t)|^2 + \int_0^T F(t,\overline{x})\,dt + \int_0^T (\nabla F(t,\overline{x}), u_k(t) - \overline{x})\,dt \\ &= (1/2)\int_0^T |\dot{u}_k(t)|^2 dt + \int_0^T F(t,\overline{x})\,dt + \int_0^T (\nabla F(t,\overline{x}), \tilde{u}_k(t))\,dt\end{aligned}$$

where $u_k = \overline{u}_k + \tilde{u}_k$ with $\overline{u}_k = (1/T)\int_0^T u_k(t)\,dt$. We obtain, using Sobolev's inequality,

$$\varphi(u_k) \geq (1/2)\int_0^T |\dot{u}_k(t)|^2 dt + \int_0^T F(t,\overline{x})\,dt$$

$$- \left(\int_0^T |\nabla F(t, \overline{x})| \, dt \right) \|\tilde{u}_k\|_\infty$$
$$\geq (1/2) \int_0^T |\dot{u}_k(t)|^2 dt - c_1 - c_2 \left(\int_0^T |\dot{u}_k(t)|^2 dt \right)^{1/2}$$

for some constants c_1 and $c_2 > 0$. Hence, there exists a constant $c_3 > 0$ such that

$$\int_0^T |\dot{u}_k(t)|^2 dt \leq c_3.$$

By Sobolev's inequality, this implies that

$$\|\tilde{u}_k\|_\infty \leq c_4 \tag{21}$$

for some constant $c_4 > 0$. Now we have, by convexity,

$$\begin{aligned} F(t, \overline{u}_k/2) &= F(t, (1/2)(u_k(t) - \tilde{u}_k(t))) \\ &\leq (1/2)F(t, u_k(t)) + (1/2)F(t, -\tilde{u}_k(t)) \end{aligned}$$

for a.e. $t \in [0, T]$ and all $k \in \mathbf{N}$, hence

$$\varphi(u_k) \geq (1/2) \int_0^T |\dot{u}_k(t)|^2 dt + 2 \int_0^T F(t, \overline{u}_k/2) \, dt - \int_0^T F(t, -\tilde{u}_k(t)) \, dt.$$

This implies, by (21),

$$\varphi(u_k) \geq 2 \int_0^T F(t, \overline{u}_k/2) \, dt - c_5$$

for some $c_5 > 0$ and therefore, by (19), $(\overline{u}_k)$ is bounded, which completes the proof. □

We consider now the case where $F(t, .)$ is strictly convex.

Theorem 1.8. *Assume that F satisfies condition* (A) *and that $F(t, .)$ is strictly convex for a.e. $t \in [0, T]$. Then the following conditions are equivalent:*

α. *Problem* (5) *is solvable.*

β. *There exists $\overline{x} \in \mathbf{R}^N$ such that*

$$\int_0^T \nabla F(t, \overline{x}) \, dt = 0.$$

γ. $\int_0^T F(t, x) \, dt \to +\infty$ *when* $|x| \to \infty$.

Proof. 1. If u is a solution of (5) then, integrating the differential equation over $(0,T)$ and using the boundary conditions, we get

$$\int_0^T \nabla F(t,u(t))\,dt = 0. \tag{22}$$

Let $u = \tilde{u} + \overline{u}$ where $\overline{u} = (1/T)\int_0^T u(t)\,dt$ and define the strictly convex functions G and $\tilde{G}$ on $\mathbf{R}^N$ by

$$G(x) = \int_0^T F(t,x)\,dt;$$

$$\tilde{G}(x) = \int_0^T F(t,x+\tilde{u}(t))\,dt.$$

Since, by (22), $\nabla\tilde{G}(\overline{u}) = 0$, Proposition 1.5 implies that

$$\tilde{G}(x) \to +\infty \quad \text{as} \quad |x| \to \infty. \tag{23}$$

By the convexity of $F(t,.)$, we have

$$\begin{aligned}\tilde{G}(x) &\leq (1/2)\int_0^T F(t,2x)\,dt + (1/2)\int_0^T F(t,2\tilde{u}(t))\,dt \\ &= (1/2)G(2x) + C.\end{aligned} \tag{24}$$

It then follows from (23) and (24) that $G(x) \to +\infty$ as $|x| \to \infty$. Hence, there exists $\overline{x} \in \mathbf{R}^N$ such that $\nabla G(\overline{x}) = 0$, i.e.,

$$\int_0^T \nabla F(t,\overline{x})\,dt = 0$$

and (α) implies (β).

2. By Proposition 1.5 applied to the function G defined above, (β) implies (γ).

3. By Theorem 1.7, (γ) implies (α). □

We now return to the case of a convex $F(t,.)$ but with $N = 1$. Setting $f(t,x) = D_x F(t,x)$, we see that $f(t,.)$ is nondecreasing for almost every $t \in [0,1]$. This implies a simpler necessary condition for the existence of a solution of

$$\begin{aligned}\ddot{u}(t) &= f(t,u(t)) \\ u(0) - u(T) &= \dot{u}(0) - \dot{u}(T) = 0.\end{aligned} \tag{25}$$

Lemma 1.1. *If* (25) *has a solution, there exists* $\overline{a} \in \mathbf{R}$ *such that*

$$\int_0^T f(t,\overline{a})\,dt = 0. \tag{26}$$

In other words, the real function $a \to \int_0^T F(t,a)\,dt$ has a critical point $\bar{a}$.

Proof. If (25) has a solution u, then, integrating both members of the equation over $[0,T]$ and using the boundary conditions imply that

$$\int_0^T f(t,u(t))\,dt = 0.$$

Therefore, if $m \le u(t) \le M$ for $t \in [0,T]$, we have, by the monotonicity of $f(t,.)$,

$$\int_0^T f(t,m)\,dt \le 0 \le \int_0^T f(t,M)\,dt$$

and the result follows from the intermediate value theorem. □

We shall now prove the more striking fact that condition (26) is also sufficient to the solvability of (25).

Theorem 1.9. *If $f(t,.)$ is nondecreasing for a.e. $t \in [0,T]$, then problem (25) has at least one solution if and only if there exists some $\bar{a} \in \mathbf{R}$ satisfying (26), i.e., if and only if the real function $a \to \int_0^T F(t,a)\,dt$ has a critical point.*

Proof. The necessity is proved in Lemma 1.1. For the sufficiency, let us first assume that

$$\int_0^T f(t,a)\,dt = 0$$

whenever $a \ge \bar{a}$. Then, by (26) and the nondecreasing character of $f(t,.)$, this implies that

$$f(t,a) = f(t,\bar{a})$$

for a.e. $t \in [0,T]$ and all $a \ge \bar{a}$. Let v be a solution of the T-periodic linear problem

$$\ddot{v}(t) = f(t,\bar{a})$$
$$v(0) - v(T) = \dot{v}(0) - \dot{v}(T) = 0$$

(such a solution exists because of (26)) and let $b \in \mathbf{R}$ sufficiently large so that

$$v(t) + b \ge \bar{a}, \quad t \in [0,T].$$

If we set $u(t) = v(t) + b$, then $u(0) - u(T) = \dot{u}(0) - \dot{u}(T) = 0$ and

$$\ddot{u}(t) = \ddot{v}(t) = f(t,\bar{a}) = f(t,v(t)+b) = f(t,u(t))$$

so that u is a solution to (25). Similarly if

$$\int_0^T f(t,a)\,dt = 0$$

whenever $a \leq \overline{a}$. It remains, therefore, to consider the case where there exists $a_1 < \overline{a} < a_2$ such that

$$c_1 \equiv \int_0^T f(t, a_1)\, dt < 0 < \int_0^T f(t, a_2)\, dt \equiv c_2.$$

Then, for $a \geq a_2$,

$$\begin{aligned} \int_0^T F(t,a)\, dt &= \int_0^T \left[F(t, a_2) + \int_0^1 f(t, (1-s)a_2 + sa)(a - a_2)\, ds \right] dt \\ &= \int_0^T F(t, a_2)\, dt + c_2(a - a_2) \end{aligned}$$

and $\int_0^T F(t,a)\, dt \to +\infty$ if $a \to +\infty$. Similarly, if $a \to -\infty$, the existence of a solution follows from Theorem 1.7. □

Remark 1.2. When f is independent of x, Theorem 1.9 reduces to the usual Fredholm necessary and sufficient condition

$$\int_0^T f(t)\, dt = 0$$

for the solvability of (25).

In the special case of a problem of the form

$$\ddot{u}(t) = g(u(t)) - h(t)$$
$$u(0) - u(T) = \dot{u}(0) - \dot{u}(T) = 0$$

with $g : \mathbf{R} \to \mathbf{R}$ continuous and nondecreasing and $h \in L^1(0,T)$, Theorem 1.9 implies existence if and only if $\overline{h} \equiv (1/T) \int_0^T h(t)\, dt$ belongs to the range of g.

In particular, the range of the nonlinear operator $\frac{d^2}{dt^2} - g(.)$ acting on T-periodic functions will be open (resp. closed) if the range of g is open (resp. closed) in $\mathbf{R}$.

Historical and Bibliographical Notes

The direct method of the calculus of variations has its origin in the Dirichlet principle, which consists in connecting the existence of a solution of the Dirichlet problem

$$\begin{aligned} \Delta u(x) &= 0, & x &\in \Omega \\ u(x) &= h(x), & x &\in \partial\Omega \end{aligned} \tag{27}$$

where $\Omega \subset \mathbf{R}^N$ is an open bounded set, $\Delta = \sum_{i=1}^n D_i^2$ is the Laplacian, $h : \partial\Omega \to \mathbf{R}$ is a given function, to the existence of a minimum of the Dirichlet integral

$$D(u) = \int_\Omega |\nabla u(x)|^2 dx, \tag{28}$$

over a class of sufficiently regular real functions on Ω which are equal to h on $\partial\Omega$. If, with Gauss (1839), Lord Kelvin (1847), Dirichlet (1850), Riemann (1851, 1857) and others, we consider the existence of this minimum as obvious, the existence of a solution for (27) follows, as (27) is the Euler–Lagrange equation associated to the extremums of D. In 1870, Weierstrass pointed out the important distinction between the notions of infimum and of minimum by producing a counterexample making doubtful the validity of the Dirichlet principle. It was only at the turn of the twentieth century that the partial results of Arzela [Arz_1] and the definitive work of Hilbert [$\text{Hil}_{1,2}$] rehabilitated the Dirichlet principle by giving conditions upon h and Ω which insure its validity. This seminal work was immediately followed by a number of significant contributions by Levi, Fubini, Lebesgue, Zaremba, Courant, Lichtenstein, and Tonelli, establishing the direct method of the calculus of variations, i.e., the extension of the Dirichlet principle to more general functionals of the type

$$F(u) = \int_\Omega f(x, u(x), \nabla u(x))\, dx, \tag{29}$$

as a powerful tool for proving the existence of solutions to linear and nonlinear boundary value problems.

Although the concepts of lower and upper semi-continuity had been introduced by Baire [Bai_1] in 1897, it was Lebesgue [$\text{Leb}_{1,2}$] who first emphasized that lower semi-continuity was the type of continuity naturally satisfied by functionals of type (29), and Tonelli [Ton_1] extensively and systematically developed the concept. One can consult Cesari's book [Ces_1] for a modern treatment and subsequent contributions.

It was soon realized, after Volterra's creation of the theory of functionals and its development into functional analysis, that this discipline would allow an elegant and general formulation of the direct method of calculus of variations. Theorem 1.1 illustrates this fact strikingly and is the result of successive refinements by a number of mathematicians among which Golomb [Gol_1], Mazur and Schauder [MaS_1], Morrey [Moy_1], Rothe [$\text{Rot}_{1,2,3}$], and Vainberg [$\text{Vai}_{1,2}$]. More complete references can be found in the survey papers and books of these last three authors.

Convexity is a very old concept and convex functions are present from the very beginning of the calculus. The systematic study of convex functions on $\mathbf{R}$ can be traced by Jensen [Jen_1] and detailed treatments of convex functions on $\mathbf{R}^N$ can be found in the books of Fenchel [Fen_1] and Rockafellar [Roc_1]. Convex sets were important in functional analysis, too, and Mazur's theorem was proved by Mazur [Maz_1] in 1933. The link between convexity and the direct method of the calculus of variations, already present in Tonelli's work, was very closely analyzed by Mazur and Schauder [MaS_1] in 1936.

The argument of Theorem 1.3 for obtaining the Euler equation can be traced, in special situations, to Euler [Eul_1], as early as 1771.

The fundamental lemma is a variant, for weak derivatives and periodic boundary conditions, of the du Bois-Reymond version of the fundamental lemma of the calculus of variations, generally stated and proved for more regular u and v and two-point boundary conditions. The pioneering rigorous approach of this lemma is that of du Bois-Reymond [DuB_1] in 1879, which can be considered as a very early contribution to the theory of distributions, containing in particular the first use of what we now call the "test functions."

The inequalities in Proposition 1.1 are generally referred as Sobolev inequalities and Wirtinger's inequality in Proposition 1.3 corresponds, for periodic functions, to Poincaré's inequality for Dirichlet boundary conditions. Results in the line of Theorem 1.4 are due to Morrey [Moy_1].

Although Poincaré [Poi_1] initiated the use of Jacobi's least action principle to the study of periodic solutions of a mechanical system with two degrees of freedom, and was followed by Whittaker, Signorini, Tonelli, and Birkhoff, the first treatment of a periodic boundary value problem for a non-autonomous second-order equation seems to be due to Lichtenstein [Lic_1] in 1915, where a problem of type (25) is considered under the assumption that F is bounded from below and $f \geq \alpha > 0$ for $u \geq R$ and $f \leq -\alpha < 0$ for $u \leq -R$; he mentions the example

$$u'' = u^{2n+1} + 1 + a_1(x)u^{2n} + \ldots + a_{2n+1}(x).$$

Condition (6) in Theorem 1.5 (coercivity on the kernel) was first introduced by Ahmad–Lazer–Paul [ALP_1] for Dirichlet problems and in the frame of a minimax method. Theorem 1.6 is due to Willem [Wil_1] and the method of proof, independently used by Dancer [Dan_1] in the case of the forced pendulum equation, had essentially been found already by Hamel [Ham_1] in 1922. Theorems 1.7, 1.8, and 1.9 are motivated by results of Mawhin–Willem [MaW_1] described in Chapter III. They generalize and improve earlier results of Berger–Schechter [BeS_1] and Gossez [Gos_1] obtained by other methods. For an abstract version of Theorems 1.7, 1.8, and 1.9, see Mawhin [Maw_1].

Further historical references on the Dirichlet principle are given in [Maw_2], as well in [Ber_2], [Bol_1], [Cou_1], [Fun_1] (which contains Hamel's treatment of the forced pendulum equation), [Had_1], [Ler_1], [$\text{McS}_{1,2}$], [San_1], [$\text{Vol}_{1,2}$], [VoP_1].

For recent contributions to convexity and semi-continuity questions in the calculus of variations, see [Ces_1], [$\text{Cla}_{3,4,5}$], and see [Bre_1], [EkT_1], [ScL_1] for elegant treatments of the foundations of convex analysis.

One can find a study of the forced periodic pendulum equation with further restrictions on A in [Cas_1] and an extension of Theorem 1.6 to a more general situation which covers the forced double pendulum in [CFS_1].

Other applications of the direct method to the existence of odd periodic solutions can be found in [Ber_5] and [Wil_{10}]. For a study of periodic solutions of autonomous systems as contrained minima, see [Ber$_6$], [CaM$_1$], and [Gor$_1$].

Exercises

1. If $\varphi : X \to]-\infty, \infty]$ is convex and bounded above by a real constant on a neighborhood of $a \in X$, then φ is continuous at a. Consequently, if X is complete, such a φ is continuous on int $D(\varphi)$ where $D(\varphi) = \{u \in X : \varphi(u) < \infty\}$ (effective domain of φ).

2. A function $\varphi : X \to]-\infty, \infty]$ is called strictly convex if

$$\varphi(1-\lambda)u + \lambda v < (1-\lambda)\varphi(u) + \lambda\varphi(v)$$

   whenever $u \neq v$ and $\lambda \in]0,1[$. Such a function achieves its infimum at one point at most.

3. If $\varphi : X \to]-\infty, \infty]$ is convex and l.s.c., then $\varphi(u) \to +\infty$, as $|u| \to \infty$, if and only if there exists $\alpha > 0$ and $\beta \geq 0$ such that

$$\varphi(u) \geq \alpha|u| - \beta$$

   for all $u \in X$.

4. The function $\varphi : X \to]-\infty, \infty]$ is l.s.c. if and only if $\varphi^c = \{u \in X : \varphi(u) \leq c\}$ is closed, for all $c \in \mathbf{R}$.

5. If the function $\varphi : X \to]-\infty, \infty]$ is convex; then for each $c \in \mathbf{R}$, the set $\varphi^c = \{u \in X : \varphi(u) \leq c\}$ is convex. Show by an example that φ^c can be convex for each $c \in \mathbf{R}$ without φ being convex.

6. If $\varphi : X \to]-\infty, \infty]$ is convex (resp. l.s.c.) the set of points at which φ achieves its infimum is convex (resp. closed).

7. If $\varphi : X \to]-\infty, \infty]$ is convex, each local minimum of φ is a global minimum of φ.

8. Under the assumptions and notations of Theorem 1.4, if $\varphi'(u) = 0$, then

$$\int_0^T D_x L(t, u(t), \dot{u}(t))\, dt = 0$$

   and there is some $c \in \mathbf{R}^N$ such that

$$D_y L(t, u(t), \dot{u}(t)) = \int_0^t D_x L(\tau, u(\tau), \dot{u}(\tau))\, d\tau + c$$

   a.e. on $[0, T]$.

9. Assume that F satisfies assumption (A) of Section 1.5 and that

$$F(t,x) \to +\infty$$

as $|x| \to \infty$ uniformly for a.e. $t \in [0,T]$. Show that the system

$$\ddot{u}(t) = \nabla F(t,u(t)),$$

$$u(0) - u(T) = \dot{u}(0) - \dot{u}(T) = 0,$$

has at least one solution which minimizes φ on H_T^1. ([BeS$_1$], [Ber$_2$]). Show that the same is true for

$$\ddot{u}(t) = \nabla F(t,u(t)) + e(t),$$

$$u(0) - u(T) = \dot{u}(0) - \dot{u}(T) = 0,$$

where $e \in L^1(0,T;\mathbf{R}^N)$ and $\int_0^T e(t)\,dt = 0$.

10. Let $g : \mathbf{R} \to \mathbf{R}$ be continuous, 2π-periodic, and such that $\int_0^{2\pi} g(u)\,du = 0$. Show that the problem

$$\ddot{u}(t) + g(t + u(t)) = 0,$$

$$u(0) - u(2\pi) = 0 = \dot{u}(0) - \dot{u}(2\pi),$$

has at least one solution ([Bat$_2$]) (and hence a continuum of solution as $u(t+c)+c$ is a solution for each $c \in \mathbf{R}$ when u is a solution).

11. Show that the equation of the compass in a rotating magnetic field

$$u''(t) + A\sin u(t) + B\sin(u(t) - t) = 0$$

always has a 2π-periodic solution.

12. Generalize Theorem 1.6 to the case where

$$\varphi(u) = \int_0^T \left[(1/2) \sum_{i,j=1}^{n} a_{ij}(u(t))\dot{u}_i(t)\dot{u}_j(t) - V(u(t)) - (u(t)|e(t)) \right] dt$$

where a_{ij} and V are in $C^1(\mathbf{R}^N, \mathbf{R})$, T_i-periodic in u_i for some $T_i > 0$ $(1 \le i \le N)$, and such that

$$\mu|\xi|^2 \le \sum_{i,j=1}^{n} a_{ij}(u)\xi_i\xi_j$$

for some $\mu > 0$ and all $\xi \in \mathbf{R}^N$, and $e \in L^1(0,T;\mathbf{R}^N)$ with $\int_0^T e(t)\,dt = 0$. ([CFS$_1$]). This applies in particular to the forced double-pendulum system.

13. Show that, in the conditions of Theorem 1.9, the set of solutions of (25) has the form $I + \tilde{u}_0$ where $\tilde{u}_0$ is a fixed T-periodic function with mean value zero and I is a closed interval (possibly empty). Relate the structure of I to the form of the function $a \to \int_0^T F(t,a)\,dt$. ([Maw$_1$]).

    Hint. Use exercise 1.6.

2

The Fenchel Transform and Duality

Introduction

The *Legendre transform* F^* of a function $F \in C^1(\mathbf{R}^N, \mathbf{R})$ is defined by the implicit formula

$$F^*(v) = (v, u) - F(u)$$
$$v = \nabla F(u)$$

when ∇F is invertible. It has the remarkable property that

$$\begin{aligned}\sum_{i=1}^{N} D_i F^*(v) dv_i &= dF^*(v) \\ &= \sum_{i=1}^{N} (v_i du_i + u_i dv_i - D_i F(u) du_i) = \sum_{i=1}^{N} u_i \, dv_i,\end{aligned}$$

or,

$$u = \nabla F^*(v),$$

so that F^* is such that

$$(\nabla F)^{-1} = \nabla F^*.$$

Its geometrical meaning is the following: the tangent hyperplane to the graph of F with normal $[v, -1]$ is given by

$$\{[w, s] \in \mathbf{R}^{N+1} \ : \ s = (w, v) - F^*(v)\}.$$

Thus, the graph of F can be described in a dual way, either as a set of points or as an envelope of tangent hyperplanes.

The *Fenchel transform* extends the Legendre transform to not necessarily smooth convex functions by using affine minorants instead of tangent hyperplanes. To motivate the analytical definition of the Fenchel transform of F we can notice that, when F is convex, the function $\tilde{F}_v \ : \ u \rightarrow (v, u) - F(u)$ is concave and the definition of the Legendre transform just expresses that u is a critical point of $\tilde{F}_v$, and hence the global maximum of $\tilde{F}_v$ is achieved at u. Consequently,

$$F^*(v) = \sup_{w \in \mathbf{R}^n} [(v, w) - F(w)]$$

and the right-hand member of this equality, which is defined as an element of $]-\infty, +\infty]$ without the smoothness and invertibility conditions required by the Legendre transform is, by definition, the Fenchel transform of the convex function F. The reciprocity property between ∇F and ∇F^*, which loses its meaning for a non-smooth convex F or a non-smooth F^*, can be recovered in terms of the *subdifferential* of a convex function G, i.e., a subset of $\mathbf{R}^N$ associated to G at u and which reduces to $\{\nabla G(u)\}$ when G is differentiable at u.

The role of the Legendre transform in classical Hamiltonian mechanics is well known. If the Lagrangian $L = L(t, q, r)$ is given, the corresponding Hamiltonian $H = H(t, q, p)$ is nothing but the Legendre transform of $L(t, q, .)$, namely

$$H(t, q, p) = (p, q) - L(t, q, r)$$

where r is expressed in terms of (t, q, p) through the relation

$$p = D_r L(t, q, r).$$

Besides this classical *Hamiltonian duality*, it is interesting to introduce, in the study of Hamiltonian systems, another duality based on the Legendre transform of $H(t, ., .)$. Indeed, if we write $u = (q, p)$, the Hamiltonian equations can be written in the compact form

$$-J\dot{u}(t) = \nabla H(t, u(t))$$

where

$$J = \begin{pmatrix} 0_N & I_N \\ -I_N & 0_N \end{pmatrix}$$

is the symplectic matrix. Setting

$$\dot{v} = -J\dot{u},$$

so that

$$u = Jv - c$$

where c is a constant, we obtain

$$\dot{v} = \nabla H(t, u)$$

or equivalently

$$u = \nabla H^*(t, \dot{v})$$

if the Legendre transform $H^*(t, .)$ of $H(t, .)$ exists. Therefore, our Hamiltonian equations expressed in terms of v become

$$Jv - \nabla H^*(t, \dot{v}) = c,$$

the integrated Euler–Lagrange equations corresponding to the critical points of the function χ defined on a suitable space of T-periodic functions by

$$\chi(v) = \int_0^T [(1/2)(J\dot{v}(t), v(t)) + H^*(t, \dot{v}(t))]\, dt.$$

This *dual action* χ can, therefore, be used as well as the Hamiltonian action

$$\psi(u) = \int_0^T [(1/2)(J\dot{u}(t), u(t)) + H(t, u(t))]\, dt$$

to prove the existence of T-periodic solutions of our Hamiltonian system because the critical points of χ are, in many situations, more easy to find than those of ψ. This observation is at the basis of the use of this *Clarke duality* in the study of Hamiltonian systems.

2.1 Definition of the Fenchel Transform

Let us first recall a basic tool in convex analysis.

Separation Theorem. *Let C and D be nonempty disjoint convex subsets of a normed vector space V. If C is closed and D is compact, there exists a closed affine hyperplane P which strictly separates C and D, i.e., there exists $l \in V^*$ and $\alpha \in \mathbf{R}$ such that*

$$l(u) > \alpha \quad \text{if } u \in C \quad \text{and} \quad l(u) < \alpha \quad \text{if } u \in D.$$

Let us recall that the epigraph of a function $F : V \to]-\infty, \infty]$, with V a normed vector space, is the set

$$\operatorname{epi} F = \{[u, t] \in V \times \mathbf{R} : F(u) \leq t\}.$$

The easy proof of the following lemma is left to the reader.

Lemma 2.1. *The function $F : V \to]-\infty, \infty]$ is convex (resp. l.s.c.) if and only if* epi F *is convex (resp. closed).*

We shall now show that a convex l.s.c. function $F : V \to]-\infty, \infty]$ can be entirely characterized by the affine functions that F dominates.

Lemma 2.2. *Let $F : V \to]-\infty, +\infty]$. The following statements are equivalent.*

a) *F is convex and l.s.c.*

b) *F is the supremum of all the continuous affine functions which are everywhere smaller than F.*

Proof. b) $\Rightarrow$ a). If b) holds, F is convex, and l.s.c. as the supremum of convex continuous functions.

a) $\Rightarrow$ b). If $F \equiv \infty$, then b) is satisfied. Assume now that F is finite at some u_0, so that epi $F \neq \phi$. To prove that a) $\Rightarrow$ b) we must show that for each $w \in V$ and $t < F(w)$ we can find an affine continuous function G such that $G(w) \geq t$ and $G \leq F$ on V. Let $t < F(w)$, so that $(w,t) \notin \operatorname{epi} F$. Since, by Lemma 2.1, epi F is closed and convex, the separation theorem implies the existence of $v \in V^*$, $c \in \mathbf{R}$, and $d \in \mathbf{R}$ such that

$$v(w) + ct < d < v(u) + cs \tag{1}$$

whenever $(u,s) \in \operatorname{epi} F$. Since $s \geq F(u_0)$ implies $(u_0, s) \in \operatorname{epi} F$ and thus

$$d - v(u_0) < cs,$$

the function $s \to sc$ is bounded below for $s \geq F(u_0)$ so that, necessarily, $c \geq 0$.

Assume first that $c > 0$. If we define G by

$$G(u) = c^{-1} v(w - u) + t,$$

then G is affine, $G(w) = t$ and, by (1), $G(u) < s$ if $s \geq F(u)$ so that $G \leq F$ on V. Assume now that $c = 0$. It follows from (1) that

$$v(w) < d < v(u) \tag{2}$$

when $F(u) < +\infty$. Thus $F(w) = +\infty$. Since $F(u_0)$ is finite, the preceding part of the proof, applied to u_0 and $F(u_0) - 1$, implies the existence of a continuous affine function G such that $G \leq F$ on V. Define, for $\lambda > 0$, the affine continuous function G_λ by

$$G_\lambda(u) = G(u) + \lambda(d - v(u)).$$

It follows from (2) that $F \geq G_\lambda$ and that, for sufficiently large λ, $G_\lambda(w) \geq t$. □

We shall denote by $\Gamma_0(\mathbf{R}^N)$ the set of all convex l.s.c. functions $F : \mathbf{R}^N \to]-\infty, +\infty]$ whose effective domain $D(F) = \{u \in \mathbf{R}^N : F(u) < +\infty\}$ is non-empty.

The *Fenchel transform* F^* of a function $F \in \Gamma_0(\mathbf{R}^N)$ is the function $F^* : \mathbf{R}^N \to]-\infty, +\infty]$ defined by

$$F^*(v) = \sup_{u \in D(F)} ((v,u) - F(u)).$$

Remarks. 1. The continuous affine function $(v, .) - \alpha$ is everywhere less than F if and only if

$$\alpha \geq (v,u) - F(u)$$

for all $u \in \mathbf{R}^N$, i.e. if and only if

$$\alpha \geq F^*(v).$$

2. It follows from the definition that F^* is convex and lower semi-continuous. On the other hand, by Lemma 2.2, there exists $(v, \alpha) \in \mathbf{R}^N \times \mathbf{R}$ such that

$$F \geq (v, .) - \alpha.$$

Thus, by definition,

$$F^*(v) \leq \alpha$$

and $D(F^*) \neq \phi$, which shows that $F^* \in \Gamma_0(\mathbf{R}^N)$.

3. An immediate consequence of the definition is the *Fenchel inequality*

$$F(u) + F^*(v) \geq (v, u)$$

for all $u \in \mathbf{R}^N$ and $v \in \mathbf{R}^N$.

4. Another immediate consequence of the definition is that if $F_1 \in \Gamma_0(\mathbf{R}^N)$, $F_2 \in \Gamma_0(\mathbf{R}^N)$, and $F_1 \leq F_2$, then

$$F_1^* \geq F_2^*. \tag{3}$$

Theorem 2.1. *If* $F \in \Gamma_0(\mathbf{R}^N)$, *then* $(F^*)^* = F$.

Proof. By Lemma 2.2 and Remark 2.1, we have, for each $u \in \mathbf{R}^N$,

$$\begin{aligned} F(u) &= \sup_{\substack{[v,\alpha] \in \mathbf{R}^N \times \mathbf{R} \\ (v,.)-\alpha \leq F}} ((v,u) - \alpha) = \sup_{\substack{v \in \mathbf{R}^N \\ \alpha \geq F^*(v)}} ((v,u) - \alpha) \\ &= \sup_{v \in D(F^*)} ((v,u) - F^*(v)) = (F^*)^*(u). \qquad \square \end{aligned}$$

We define the *subdifferential* of a function $F \in \Gamma_0(\mathbf{R}^N)$ at a point $u \in \mathbf{R}^N$ to be the set

$$\partial F(u) = \{v \in \mathbf{R}^N \; : \; F(w) \geq F(u) + (v, w - u) \text{ for all } w \in \mathbf{R}^N\}.$$

We shall say that F is *subdifferentiable at* u if $\partial F(u) \neq \phi$.

Remarks. 1. F is subdifferentiable at u if and only if $u \in D(F)$ and there is an affine continuous function everywhere less than F and *equal* to $F(u)$ at u.

2. $F(u) = \inf F$ if and only if $0 \in \partial F(u)$.

3. If $v_i \in \partial F(u_i)$ $(i = 1, 2)$, then

$$F(u_2) \geq F(u_1) + (v_1, u_2 - u_1)$$

$$F(u_1) \geq F(u_2) + (v_2, u_1 - u_2)$$

so that

$$0 \geq (v_2 - v_1, u_1 - u_2),$$

i.e.

$$(v_1 - v_2, u_1 - u_2) \geq 0,$$

which we express by saying that $\partial F(u)$ is a monotone multivalued mapping.

4. It is easy to check that $\partial F(u)$ is closed and convex.

We now state and prove a fundamental property of the Fenchel transforms which is the basis of the duality method in optimization.

Theorem 2.2. *If $F \in \Gamma_0(\mathbf{R}^N)$, the following statements are equivalent*

a. $v \in \partial F(u)$

b. $F(u) + F^*(v) = (v, u)$

c. $u \in \partial F^*(v)$.

Proof. By definition

$$\begin{aligned} v \in \partial F(u) \quad &\Leftrightarrow \quad (v,u) - F(u) \geq (v,w) - F(w) \text{ for all } w \in \mathbf{R}^N \\ &\Leftrightarrow \quad (v,u) - F(u) = \sup_{v \in \mathbf{R}^N} ((v,w) - F(w)) \\ &\Leftrightarrow \quad (v,u) - F(u) = F^*(v) \end{aligned}$$

so that (a) $\Leftrightarrow$ (b). By Theorem 2.1 and the first equivalence

$$F(u) + F^*(v) = (v,u) \Leftrightarrow (F^*)^*(u) + F^*(v) = (v,u)$$

$$\Leftrightarrow \; F^*(v) + (F^*)^*(u) = (u,v) \Leftrightarrow u \in \partial F^*(v)$$

so that (b) $\Leftrightarrow$ (c), and the proof is complete. □

Proposition 2.1. *If $F \in \Gamma_0(\mathbf{R}^N)$, the graph $\{[u,v] \in \mathbf{R}^N \times \mathbf{R}^N : v \in \partial F(u)\}$ of ∂F is closed.*

Proof. Let $([u_k, v_k])$ be a sequence in ∂F such that $u_k \to u$ and $v_k \to v$ as $k \to \infty$. For every $w \in \mathbf{R}^N$, we have

$$F(w) \geq F(u_k) + (v_k, w - u_k), \quad k \in \mathbf{N}^*.$$

and hence, by lower semi-continuity

$$F(w) \geq \liminf_{k \to \infty} +(v, w-u) \geq F(u) + (v, w-u).$$

Thus, $v \in \partial F(u)$ and the proof is complete. □

Example. Let $G : \mathbf{R}^N \to \mathbf{R}$ be defined by

$$G(u) = \alpha q^{-1} |u|^q + \gamma$$

where $\alpha > 0$, $q > 1$, $\gamma \in \mathbf{R}$. Then

$$G^*(v) = \sup_{u \in \mathbf{R}^N} ((v,u) - \alpha q^{-1}|u|^q - \gamma) = \alpha^{-p/q} p^{-1}|v|^p - \gamma, \tag{4}$$

where $q^{-1} + p^{-1} = 1$.

Proposition 2.2. *Let $F \in \Gamma_0(\mathbf{R}^N)$ be such that, for some $\alpha > 0$, $q > 1$, $\beta \geq 0$, $\gamma \geq 0$, one has*

$$-\beta \leq F(u) \leq \alpha q^{-1}|u|^q + \gamma \tag{5}$$

whenever $u \in \mathbf{R}^N$. Then, if $v \in \partial F(u)$, one has

$$\alpha^{-p/q} p^{-1}|v|^p \leq (v,u) + \beta + \gamma \tag{6}$$

and

$$|v| \leq \{p\alpha^{p/q}[|u| + \beta + \gamma] + 1\}^{q-1}. \tag{7}$$

Proof. By Theorem 2.2, $v \in \partial F(u) \Leftrightarrow F^*(v) = (v,u) - F(u)$ and hence, by (3), (4), and (5),

$$\alpha^{-p/q} p^{-1}|v|^p - \gamma \leq F^*(v) \leq (v,u) + \beta$$

which directly gives (6). If $|v| \leq 1$, (7) is obvious. If we now assume that $|v| \geq 1$, then, by (6),

$$|v|^{p-1} \leq \alpha^{p/q} p[|u| + \beta + \gamma]$$

and the proof is complete. □

2.2 Differentiable Convex Functions

We shall study the regularity of the Fenchel transform of a convex function.

Proposition 2.3. *If $F : \mathbf{R}^N \to \mathbf{R}$ is convex and differentiable at u, then $\partial F(u) = \{\nabla F(u)\}$.*

Proof. By the convexity of F and Proposition 1.2, $\nabla F(u) \in \partial F(u)$. Now, if $v \in \partial F(u)$, then

$$F(w) - (v,w) \geq F(u) - (v,u)$$

for all $w \in \mathbf{R}^N$, i.e., $F(.) - (v,.)$ has a minimum at u. As F is differentiable, this implies that

$$\nabla F(u) - v = 0$$

and the proof is complete. □

Recall that a function $F : \mathbf{R}^N \to [-\infty, +\infty[$ is *strictly concave* if $-F$ is strictly convex.

Proposition 2.4. *If $F \in \Gamma_0(\mathbf{R}^N)$ is strictly convex and such that*

$$F(u)/|u| \to +\infty \tag{8}$$

if $|u| \to \infty$, then $F^* \in C^1(\mathbf{R}^N, \mathbf{R})$.

Proof. Without loss of generality, we can assume that $0 \in D(F)$.

1. By assumption, for $v \in \mathbf{R}^N$ fixed, the function G_v defined by

$$G_v(w) = (v, w) - F(w)$$

is strictly concave and $G_v(w) \to -\infty$ as $|w| \to \infty$ by (8). Thus, G_v has exactly one maximum point u. By Theorem 2.2,

$$\partial F^*(v) = \{u\}.$$

2. Let us show that $\partial F^* : \mathbf{R}^N \to \mathbf{R}^N$, $v \to u$ where u is such that $\partial F^*(v) = \{u\}$, is continuous. By Proposition 2.1, the graph of ∂F^* is closed, hence it suffices to prove that ∂F^* takes bounded sets into bounded sets. Let $|v| \le \rho$ for some ρ and $\{u\} = \partial F^*(v)$; then $v \in \partial F(u)$ hence

$$F(0) \ge F(u) - (v, u) \tag{9}$$

so that, by the Cauchy–Schwarz inequality

$$\rho \ge |v| \ge (F(u) - F(0))/|u|.$$

Relations (8) and (9) imply that $|u|$ is bounded.

3. Finally, if $\{u\} = \partial F^*(v)$ and $\{u_h\} = \partial F^*(v + h)$ for some $v \in \mathbf{R}^N$, $h \in \mathbf{R}^N \setminus \{0\}$, then, by definition of the subdifferential, we have

$$0 \le \frac{F^*(v+h) - F^*(v) - (h, u)}{|h|} \le \frac{(h, u_h - u)}{|h|} \le |u_h - u|.$$

By the continuity of ∂F^*, $u_h \to u$ if $h \to 0$ and F^* is differentiable at v with $\{\nabla F^*(v)\} = \{u\} = \partial F^*(v)$. Hence $F^* \in C^1(\mathbf{R}^N, \mathbf{R})$. □

2.3 Hamiltonian Duality

Let $L : [0, T] \times \mathbf{R}^N \times \mathbf{R}^N \to \mathbf{R}$, $(t, x, y) \to L(t, x, y)$ be a smooth function such that, for each $(t, x) \in [0, T] \times \mathbf{R}^N$, $L(t, x, .)$ satisfies the assumptions of Proposition 2.4. The Fenchel (or Legendre) transform $H(t, x, .)$ of $L(t, x, .)$ is defined by

$$H(t, x, z) = \sup_{y \in \mathbf{R}^N} [(z, y) - L(t, x, y)]$$

or

$$\begin{aligned} H(t,x,z) &= (z,y) - L(t,x,y) \\ z = D_y L(t,x,y), &\quad y = D_z H(t,x,z). \end{aligned} \tag{10}$$

The Lagrangian action is defined on a suitable space of T-periodic functions by

$$\varphi(q) = \int_0^T L(t,q(t),\dot{q}(t))\,dt.$$

The corresponding Euler equations are

$$\frac{d}{dt}[D_y L(t,q(t),\dot{q}(t))] = D_x L(t,q(t),\dot{q}(t)). \tag{11}$$

Formula (10) suggests replacing $L(t,q,\dot{q})$ in φ by $(\dot{q},p) - H(t,q,p)$. The Hamiltonian action is thus defined on a suitable space of T-periodic functions by

$$\psi(q,p) = \int_0^T [(\dot{q}(t),p(t)) - H(t,q(t),p(t))]\,dt.$$

The corresponding Euler equations are the Hamilton equations

$$\dot{q}(t) = D_z H(t,q(t),p(t)) \tag{12}$$

$$\dot{p}(t) = -D_x H(t,q(t),p(t)). \tag{13}$$

By duality, (12) is equivalent to

$$p(t) = D_y L(t,q(t),\dot{q}(t)). \tag{14}$$

We obtain, at least formally, from (10)

$$D_x H = -D_x L - (D_x y)(D_y L) + (D_x y)(z) = -D_x L,$$

so that (13) is equivalent to

$$\dot{p}(t) = D_x L(t,q(t),\dot{q}(t)). \tag{15}$$

The Euler equation follows directly from (14) and (15). Let J be the symplectic matrix, so that $J^2 = -J$ and $(Ju,v) = -(u,Jv)$ for all $u,v \in \mathbf{R}^{2N}$. If $u = [q,p]$, the system (12)–(13) becomes

$$\dot{u}(t) = J\nabla H(t,u(t)) \tag{16}$$

or

$$J\dot{u}(t) + \nabla H(t,u(t)) = 0,$$

where ∇H denotes the gradient of H with respect to u. We have, by T-periodicity

$$\int_0^T (\dot{q}(t),p(t))\,dt = \frac{1}{2}\int_0^T \left[(\dot{q}(t),p(t)) + \frac{d}{dt}(q(t),p(t)) - (q(t),\dot{p}(t))\right] dt$$

$$= -\frac{1}{2}\int_0^T [(\dot{p}(t), q(t)) + (-\dot{q}(t), p(t))]\, dt = -\frac{1}{2}\int_0^T (J\dot{u}(t), u(t))\, dt.$$

Consequently, the Hamiltonian action ψ can be written

$$\psi(u) = -\int_0^T \left[\frac{1}{2}(J\dot{u}(t), u(t)) + H(t, u(t))\right] dt.$$

Now the quadratic form

$$u \to -\frac{1}{2}\int_0^T (J\dot{u}(t), u(t))\, dt$$

is "strongly indefinite" (see Section 3.1) and "dominates" the functional

$$u \to -\int_0^T H(t, u(t))\, dt.$$

Consequently, the function ψ will be neither bounded from above nor from below and critical points will be rather difficult to obtain as global minimums or maximums do not exist.

2.4 Clarke Duality

Let $H : [0,T] \times \mathbf{R}^{2N} \to \mathbf{R}$, $(t,u) \to H(t,u)$ be a smooth Hamiltonian such that, for each $t \in [0,T]$, $H(t,.)$ satisfies the assumptions of Proposition 2.4. The Fenchel (or Legendre) transform $H^*(t,.)$ of $H(t,.)$ is defined by

$$H^*(t,v) = \sup_{u \in \mathbf{R}^{2N}} [(v,u) - H(t,u)]$$

or

$$\begin{aligned} H^*(t,v) &= (v,u) - H(t,u) \\ v = \nabla H(t,u), &\quad u = \nabla H^*(t,v). \end{aligned} \tag{17}$$

If

$$v = -Ju \quad \text{or} \quad u = Jv,$$

we obtain

$$\begin{aligned} \psi(u) &= \int_0^T \left[\frac{1}{2}(\dot{v}(t), u(t)) - H(t, u(t))\right] dt \\ &= \int_0^T \left[-\frac{1}{2}(\dot{v}(t), u(t)) + (\dot{v}(t), u(t)) - H(t, u(t))\right] dt \\ &= \int_0^T \left[\frac{1}{2}(J\dot{v}(t), v(t)) + (\dot{v}(t), u(t)) - H(t, u(t))\right] dt. \end{aligned}$$

Formula (17) suggests replacing $(\dot{v}, u) - H(t,u)$ by $H^*(t,\dot{v})$. The dual action is thus defined on a suitable space of T-periodic functions by

$$\chi(v) = \int_0^T \left[\frac{1}{2}(J\dot{v}(t), v(t)) + H^*(t,\dot{v}(t))\right] dt.$$

We shall see in the next chapter that χ can be bounded below under reasonable assumptions upon H.

Another useful property of the dual action is that

$$\chi(v+c) = \chi(v)$$

for all $c \in \mathbf{R}^{2N}$. Thus, it suffices to find critical points of χ restricted to the space

$$\tilde{W}_T^{1,p} = \left\{ v \in W_T^{1,p} : \int_0^T v(t)\,dt = 0 \right\}.$$

Theorem 2.3. *Let*

$$H : [0,T] \times \mathbf{R}^{2N} \to \mathbf{R}, \quad (t,u) \to H(t,u)$$

be measurable in t for each $u \in \mathbf{R}^{2N}$ and strictly convex and continuously differentiable in u for almost every $t \in [0,T]$. Assume that there exists $q \in]1,+\infty[$, $\alpha > 0$, $\delta > 0$, $\beta, \gamma \in L^p(0,T;\mathbf{R}^+)$, with $\frac{1}{p} + \frac{1}{q} = 1$, such that, for all $u \in \mathbf{R}^{2N}$ and a.e. $t \in [0,T]$, one has

$$\delta(|u|^q/q) - \beta(t) \le H(t,u) \le \alpha(|u|^q/q) + \gamma(t). \tag{18}$$

Then the dual action χ is a continuously differentiable on $\tilde{W}_T^{1,p}$ and, if $v \in \tilde{W}_T^{1,p}$ is a critical point of χ, the function u defined by

$$u(t) = \nabla H^*(t,\dot{v}(t))$$

satisfies (16) *and* $u(0) = u(T)$.

Proof. It follows directly from Proposition 2.4 that $H^*(t,u)$ is continuously differentiable in u for a.e. $t \in [0,T]$. By assumption (18) and relation (4), we obtain, for all $u \in \mathbf{R}^{2N}$ and a.e. $t \in [0,T]$,

$$\alpha^{-p/q}(|v|^p/p) - \gamma(t) \le H^*(t,v) \le \delta^{-p/q}(|v|^p/p) + \beta(t). \tag{19}$$

Proposition 2.2 implies that

$$\begin{aligned} |\nabla H^*(t,v)| &\le [(q/\delta)(|v| + \beta(t) + \gamma(t)) + 1]^{p-1} \\ &\le c_1|v|^{p-1} + c_2(\beta(t) + \gamma(t) + 1)^{p-1}, \end{aligned} \tag{20}$$

for some positive constants c_1 and c_2. Let us note that $(\beta+\gamma+1)^{p-1} \in L^q$ since $(\beta+\gamma+1) \in L^p$. By (19) and (20), the function L defined by

$$L(t,x,y) = (1/2)(Jy,x) + H^*(t,y)$$

satisfies the assumptions of Theorem 1.4. Consequently, the dual action is continuously differentiable on $W_T^{1,p}$, and, hence, on $\tilde{W}_T^{1,p}$.

Finally, if $v \in \tilde{W}_T^{1,p}$ is a critical point of χ, Theorem 1.4 implies that, for all $h \in \tilde{W}_T^{1,p}$, one has

$$0 = \int_0^T \left[\frac{1}{2}(J\dot{v}(t), h(t)) + (\nabla H^*(t, \dot{v}(t)) - \frac{1}{2}Jv(t), \dot{h}(t))\right] dt. \tag{21}$$

It is then easy to verify the preceding relation for all $h \in W_T^{1,p}$, and hence for all $h \in C_T^\infty$. By (21), the fundamental lemma is applicable, so that

$$\nabla H^*(t, \dot{v}(t)) - (1/2)Jv(t) = \int_0^t (1/2)J\dot{v}(s)\, ds + c$$

a.e. on $[0, T]$, i.e.,

$$Jv(t) = \nabla H^*(t, \dot{v}(t)) + \tilde{c}$$

a.e. on $[0, T]$. Setting

$$u(t) = \nabla H^*(t, \dot{v}(t)) = Jv(t) - \tilde{c},$$

we obtain $u \in W_T^{1,p}$, $\dot{u} = J\dot{v}$ and, by duality

$$\dot{v}(t) = \nabla H(t, u(t)).$$

Thus,

$$\dot{u}(t) = J\dot{v}(t) = J\nabla H(t, u(t))$$

a.e. on $[0, T]$. Moreover, $u(0) = u(T)$ since $u \in W_T^{1,p}$. □

Historical and Bibliographical Notes

The Fenchel transform first appeared in 1939 for convex functions on **R** in a paper of S. Mandelbrojt [Man$_1$], which motivated an improved and more general definition by Fenchel [Fen$_2$] in 1949 for convex functions in $\mathbf{R}^N$. A special case of the Fenchel inequality (in **R**) was already given and used by W.H. Young [You$_1$] in 1917. The Fenchel transform is an extension of the Legendre transform [Leg$_1$] introduced in 1787. The Fenchel transform was extended to topological vector spaces by Bronsted [Bro$_1$], Moreau [Mor$_{1,2,3}$], and Rockafellar [Roc$_{1,2,3}$], as well as the concept of subdifferential.

The Hamiltonian duality is basic in analytic mechanics and in the calculus of variations. The Hamilton's equations appear for the first time in a paper of Lagrange (1809) on perturbation theory, but it was Cauchy (1831) who first gave the true significance of those equations. In 1835, Hamilton put those equations at the basis of his analytical mechanics and gave the first exact formulation of the least action principle. That the Hamiltonian with periodic boundary conditions is indefinite was already noticed by Birkhoff [Bir$_1$], who developed the first minimax approach to handle it.

The Clarke duality was introduced in 1978 by Clarke [Cla$_1$] and developed by Clarke–Ekeland [Cla$_{1,2}$, ClE$_{1,2}$, Eke$_{1,2,3}$] to overcome the above mentioned difficulty by replacing the Hamiltonian action by a dual action which can be bounded from below. A heuristic exposition of duality in the calculus of variations is given in Courant–Hilbert [CoH$_1$].

See also [You$_2$], [EkT$_1$], [EkTu$_1$], [Fen$_3$], [Wil$_5$] for various aspects of the role of convexity and duality in the calculus of variations. Duality methods for first and second order evolution equations are developed in [AuE$_2$], [BrE$_1$], and [BrE$_2$]. Non-convex optimization problems are considered, using duality, in [Tol$_{1,2}$], [Eke$_{9,10}$], [Mas$_1$]. On the relations between optimization and periodic orbits, see also [Cla$_{7,8,9}$].

Exercises

1. Let $F \in \Gamma_0(\mathbf{R}^N)$, $u_0 \in \mathbf{R}^N$, $\lambda \in \mathbf{R}$. If

$$\begin{aligned} G(u) &= F(u) + \lambda, && \text{then } G^*(v) = F^*(v) - \lambda; \\ G(u) &= F(u - u_0), && \text{then } G^*(v) = F^*(v) + (u_0, v); \\ G(u) &= F(u) + (u, u_0), && \text{then } G^*(v) = F^*(u - u_0). \end{aligned}$$

2. A function $F \in \Gamma_0(\mathbf{R}^N)$ has a global minimum at u if and only if $u \in \partial F^*(0)$, in which case

$$\min_{\mathbf{R}^N} F = -F^*(0).$$

3. Let (v_n) be a sequence in $\mathbf{R}^N$, $v \in \mathbf{R}^N$, (F_n) be a sequence in $\Gamma_0(\mathbf{R}^N)$ and $F \in \Gamma_0(\mathbf{R}^N)$. If

$$v_n \longrightarrow v$$

and

$$\underline{\lim}\, F_n(u) \geq F(u) \quad \text{for each} \quad u \in \mathbf{R}^N,$$

then

$$F^*(v) \leq \underline{\lim}\, F_n^*(v_n).$$

4. If $F : \mathbf{R}^N \to \mathbf{R}^N$ is convex and differentiable, then

$$F(w) = \sup_{u \in \mathbf{R}^N} [F(u) + (\nabla F(u), w - u)].$$

5. Let $F \in \Gamma_0(\mathbf{R}^N)$ be such that $F(u) \geq a|u| - \beta$ for all $u \in \mathbf{R}^N$ and some $\alpha > 0$, $\beta \geq 0$. Then

$$F^*(v) \leq \beta \quad \text{whenever} \quad |v| \leq \alpha.$$

6. Compute F^* if $F(u) = (\ell, u) - \alpha$ for some $\ell \in \mathbf{R}^N$ and $\alpha \in \mathbf{R}$.

7. Let $F : \mathbf{R}^N \to]-\infty, +\infty]$ be convex and $C \subset \mathbf{R}^N$ a convex set. Assume that F is finite and continuous at $u_0 \in C$ or finite at $u_0 \in \operatorname{int} C$. Then u minimizes F on C if and only if there is some $v \in \partial F(u)$ such that
$$(v, w - u) \geq 0 \text{ for all } w \in C.$$

8. If $F \in \Gamma_0(\mathbf{R}^N)$ is continuous at $u_0 \in D(F)$, then $\partial F(u_0)$ is compact.

3

Minimization of the Dual Action

Introduction

A basic problem in mechanics (classical and celestial) is the study of the periodic solutions of Hamiltonian systems

$$J\dot{u}(t) + \nabla H(t, u(t)) = 0.$$

Although the variational structure of the problem suggests that the best results should be obtained through a variational approach, progress in this direction has been rather slow. This is due to the fact that the associated Hamiltonian action ψ given by

$$\psi(u) = \int_0^T [(1/2)(J\dot{u}(t), u(t)) + H(t, u(t))]\, dt$$

is indefinite. This is easily shown by substituting

$$u_k(t) = (\cos \lambda_k t)c - (\sin \lambda_k t)Jc$$

with $\lambda_k = 2k\pi/T$, $k \in \mathbf{Z}$, $c \in \mathbf{R}^{2N}$, $|c| = 1$, so that $|u_k(t)| = 1$ and $(J\dot{u}_k(t), u_k(t)) = \lambda_k$ for all $t \in \mathbf{R}$ and $k \in \mathbf{Z}$. Consequently,

$$\psi(u_k) = k\pi + \int_0^T H(t, u_k(t))\, dt \to +\infty \text{ or } -\infty$$

according to $k \to \infty$ or $-\infty$. Therefore, the direct method of the calculus of variations cannot be applied in a straightforward way and more sophisticated approaches like minimax methods, isoperimetric natural constraints, or dual least action principles have to be used.

In this chapter, we shall concentrate on situations where the Hamiltonian $H(t, u)$ is convex in u, in which case the dual least action principle seems to provide the best results in the simplest way. We shall base the various type of existence results (which deal with subharmonic solutions, periodic solutions of autonomous systems with fixed period or with fixed energy as shown in Sections 3.3 to 3.5) on a single basic existence theorem given in Section 3.2. This theorem only requires a suitable quadratic growth restriction on $H(t,.)$ and a coercivity condition on $\mathbf{R}^{2N}$ for the averaged Hamiltonian

$$T^{-1}\int_0^T H(t,v)\,dt.$$

The special case of second order systems, which is particularly important for the applications, deserves a special study made in Sections 3.6 to 3.8 where, in particular, nonlinear extensions of linear problems of the type

$$\ddot{u}(t) = \lambda u - h(t) \quad (\lambda < 0)$$

$$u(0) - u(T) = \dot{u}(0) - \dot{u}(T) = 0$$

are given, which complete the study initiated in Chapter 1 for $\lambda \geq 0$. For example, some necessary and sufficient conditions for the solvability of the scalar problem

$$\ddot{u}(t) = g(u(t)) - h(t)$$

$$u(0) - u(T) = \dot{u}(0) - \dot{u}(T) = 0,$$

with g continuous and non-increasing, are given and connected to the Landesman–Lazer conditions.

3.1 Eigenvalues and Eigenfunctions of $J(d/dt)$ with Periodic Boundary Conditions

Before going to nonlinear problems, it is of interest to discuss the simple linear periodic problem

$$J\dot{u}(t) = \lambda u(t), \quad u(0) = u(T) \tag{1}$$

where $\lambda \in \mathbf{R}$. The differential equation in (1) is equivalent to

$$\dot{u}(t) = -\lambda J u(t)$$

and its solutions are of the form

$$u(t) = \exp(-\lambda t J)c$$

with arbitrary $c \in \mathbf{R}^{2N}$. Now, as $J^2 = -I$,

$$\begin{aligned}
\exp(-\lambda t J) &= \sum_{k=0}^{\infty}(-1)^k(\lambda t)^k J^k k! \\
&= \sum_{k=0}^{\infty}(\lambda t)^{2k}(-1)^k I(2k)! + \sum_{k=0}^{\infty}(-1)(\lambda t)^{2k+1}(-1)^k J(2k+1)! \\
&= (\cos \lambda t)I - \sin(\lambda t)\, J
\end{aligned}$$

hence

$$u(t) = (\cos \lambda t)c - (\sin \lambda t)Jc.$$

This solution will satisfy the T-periodicity condition if and only if c satisfies the equation

$$[(1 - \cos \lambda T)I + (\sin \lambda T)J]c = 0. \tag{2}$$

Taking the inner product with c and with Jc, we obtain, since $(Jc, c) = 0$,

$$(1 - \cos \lambda T)|c|^2 = (\sin \lambda T)|Jc|^2 = 0.$$

Thus (2) has a nontrivial solution if and only if

$$\lambda = \lambda_k \equiv 2k\pi/T, \quad k \in \mathbf{Z}.$$

Now, if $\lambda = \lambda_k$, equation (2) becomes

$$Oc = 0$$

so that $c \in \mathbf{R}^{2N}$ is arbitrary. We have, therefore, proved the following.

Proposition 3.1. *The periodic eigenvalue problem* (1) *has a nontrivial solution if and only if*

$$\lambda = \lambda_k \equiv 2k\pi/T$$

for some $k \in \mathbf{Z}$, *in which case* (1) *possesses the* $2N$*-dimensional vector space of solutions*

$$u(t) = (\cos \lambda_k t)c - (\sin \lambda_k t)\, Jc \tag{3}$$

where $c \in \mathbf{R}^{2N}$ *is arbitrary.*

As the set of eigenvalues $\{\lambda_k : k \in \mathbf{Z}\}$ is unbounded from below and from above, the quadratic form

$$u \to \frac{1}{2}\int_0^T (J\dot{u}(t), u(t))\, dt$$

will be indefinite on the space

$$H_T^1 = \{u : [0,T] \to \mathbf{R}^{2N} : u \text{ is absolutely continuous, } u(0) = u(T) \text{ and } \dot{u} \in L^2(0,T)\}.$$

Indeed, with $u(t)$ defined in (3),

$$\frac{1}{2}\int_0^T (J\dot{u}(t), u(t))\, dt = \frac{\lambda_k}{2}\int_0^T |u(t)|^2 dt = \frac{\lambda_k}{2} T\,|c|^2 = k\pi\,|c|^2.$$

The following estimate is useful.

Proposition 3.2. *For every* $u \in H^1_T$,

$$\int_0^T (J\dot{u}(t), u(t))\, dt \geq -\frac{T}{2\pi}\int_0^T |\dot{u}(t)|^2 dt.$$

Proof. Let us write $\tilde{u} = u(t) - \frac{1}{T}\int_0^T u(s)\, ds$. Cauchy–Schwarz and Wirtinger inequalities imply that

$$\begin{aligned}
\int_0^T (J\dot{u}(t), u(t))\, dt &= \int_0^T (J\dot{u}(t), \tilde{u}(t))\, dt \\
&\geq -\left(\int_0^T |J\dot{u}(t)|^2 dt\right)^{1/2}\left(\int_0^T |\tilde{u}(t)|^2 dt\right)^{1/2} \\
&\geq -\frac{T}{2\pi}\left(\int_0^T |J\dot{u}(t)|^2 dt\right)^{1/2}\left(\int_0^T |\dot{\tilde{u}}(t)|^2 dt\right)^{1/2} \\
&= -\frac{T}{2\pi}\int_0^T |\dot{u}(t)|^2 dt. \quad \square
\end{aligned}$$

3.2 A Basic Existence Theorem for Periodic Solutions of Convex Hamiltonian Systems

We consider the periodic boundary value problem

$$\begin{gathered}
J\dot{u}(t) + \nabla H(t, u(t)) = 0 \quad \text{a.e. on} \quad [0, T] \\
u(0) = u(T)
\end{gathered} \tag{4}$$

where $H : [0, T] \times \mathbf{R}^{2N} \to \mathbf{R}$, $(t, u) \to H(t, u)$ is measurable for t for each $u \in \mathbf{R}^{2N}$ and continuously differentiable and convex in u for almost every $t \in [0, T]$.

Theorem 3.1. *Assume that the following conditions are satisfied.*

A_1. *There exists* $l \in L^4(0, T; \mathbf{R}^{2N})$ *such that for all* $u \in \mathbf{R}^{2N}$ *and a.e.* $t \in [0, T]$ *one has*

$$H(t, u) \geq (l(t), u). \tag{5}$$

A_2. *There exists* $\alpha \in\,]0, 2\pi/T[$ *and* $\gamma \in L^2(0, T; \mathbf{R}^+)$ *such that, for every* $u \in \mathbf{R}^{2N}$ *and a.e.* $t \in [0, T]$ *one has*

$$H(t, u) \leq \frac{\alpha}{2}|u|^2 + \gamma(t). \tag{6}$$

A_3.

$$\int_0^T H(t, u)\, dt \to +\infty \quad \text{as} \quad |u| \to \infty, u \in \mathbf{R}^{2N}. \tag{7}$$

Then problem (4) *has at least one solution u such that*

$$v(t) = -J\left[u(t) - \frac{1}{T}\int_0^T u(s)\,ds\right]$$

minimizes the dual action

$$\chi : H_T^1 \to]-\infty,\infty], \quad v \to \int_0^T \left[\frac{1}{2}(J\dot{v}(t), v(t)) + H^*(t,\dot{v}(t))\right] dt.$$

Proof. a) Existence of a solution for a perturbed problem. Let $\epsilon_0 > 0$ be such that

$$0 < \alpha + \epsilon_0 < 2\pi/T \tag{8}$$

and let

$$H_\epsilon : [0,T] \times \mathbf{R}^{2N} \to \mathbf{R}, \quad (t,u) \to \frac{\epsilon|u|^2}{2} + H(t,u)$$

where $0 < \epsilon < \epsilon_0$. Clearly $H_\epsilon(t,.)$ is strictly convex and continuously differentiable for a.e. $t \in [0,T]$ and $H_\epsilon(.,u)$ is measurable on $[0,T]$ for every $u \in \mathbf{R}^{2N}$. We obtain, from (A_1) and (A_2),

$$-|l(t)|\,|u| + \frac{\epsilon|u|^2}{2} \le H_\epsilon(t,u) \le (\alpha+\epsilon_0)\frac{|u|^2}{2} + \gamma(t),$$

hence

$$\frac{\epsilon|u|^2}{4} - \frac{|l(t)|^2}{\epsilon} \le H_\epsilon(t,u) \le (\alpha+\epsilon_0)\frac{|u|^2}{2} + \gamma(t), \tag{9}$$

so that by Theorem 2.3, the perturbed dual action

$$\chi_\epsilon(v) = \int_0^T \left[\frac{1}{2}(J\dot{v}(t), v(t)) + H_\epsilon^*(t,\dot{v}(t))\right] dt$$

is continuously differentiable on $\tilde{H}_T^1 = \{u \in H_T^1 : \int_0^T u(t)\,dt = 0\}$ and if $v_\epsilon \in \tilde{H}_T^1$ is a critical point of χ_ϵ, the function u_ϵ defined by

$$u_\epsilon(t) = \nabla H_\epsilon^*(t,\dot{v}_\epsilon(t))$$

is a solution of

$$\begin{gathered} J\dot{u}(t) + \epsilon u(t) + \nabla H(t,u(t)) = 0, \\ u(0) = u(T), \end{gathered} \tag{10}$$

and the relation

$$J\dot{v}_\epsilon = \dot{u}_\epsilon$$

holds. We have, by (9) and Propositions 3.2 and 2.2,

$$\chi_\epsilon(v) \ge \frac{1}{2}\left(\frac{1}{\alpha+\epsilon_0} - \frac{T}{2\pi}\right)\int_0^T |\dot{v}(t)|^2 dt - \int_0^T \gamma(t)\,dt$$

$$= \delta_0 \int_0^T (\dot{v}(t))^2 dt - \gamma_0 \tag{11}$$

where $\delta_0 > 0$ by (8). Let (v_k) be a minimizing sequence for χ_ϵ. By (11), $(|\dot{v}_k|)_{L^2})$ is bounded, and hence, by Wirtinger inequality, (v_k) is bounded in H_T^1. Now $\chi_{\epsilon,1}(v) = \int_0^T H_\epsilon^*(t, \dot{v}(t))\, dt$ is weakly lower semi-continuous on $\tilde{H}_T^1$ by Theorem 1.2 and $\chi_{\epsilon,2}(v) = \frac{1}{2}\int_0^T (J\dot{v}(t), v(t))\, dt$ is w.l.s.c. (even weakly continuous) by Proposition 1.2. Thus $\chi_\epsilon = \chi_{\epsilon,1} + \chi_{\epsilon,2}$ is w.l.s.c. and, by Theorem 1.1, has a minimum at some point $v_\epsilon \in \tilde{H}_T^1$.

b) A posteriori estimates on u_ϵ. It follows from (A_1), (A_2), and Proposition 2.2 that

$$|\nabla H(t,u)| \le 2(\alpha+1)[|u| + |l(t)|^2/2) + \gamma(t)] + 1 + |u|.$$

It is then easy to verify that the function

$$\overline{H} : \mathbf{R}^{2N} \to \mathbf{R}, \quad u \to \int_0^T H(t,u)\, dt$$

is continuously differentiable. Now, by assumption (A_3), $\overline{H}$ has a minimum at some point $\overline{u} \in \mathbf{R}^{2N}$ for which

$$\int_0^T \nabla H(t,\overline{u})\, dt = 0$$

so that the problem

$$\dot{v}(t) = \nabla H(t,\overline{u}) \tag{12}$$

has a unique solution w in H_T^1 such that $\int_0^T w(s)\, ds = 0$. By (12), $H^*(t, \dot{w}(t)) = (\dot{w}(t), \overline{u}) - H(t,\overline{u})$ so that $H^*(.,\dot{w}(.)) \in L^1(0,T;\mathbf{R})$. From the obvious inequality $H(t,u) \le H_\epsilon(t,u)$ we deduce $H_\epsilon^*(t,v) \le H^*(t,v)$ and from (11) we obtain

$$\begin{aligned} \delta_0 \int_0^T |\dot{v}_\epsilon(t)|^2 dt - \gamma_0 &\le \chi_\epsilon(v_\epsilon) \le \chi_\epsilon(w) \\ &\le \int_0^T \left[\frac{1}{2}(J\dot{w}(t), w(t)) + H^*(t,\dot{w}(t))\right] dt = c_1 < \infty. \end{aligned}$$

Therefore, $|\dot{v}_\epsilon|_{L^2} \le c_2$ and from $J\dot{v}_\epsilon = \dot{u}_\epsilon$, we have

$$|\dot{\tilde{u}}_\epsilon|_{L^2} = |\dot{u}_\epsilon|_{L^2} \le c_2$$

where $\tilde{u}_\epsilon = u_\epsilon - \overline{u}_\epsilon$, $\overline{u}_\epsilon = \frac{1}{T}\int_0^T u_\epsilon(t)\, dt$. Wirtinger's inequality implies that $\|\tilde{u}_\epsilon\| \le c_3$. By the convexity of $H(t,.)$ and (10) we obtain

$$\begin{aligned} H\left(t, \frac{\overline{u}_\epsilon}{2}\right) &= H\left(t, \frac{u_\epsilon(t)}{2} - \frac{\tilde{u}_\epsilon(t)}{2}\right) \le \frac{1}{2}H(t,u_\epsilon(t)) + \frac{1}{2}H(t,-\tilde{u}_\epsilon(t)) \\ &\le \frac{1}{2}(\nabla H(t,u_\epsilon(t)), u_\epsilon(t)) + \frac{1}{2}H(t,0) + \frac{\alpha}{4}|\tilde{u}_\epsilon(t)|^2 + \frac{\gamma(t)}{2} \\ &\le \frac{1}{2}(-J\dot{u}_\epsilon(t), u_\epsilon(t)) - \frac{\epsilon}{2}|u_\epsilon(t)|^2 + \frac{\alpha}{4}|\tilde{u}_\epsilon(t)|^2 + \gamma(t). \end{aligned}$$

Using Proposition 3.2, we have

$$\begin{aligned}\int_0^T H(t,\overline{u}_\epsilon)dt &\leq -\frac{1}{2}\int_0^T (J\dot{u}_\epsilon(t),u_\epsilon(t))\,dt+\frac{\alpha}{4}|\tilde{u}_\epsilon|_{L^2}^2+\gamma_0\\ &\leq \frac{T}{4\pi}|\dot{u}_\epsilon|_{L^2}^2+\frac{\alpha}{4}|\tilde{u}_\epsilon|_{L^2}^2+\gamma_0\leq\frac{T}{4\pi}c_2^2+\frac{\alpha}{4}c_3^2+\gamma_0=c_4.\end{aligned}$$

By assumption (A_3), $|\overline{u}_\epsilon|\leq c_5$. Finally,

$$\|u_\epsilon\|\leq\|\tilde{u}_\epsilon\|+\|\overline{u}_\epsilon\|\leq c_3+\sqrt{T}c_5=c_6.$$

c) Existence of a solution for the original problem. Since $\|u_\epsilon\|\leq c_6$, there is a sequence (ϵ_n) in $]0,\epsilon_0]$ tending to 0 and some $u\in H_T^1$ such that u_{ϵ_n} converge weakly to u in H_T^1. Moreover, as $\dot{v}_\epsilon=-J\dot{u}_\epsilon$, we have

$$v_\epsilon(t)=-J(u_\epsilon(t)-\overline{u}_\epsilon),$$

so that (v_{ϵ_n}) converges weakly to

$$v=-J(u-\overline{u}). \tag{13}$$

By Proposition 1.2, u_{ϵ_n} (resp. v_{ϵ_n}) converges uniformly to u (resp. v) on $[0,T]$. From (10) in integrated form

$$Ju_{\epsilon_n}(t)-Ju_{\epsilon_n}(0)+\int_0^T[\epsilon_n u_{\epsilon_n}(s)+\nabla H(s,u_{\epsilon_n}(s))]\,ds=0$$

we deduce

$$Ju(t)-Ju(0)+\int_0^T\nabla H(s,u(s))\,ds=0,$$

i.e. $u\in H_T^1$ is a solution of (4).

Finally, as $H_\epsilon^*(t,v)\leq H^*(t,v)$, we have, for all $h\in H_T^1$,

$$\chi_{\epsilon_n}(v_{\epsilon_n})\leq\chi_{\epsilon_n}(h)\leq\chi(h).$$

Now, by the duality between u_{ϵ_n} and $\dot{v}_{\epsilon_n}$, we have

$$\begin{aligned}\chi_{\epsilon_n}(v_{\epsilon_n}) &= \int_0^T\left[\frac{1}{2}(J\dot{v}_{\epsilon_n}(t),v_{\epsilon_n}(t))+(u_{\epsilon_n}(t),\dot{v}_{\epsilon_n}(t))-H_{\epsilon_n}(t,u_{\epsilon_n}(t))\right]dt\\ &= \int_0^T\left[\frac{1}{2}(J\dot{v}_{\epsilon_n}(t),v_{\epsilon_n}(t))+(u_{\epsilon_n}(t),\dot{v}_{\epsilon_n}(t))\right.\\ &\quad\left.-\ H(t,u_{\epsilon_n}(t))-\frac{\epsilon_n}{2}|u_{\epsilon_n}(t)|^2\right]dt.\end{aligned} \tag{14}$$

It follows from (4) and (13) that

$$\dot{v}(t)=\nabla H(t,u(t))\quad\text{a.e. on }[0,T]. \tag{15}$$

Letting $n \to \infty$ in (14), we obtain, by (15)

$$\lim_{n\to\infty} \chi_{\epsilon_n}(v_{\epsilon_n}) = \int_0^T \left[\frac{1}{2}(J\dot{v}(t), v(t)) + H^*(t, \dot{v}(t)) - H(t, u(t))\right] dt$$

$$= \int_0^T \left[\frac{1}{2}(J\dot{v}(t), v(t)) + H^*(t, \dot{v}(t))\right] dt = \chi(v).$$

Thus $\chi(v) \le \chi(h)$ for all $h \in H_T^1$ and the proof is complete. □

Under stronger assumptions, it is possible to obtain a priori bounds for all the solutions of (4).

Proposition 3.3. *If there exists $\alpha \in]0, \pi/T[$, $\beta \ge 0$, $\gamma \ge 0$ and $\delta > 0$ such that*

$$\delta|u| - \beta \le H(t,u) \le \frac{\alpha}{2}|u|^2 + \gamma$$

for all $t \in [0,T]$ and $u \in \mathbf{R}^{2N}$, then each solution of (4) satisfies the inequalities

$$\int_0^T |\dot{u}(t)|^2 dt \le \frac{2\alpha(\beta+\gamma)\pi T}{\pi - \alpha T} \tag{16}$$

$$\int_0^T |u(t)|\, dt \le \frac{\pi T(\beta+\gamma)}{\delta(\pi - \alpha T)}. \tag{17}$$

Proof. By Proposition 2.2, we have

$$\frac{1}{2\alpha}|H(t,u(t))|^2 \le (\nabla H(t,u(t)), u(t)) + \beta + \gamma.$$

It follows from (4) that

$$\frac{1}{2\alpha}\int_0^T |\dot{u}(t)|^2 dt + \int_0^T (J\dot{u}(t), u(t))\, dt \le (\beta+\gamma)T$$

and, by Proposition 3.2,

$$\frac{1}{2\alpha} - \frac{T}{2\pi}\int_0^T |\dot{u}(t)|^2 dt \le (\beta+\gamma)T,$$

which gives (16). Now, by convexity, (4), Proposition 3.2, and (16), we have

$$\begin{aligned}
\delta\int_0^T |u(t)|\, dt - \beta T &\le \int_0^T H(t,u(t))\, dt \\
&\le \int_0^T [H(t,0) + (\nabla H(t,u(t)), u(t))]\, dt \\
&\le \gamma T - \int_0^T (J\dot{u}(t), u(t))\, dt \le \gamma T + \frac{T}{2\pi}\int_0^T |\dot{u}(t)|^2 dt \\
&\le \gamma T + \frac{T}{2\pi}\left(\frac{2\alpha(\beta+\gamma)\pi T}{\pi - \alpha T}\right)
\end{aligned}$$

which gives (17). □

When $H(t,.)$ is strictly convex we can proceed exactly like in Theorem 8 of Chapter 1 (with F replaced by H) to deduce from Theorem 3.1 a necessary and sufficient condition for the solvability of (4).

Corollary 3.1. *Assume that $H(t,.)$ is strictly convex for a.e. $t \in [0,T]$ and satisfies the conditions* (A_1) *and* (A_2) *of Theorem* 3.1. *Then the following conditions are equivalent.*

α. *Problem* (4) *is solvable.*

β. *There exists $\overline{x} \in \mathbf{R}^{2N}$ such that*

$$\int_0^T \nabla H(t,\overline{x})\,dt = 0.$$

γ. $\int_0^T H(t,x)\,dt \to +\infty$ *when* $|x| \to \infty$.

3.3 Subharmonics of Non-Autonomous Convex Hamiltonian Systems

Now let $H : \mathbf{R} \times \mathbf{R}^{2N} \to \mathbf{R}$ be continuous with $H(t,.)$ convex and differentiable on $\mathbf{R}^{2N}$ for each $t \in \mathbf{R}$, $\nabla H : \mathbf{R} \times \mathbf{R}^{2N} \to \mathbf{R}^{2N}$ continuous and $H(.,u)$ T-periodic for each $u \in \mathbf{R}^{2N}$, i.e.,

$$H(t,u) = H(t+T,u)$$

for all $(t,u) \in \mathbf{R} \times \mathbf{R}^{2N}$. We still consider the corresponding system

$$J\dot{u} + \nabla H(t,u) = 0. \tag{18}$$

Clearly a solution u of (18) over $[0,T]$ verifying

$$u(0) = u(T)$$

can be extended by T-perodicity over $\mathbf{R}$ to give a *T-periodic* or *harmonic solution* of (18), i.e. a solution satisfying

$$u(t+T) = u(t), \quad t \in \mathbf{R}, \tag{19}$$

and Theorem 3.1 gives conditions for the existence of such a solution. If (19) is not satisfied with T replaced by T/k ($k \in \mathbf{R}$, k$\geq$ 2), T is called the *minimal period* of u. We shall show that systems like (18) may admit solutions u such that

$$u(t) = u(t+kT), \quad t \in \mathbf{R},$$

for some $k \geq 2$ where minimal period is strictly greater than T. Such solutions are called *subharmonic solutions* or simply *subharmonics*. Their existence will follow from Theorem 3.1 and Proposition 3.3.

Theorem 3.2. *Assume that*

$$H(t,u)/|u|^2 \to 0 \tag{20}$$

and

$$H(t,u) \to +\infty \tag{21}$$

as $|u| \to \infty$ *uniformly in* $t \in \mathbf{R}$.

Then, for each $k \in \mathbf{N} \setminus \{0\}$, *there exists a* kT*-periodic solution* u_k *of* (18), *such that*

$$\|u_k\|_\infty \to \infty \tag{22}$$

and such that the minimal period T_k *of* u_k *tends to* $+\infty$ *when* $k \to \infty$.

Proof. Let $c_1 = \max_{t\in\mathbf{R}} |H(t,)|$. By condition (21) there exists $R > 0$ such that

$$H(t,u) \geq 1 + c_1$$

for all $t \in \mathbf{R}$ and u with $|u| \geq R$. By convexity we have, for all $(t,u) \in \mathbf{R} \times \mathbf{R}^{2N}$ with $|u| \geq R$,

$$\begin{aligned} 1 + c_1 &\leq H\left(t, \frac{R}{|u|}u\right) \leq \frac{R}{|u|}H(t,u) + \left(1 - \frac{R}{|u|}\right)H(t,0) \\ &\leq \frac{R}{|u|}H(t,u) + c_1 \end{aligned}$$

and hence there is $\beta > 0$ and $\delta > 0$ such that

$$H(t,u) \geq \delta|u| - \beta \tag{23}$$

for all $(t,u) \in \mathbf{R} \times \mathbf{R}^{2N}$.

If $k \in \mathbf{N} \setminus \{0\}$ is fixed, condition (20) implies that there exists $\alpha \in]0, 2\pi/kT[$ and $\gamma > 0$ such that

$$H(t,u) \leq (\alpha/2)|u|^2 + \gamma \tag{24}$$

for all $(t,u) \in \mathbf{R} \times \mathbf{R}^{2N}$. By (23), (24), and Theorem 3.1 with T replaced by kT, the system (18) will have a kT-periodic solution u_k such that

$$v_k = -J\left[u_k - \frac{1}{kT}\int_0^{kT} u_k(s)\,ds\right]$$

minimizes the dual action

$$\chi_k : v \to \int_0^{kT} [(1/2)(J\dot{v}(t), v(t)) + H^*(t, \dot{v}(t))]\,dt$$

on H^1_{kT}.

The remainder of the proof depends upon the obtention of an upper estimate for $c_k = \chi_k(v_k)$. Let us first notice that by (23)

$$(v,u) - H(t,u) \leq (v,u) - \delta|u| + \beta \leq \beta$$

if $|v| \leq \delta$, so that

$$H^*(t,v) \leq \beta \ \text{ whenever } \ |v| \leq \delta.$$

If $\rho \in \mathbf{R}^{2N}$ is such that $|\rho| = 1$, then the eigenfunction h_k associated to the eigenvalue $\lambda_{-1} = -2\pi/kT$ of the kT-periodic problem for (1) given by

$$h_k(t) = \frac{\delta Tk}{2\pi}\left(\cos\frac{2\pi t}{kT}I + \sin\frac{2\pi t}{kT}J\right)\rho$$

belongs to H^1_{kT} and is such that $|\dot{h}_k| = \delta$. Thus,

$$c_k \leq \chi_k(h_k) \leq \int_0^{kT} [(1/2)(J\dot{h}_k(t), h_k(t)) + \beta]\, dt = -\frac{\delta^2}{4\pi}k^2T^2 + \beta kT. \quad (25)$$

If (22) does not hold, there exists $c_2 > 0$ and a subsequence (k_n) such that

$$||u_{k_n}||_\infty \leq c_2.$$

By (18), this implies that

$$||\dot{u}_{k_n}||_\infty \leq c_3$$

for some $c_3 > 0$ and hence

$$||v_{k_n}||_\infty \leq 2c_2, \quad ||\dot{v}_{k_n}||_\infty \leq c_3.$$

Consequently, as

$$H^*(t,v) \geq -H(t,0) \geq -c_1,$$

we have

$$c_{k_n} \geq -(2c_2c_3 + c_1)\, k_nT,$$

which is impossible by (25) for n sufficiently large. Thus

$$||u_k||_\infty \to \infty \ \text{ if } \ k \to \infty.$$

It remains only to prove that the minimal period of u_k tends to $+\infty$ as $k \to \infty$. If not, there exists $\tau > 0$ and a subsequence (k_n) such that the minimal period T_{k_n} of u_{k_n} satisfies $T_{k_n} \leq \tau$ ($n \in \mathbf{N}^*$). By (20) there exists $\alpha \in]0, \pi/\tau[$ and $\gamma > 0$ such that

$$H(t,u) \leq (\alpha/2)|u|^2 + \gamma,$$

for all $(t,u) \in \mathbf{R}\times\mathbf{R}^{2N}$. Hence, by (23) and Proposition 3.3 with T replaced by T_{k_n} we get

$$\int_0^{T_{k_n}} |\dot{u}_{k_n}(t)|^2 dt \leq \frac{2\alpha(\beta+\gamma)\pi T_{k_n}}{\pi - \alpha T_{k_n}} \leq \frac{2\alpha(\beta+\gamma)\pi\tau}{\pi-\alpha\tau} \tag{26}$$

$$\int_0^{T_{k_n}} |u_{k_n}(t)|\,dt \leq \frac{T_{k_n}(\beta+\gamma)\pi}{\delta(\pi-\alpha T_{k_n})} \leq \frac{T_{k_n}(\beta+\gamma)\pi}{\delta(\pi-\alpha\tau)}. \tag{27}$$

Let us write $u_{k_n} = \overline{u}_{k_n} + \tilde{u}_{k_n}$ where

$$\overline{u}_{k_n} = \frac{1}{T_{k_n}} \int_0^{T_{k_n}} u_{k_n}(t)\,dt.$$

Sobolev inequality and (26) imply that

$$\|\tilde{u}_{k_n}\|_\infty^2 \leq \frac{\tau}{12}\left(\frac{2\alpha(\beta+\gamma)\pi\tau}{\pi-\alpha\tau}\right).$$

Inequality (27) implies that

$$\|\overline{u}_{k_n}\|_\infty \leq \frac{1}{T_{k_n}} \int_0^{T_{k_n}} |u_{k_n}(t)|\,dt \leq \frac{(\beta+\gamma)\pi}{(\pi-\alpha\tau)}.$$

Thus $(\|u_{k_n}\|_\infty)$ is bounded, a contradiction with (22). □

3.4 Periodic Solutions with Prescribed Minimal Period of Autonomous Convex Hamiltonian Systems

Let $H : \mathbf{R}^{2N} \to \mathbf{R}$ be convex and of class C^1. We shall now consider the autonomous system

$$J\dot{u}(t) + \nabla H(u(t)) = 0. \tag{28}$$

In this section, we shall deal with the problem of periodic solutions with a prescribed *minimal* period. In the next one, we shall consider the case of periodic solutions with a fixed energy.

Theorem 3.3. *Assume that*

$$H(u)/|u|^2 \to 0 \tag{29}$$

and

$$H(u) \to +\infty \tag{30}$$

as $|u| \to \infty$. *Then there exists* $T_0 > 0$ *such that, for each* $T > T_0$, *the system* (28) *has a periodic solution* u_T *with minimal period* T. *Moreover,*

$$\min_{t\in\mathbf{R}} |u_T(t)| \to +\infty \text{ if } T \to +\infty. \tag{31}$$

Proof. We can always assume that $H(0) = 0$. Theorem 3.1 implies, for each $T > 0$, the existence of a T-periodic solution u_T of (28) such that

$$v_T = -J\left[u_T - \frac{1}{T}\int_0^T u_T(s)\,ds\right]$$

minimizes the dual action

$$\chi_T(v) = \int_0^T \left[\frac{1}{2}(J\dot{v}(t), v(t)) + H^*(\dot{v}(t))\right] dt$$

on H_T^1. Let us estimate $c_T = \chi_T(v_T)$ from above. Like in the proof of Theorem 3.2, there exist $\beta, \delta > 0$ such that

$$|v| \leq \delta \Rightarrow H^*(v) \leq \beta.$$

If $\rho \in \mathbf{R}^{2N}$ is such that $|\rho| = 1$, then the eigenfunction h_T associated with the eigenvalue $\lambda_{-1} = -2\pi/T$ of the T-periodic problem for (28) given by

$$h_T(t) = \frac{\delta T}{2\pi}\left[\left(\cos\frac{2\pi}{T}t\right)I + \left(\sin\frac{2\pi}{T}t\right)J\right]\rho$$

belongs to H_T^1 and is such that $|\dot{h}_T(t)| = \delta$. Thus

$$c_T \leq \chi_T(h_T) \leq \int_0^T \left[\frac{1}{2}(J\dot{h}_T(t), h_T(t)) + \beta\right] dt = -\frac{\delta^2}{4\pi}T^2 + \beta T. \qquad (32)$$

If $T > T_0 = 4\pi\beta/\delta^2$, we have $c_T < 0$. Suppose that u_T is (T/k)-periodic for some $k \geq 0$. Then $v(t) = kv_T(\frac{t}{k})$ belongs to H_T^1 and

$$\begin{aligned}
\chi_T(v) &= \frac{k}{2}\int_0^T \left(J\dot{v}_T\left(\frac{t}{k}\right), v_T\left(\frac{t}{k}\right)\right) dt + \int_0^T H^*\left(\dot{v}_T\frac{t}{k}\right) dt \\
&= \frac{k^2}{2}\int_0^{T/k} (J\dot{v}_T(t), v_T(t))\,dt + k\int_0^{T/k} H^*(\dot{v}_T(t))\,dt \\
&= \frac{k}{2}\int_0^T (J\dot{v}_T(t), v_T(t))\,dt + \int_0^T H^*(\dot{v}_T(t))\,dt \\
&= \chi_T(v_T) + \frac{k-1}{2}\int_0^T (J\dot{v}_T(t), v_T(t))\,dt. \qquad (33)
\end{aligned}$$

Since $H(0) = 0$, we have $H^*(v) \geq (v, 0) - H(0) = 0$. Hence, if $T > T_0$,

$$0 > \chi_T(v_T) \geq \int_0^T \frac{1}{2}(J\dot{v}_T(t), v_T(t))\,dt$$

and (33) implies $\chi_T(v) < \chi_T(v_T)$, a contradiction. Thus, for $T > T_0$, the minimal period of u_T is T.

As in the proof of Theorem 3.2, (32) implies that $\|u_T\|_\infty \to \infty$ if $T \to +\infty$. Let $R > 0$ and let

$$m = \max\{H(e) \,:\, e \in \mathbf{R}^{2N}, |e| \leq R\}.$$

By (30) there is $\rho > 0$ such that, for all $e \in \mathbf{R}^{2N}$,

$$|e| \geq \rho \Rightarrow H(e) \geq m + 1. \tag{34}$$

There is also some τ such that

$$T > \tau \Rightarrow \|u_T\|_\infty = |u_T(t_T)| \geq \rho. \tag{35}$$

Now, we have, for $T > \tau$ and $t \in \mathbf{R}$,

$$m + 1 \leq H(u_T(t_T)) = H(u_T(t)) \tag{36}$$

since the energy is conserved. The definition of m implies that, for $T > \tau$,

$$\min_{t \in \mathbf{R}} |u_T(t)| \geq R$$

and the proof is complete. □

Remark 3.1. There is no assumption in Theorem 3.3 about the behavior of H near zero, so that bifurcation theory is not applicable.

Remark 3.2. The following example shows that *Theorem 3.3 cannot be generalized to the case of a superquadratic Hamiltonian.* Let $F \in C^1(\mathbf{R}, \mathbf{R})$ be such that $F'(s) \geq 1$ for all $s \in \mathbf{R}$ and let $H(u) = F(|u|^2)/2$. Thus H is convex when F is convex. The solutions of the Hamiltonian systems

$$J\dot{u}(t) + F'(|u|^2)u = 0$$

are

$$u(t) = \cos(F'(|e|^2)t)e + \sin(F'(|e|^2)t)\, Je,$$

for any $e \in \mathbf{R}^{2N}$. Thus, the minimal period of u is less than 2π.

We shall now show that in Theorem 3, the existence of a non-constant T-periodic solution cannot be expected for every $T > 0$.

Proposition 3.4. *Let $f : \mathbf{R}^N \to \mathbf{R}^N$ be Lipschitz continuous on $\mathbf{R}^N$ with Lipschitz constant c. If u is a non-constant T-periodic solution of*

$$\dot{u} = f(u),$$

then

$$T \geq 2\pi/c.$$

Proof. Since $\dot{u} = f(u)$, we obtain, for a $t \in \mathbf{R}$, and $h \neq 0$

$$|\dot{u}(t+h) - \dot{u}(t)| = |f(u(t+h)) - f(u(t))| \leq c|u(t+h) - u(t)|,$$

hence

$$|\ddot{u}(t)| \leq c|\dot{u}(t)|$$

for a.e. $t \in \mathbf{R}$. Then, by Wirtinger's inequality,

$$\int_0^T |\dot{u}(t)|^2 dt \leq \frac{T^2}{4\pi^2} \int_0^T |\ddot{u}(t)|^2 dt \leq \frac{T^2 c^2}{4\pi^2} \int_0^T |\dot{u}(t)|^2 dt$$

and the result follows as u is nonconstant. □

Now Lipschitz continuity of ∇H is compatible with the assumptions of Theorem 3.3, and Proposition 3.4 shows that the conclusion of Theorem 3.3 is optimal.

3.5 Periodic Solutions with Prescribed Energy of Autonomous Hamiltonian Systems

If $u(t)$ is a solution over $[0, T]$ of an autonomous Hamiltonian system

$$J\dot{u}(t) + \nabla H(u(t)) = 0 \tag{37}$$

then scalar multiplication of both members by $\dot{u}(t)$ gives

$$(\nabla H(u(t)), \dot{u}(t)) = 0,$$

i.e. the energy

$$H(u(t)) = \text{constant}. \tag{38}$$

It is therefore natural to look for solutions, and in particular for periodic solutions, with prescribed energy. The difficulty in this case is that the period, and hence the underlying function space of solutions, is not a priori known. We shall see, however, how to reduce, under some assumptions upon H, the fixed energy case to the fixed period case.

We first prove that, under some conditions on ∇H, the orbits of (37) on an energy hypersurface S, i.e. the sets $\{u(t) \in \mathbf{R}^{2N} : t \in [0, T]\}$ with u verifying (37) and (38) are independent of H and depend only on S.

Lemma 3.1. *Let $H_i \in C^1(\mathbf{R}^{2N}, \mathbf{R})$ and $c_i \in \mathbf{R}$ $(i = 1, 2)$ be such that*

$$S = H_1^{-1}(c_1) = H_2^{-1}(c_2).$$

If

$$\nabla H_i(u) \neq 0, \quad u \in S, \quad i = 1, 2, \tag{39}$$

then the orbits of the systems

$$J\dot{u}(t) + \nabla H_i(u(t)) = 0, \quad i = 1, 2$$

on S are the same.

Proof. Let $u_1 : [0,T] \to S$ be a solution of

$$J\dot{u}(t) + \nabla H_1(u(t)) = 0.$$

Since $\nabla H_1(u_1(t))$ and $\nabla H_2(u_1(t))$ are normal to S and continuous, there is a function $\lambda : [0,T] \to \mathbf{R}$ such that

$$\nabla H_2(u_1(t)) = \lambda(t)\nabla H_1(u_1(t)), \quad t \in [0,T].$$

By (39), $\lambda(t) \neq 0$ and the relation

$$\lambda(t) = (\nabla H_2(u_1(t)), \nabla H_1(u_1(t)))/|\nabla H_1(u_1(t))|^2$$

shows that λ is continuous and hence either positive or negative. Define the strictly monotone function ψ by

$$\psi(s) = \int_0^s \frac{dt}{\lambda(t)}, \quad s \in [0,T]$$

and let $u_2 : \psi[0,T] \to \mathbf{R}^{2N}$ be given by $u_2 = u_1 \circ \psi^{-1}$. Then

$$\begin{aligned} J\dot{u}_2(t) &= J(\dot{u}_1 \circ \psi^{-1})(1/\psi' \circ \psi^{-1})(t) = -[\nabla H_1(u_1 \circ \psi^{-1})(t)](\lambda \circ \psi^{-1})(t) \\ &= \nabla H_2(u_1\psi^{-1}(t)) = -\nabla H_2(u_2(t)), \quad t \in \psi[0,T]. \end{aligned}$$

Consequently, u_2 is a solution of $J\dot{u} + \nabla H_2(u) = 0$. □

We shall now use the above lemma and some convexity properties of H to replace it by another Hamiltonian leading to the same orbits and satisfying the conditions of Theorem 3.3. To this end, let us recall that if C is a closed convex set in $\mathbf{R}^m$ and $0 \in \operatorname{int} C$, the gauge j of C is defined on $\mathbf{R}^m$ by

$$j(u) = \inf\{\lambda > 0 : u/\lambda \in C\}.$$

Clearly, j maps $\mathbf{R}^m$ onto $\mathbf{R}_+$, $j(0) = 0$ and j is *positive homogeneous.* If $u \in \mathbf{R}^m$, $v \in \mathbf{R}^m$,

$$\lambda \in \left\{\lambda > 0 : \frac{u}{\lambda} \in C\right\}, \quad \mu \in \left\{\mu > 0 : \frac{v}{\mu} \in C\right\},$$

then

$$\frac{u+v}{\lambda+\mu} = \frac{\lambda}{\lambda+\mu}\left(\frac{u}{\lambda}\right) + \frac{\mu}{\lambda+\mu}\left(\frac{v}{\mu}\right) \in C,$$

i.e.

$$\lambda + \mu \in \left\{\nu > 0 : \frac{u+v}{\nu} \in C\right\}.$$

Thus,

$$\left\{\lambda > 0 : \frac{u}{\lambda} \in C\right\} + \left\{\mu > 0 : \frac{v}{\mu} \in C\right\} \subset \left\{\nu > 0 : \frac{u+v}{\nu} \in C\right\}$$

so that

$$j(u) + j(v) \geq j(u+v).$$

Then, if $0 < \alpha < 1$, $u \in \mathbf{R}^m$, $v \in \mathbf{R}^m$, one has

$$j((1-\alpha)u + \alpha v) \leq j((1-\alpha)u + j(\alpha v) = (1-\alpha)j(u) + \alpha j(v)$$

which shows that j is a *convex function.*

On the other hand, if $u \in C$, $j(u) \leq 1$ and if $u \in \operatorname{int} C$, $(1+\epsilon)u \in C$ for $\epsilon > 0$ sufficiently small so that $j(u) \leq \frac{1}{1+\epsilon} < 1$. Conversely, if $j(u) < 1$, there is some $j(u) < \alpha < 1$ such that $\frac{u}{\alpha} \in C$; as $0 \in \operatorname{int} C$, there is $r_0 > 0$ such that $v \in C$ whenever $|v| \leq r_0$, so that, for all such v we have

$$u + (1-\alpha)v = \alpha\frac{u}{\alpha} + (1-\alpha)v \in C$$

so $B[u, (1-\alpha)r_0] \subset C$ and $u \in \operatorname{int} C$. Finally, if $j(u) = 1$, there is a sequence (λ_n) with $\lambda_n > 1$ tending to 1 such that $\frac{u}{\lambda_n} \in C$. C being closed, $u = \lim_{n\to\infty} \frac{u}{\lambda_n} \in C$. We have, therefore, proved that $u \in C$ *if and only if* $j(u) \leq 1$ and $u \in \operatorname{int} C$ *if and only if* $j(u) < 1$. As C is closed this implies that $u \in \partial C$ *if and only if* $j(u) = 1$. Thus j characterizes C.

Lemma 3.2. *Let $H \in C^1(\mathbf{R}^{2N}, \mathbf{R})$ and $c \in \mathbf{R}$ be such that $\nabla H(u) \neq 0$ for every $u \in S = H^{-1}(c)$. Assume that S is the boundary of a convex compact set C containing* 0 *as an interior point. Let j be the gauge of C and let $F = j^{3/2}$. Then*

(i) $F^{-1}(1) = S$.

(ii) *F is positively homogeneous of degree* 3/2.

(iii) *There is $\beta > 0$ and $\gamma > 0$ such that*

$$\beta|u|^{3/2} \leq F(u) \leq \gamma|u|^{3/2} \tag{40}$$

for all $u \in \mathbf{R}^{2N}$.

(iv) *$F \in C^1(\mathbf{R}^{2N}, \mathbf{R})$ and ∇F is positively homogeneous of degree* 1/2.

Proof. (i) and (ii) follow directly from the definition of F and the properties of j.

For (iii), as $F(u) > 0$ for $u \neq 0$, we have

$$0 < \beta = \min_{|u|=1} F(u) \leq \max_{|u|=1} F(u) = \gamma$$

and hence the result by (ii). To show (iv), first let $u \neq 0$, $\lambda \neq 0$ and $G(u,\lambda) = H(\frac{u}{\lambda}) - c$. Then

$$G(u,\lambda) = 0 \Leftrightarrow \frac{u}{\lambda} \in S \Leftrightarrow j\left(\frac{u}{\lambda}\right) = 1 \Leftrightarrow \lambda = j(u)$$

and

$$\frac{\partial G}{\partial \lambda}(u,\lambda) = -\frac{1}{\lambda}\left(\nabla H\left(\frac{u}{\lambda}\right), \frac{u}{\lambda}\right)$$

so that

$$\frac{\partial G}{\partial \lambda}(u, j(u)) = -\frac{1}{j(u)}\left(\nabla H\left(\frac{u}{j(u)}\right), \frac{u}{j(u)}\right) \neq 0$$

as $0 \in \operatorname{int} C$ and

$$\left\{v : \left(\nabla H\left(\frac{u}{j(u)}\right), v - \frac{u}{j(u)}\right) = 0\right\}$$

is a supporting hyperplane for C. Therefore, the implicit function theorem implies that j, and hence F, is of class C^1 on $\mathbf{R}^{2N}\setminus\{0\}$. Moreover, it follows directly from (iii) that $\nabla F(0) = 0$. Property (ii) implies trivially that ∇F is positively homogeneous of degree $1/2$. But then

$$\lim_{u\to 0} \nabla F(u) = 0 = \nabla F(0)$$

so that ∇F is also continuous at zero. □

We can now state and prove the basic theorem of this section.

Theorem 3.4. *Let $H \in C^1(\mathbf{R}^{2N}, \mathbf{R})$ and $c \in \mathbf{R}$ be such that $\nabla H(u) \neq 0$ for every $u \in S = H^{-1}(c)$. Assume that S is the boundary of a convex compact set C containing 0 as an interior point. Then there exists at least one periodic solution of* (37) *whose orbit lies on S.*

Proof. Let $F : \mathbf{R}^{2N} \to \mathbf{R}$ be given by Lemma 3.2. All the conditions of Theorem 3.3 are satisfied for

$$J\dot{u}(t) + \nabla F(u(t)) = 0 \tag{41}$$

and hence, if we fix any $T > T_0$, (41) has a periodic solution u with minimal period T. Conservation of energy and properties of F imply that

$$F(u(t)) = d > 0, \quad t \in \mathbf{R}.$$

Let us define w by

$$w(t) = d^{-2/3}u(d^{1/3}t), \quad t \in \mathbf{R}.$$

Then w is $(T/d^{1/3})$-periodic,

$$F(w(t)) = F(d^{-2/3}u(d^{1/3}t)) = d^{-1}F(u(d^{1/3}t)) = 1$$

for all $t \in \mathbf{R}$ and

$$\begin{aligned} J\dot{w}(t) &= d^{1/3}J\dot{u}(d^{1/3}t) = -d^{-1/3}\nabla F(u(d^{1/3}t)) \\ &= -\nabla F(d^{-2/3}u(d^{1/3}t)) = -\nabla F(w(t)), \end{aligned}$$

so that w is a $(T/d^{1/3})$-periodic solution of (41) whose orbit Γ lies on S. By Lemma 1, system (41) admit on S the same orbits as (37) and in particular the closed orbit Γ which corresponds to a periodic solution of (37). □

3.6 Periodic Solutions of Non-Autonomous Second Order Systems with Convex Potential

In this section we shall use Theorem 3.1 to study the periodic boundary value problem

$$\begin{aligned} \ddot{q}(t) + \nabla F(t, q(t)) = 0 \\ q(0) - q(T) = \dot{q}(0) - \dot{q}(T) = 0 \end{aligned} \tag{42}$$

where $F : [0,T] \times \mathbf{R}^N \to \mathbf{R}$, $(t,q) \to F(t,q)$ is measurable in t for every $q \in \mathbf{R}^N$ and continuously differentiable and convex in q for almost every $t \in [0,T]$.

Theorem 3.5. *Assume that the following conditions are satisfied.*

A$_1$. *There exists $l \in L^4(0,T;\mathbf{R}^N)$ such that, for all $q \in \mathbf{R}^N$ and a.e. $t \in [0,T]$ one has*

$$(l(t), q) \leq F(t,q).$$

A$_2$. *There exists $\alpha \in]0, 2\pi/T[$ and $\gamma \in L^2(0,T;\mathbf{R}^+)$ such that, for each $q \in \mathbf{R}^N$ and a.e. $t \in [0,T]$, one has*

$$F(t,q) \leq \frac{\alpha^2}{2}|q|^2 + \gamma(t).$$

A$_3$.

$$\int_0^T F(t,q)\,dt \to +\infty \quad \text{as} \quad |q| \to \infty, \quad q \in \mathbf{R}^N.$$

Then problem (42) *has at least one solution.*

Proof. Define $H : [0,T] \times \mathbf{R}^{2N} \to \mathbf{R}$ by

$$H(t,u) = \alpha\frac{|u_2|^2}{2} + \frac{1}{\alpha}F(t,u_1)$$

where $u = (u_1, u_2)$. For a.e. $t \in [0,T]$, $H(t,.)$ is convex and continuously differentiable. For every $u \in \mathbf{R}^{2N}$ and a.e. $t \in [0,T]$,

$$\left(\frac{l(t)}{\alpha}, u_1\right) \leq H(t,u) \leq \frac{\alpha}{2}(|u_2|^2 + |u_1|^2) + \frac{\gamma(t)}{\alpha} = \frac{\alpha}{2}|u|^2 + \frac{\gamma(t)}{\alpha}.$$

Moreover,

$$\int_0^T H(t,u)\,dt = \alpha T\frac{|u_2|^2}{2} + \frac{1}{\alpha}\int_0^T F(t,u_1)\,dt \to \infty$$

if $|u| \to \infty$. By Theorem 3.1, the problem (4) is solvable, i.e. u_1 and u_2 satisfy

$$\dot{u}_2(t) + \frac{1}{\alpha}\nabla F(t, u_1(t)) = 0$$

$$-\dot{u}_1(t) + \alpha u_2(t) = 0$$
$$u_1(0) - u_1(T) = u_2(0) - u_2(T) = 0$$

and hence $q(t) = u_1(t)$ is a solution of (42). □

By applying Corollary 3.1 to the Hamiltonian system introduced in the above proof, we easily get the following result.

Corollary 3.2. *If $F(t,.)$ is strictly convex for a.e. $t \in [0,T]$ and satisfies conditions (A_1) and (A_2) of Theorem 3.5, the following conditions are equivalent.*

α. Problem (42) is solvable.

β. There exists $\overline{x} \in \mathbf{R}^N$ such that $\int_0^T \nabla F(t,\overline{x})\,dt = 0$.

γ. $\int_0^T F(t,x)\,dt \to +\infty$ as $|x| \to \infty$.

In the special case of $N = 1$, we can proceed exactly as in Theorem 1.9 to deduce from Theorem 3.5 a necessary and sufficient condition for the solvability of (42).

Theorem 3.6. *If $N = 1$ and F satisfies conditions (A_1) and (A_2) of Theorem 5, then the problem (42) has a solution if and only if there exists $\overline{q} \in \mathbf{R}$ such that*

$$\int_0^T \nabla F(t,\overline{q})\,dt = 0 \tag{43}$$

(or equivalently if and only if the function $\int_0^T F(t,.)\,dt$ has a critical point).

Remark 3.3. Condition (A_2) in Theorem 3.5 is sharp as shown by the example

$$\ddot{q}(t) + \omega^2 q(t) = a\cos\omega t \tag{44}$$
$$q(0) - q(T) = \dot{q}(0) - \dot{q}(T) = 0 \tag{45}$$

where $\omega = 2\pi/T$ where $a \in \mathbf{R}^N \backslash \{0\}$ which has no solution and corresponds to

$$F(t,q) = (\omega^2/2)\,|q|^2 - \cos(\omega t)(a,q) \tag{46}$$

which satisfies the regularity and convexity assumptions of Theorem 3.5 as well as conditions (A_1) and (A_3).

3.7 A Variant of the Dual Least Action Principle for Non-Autonomous Second Order Systems

Let us consider the following generalization of the problem considered in Section 3.6,

$$\begin{aligned} \ddot{q}(t) + m^2\omega^2 q(t) + \nabla F(t,q(t)) = 0 \\ q(0) - q(T) = \dot{q}(0) - \dot{q}(T) = 0 \end{aligned} \tag{47}$$

where $F : [0,T] \times \mathbf{R}^N \to \mathbf{R}$ satisfies the regularity and convexity assumptions listed at the beginning of Section 3.6 as well as condition (A_1) of Theorem 3.5, and where $m \in \mathbf{N}$ and $\omega = 2\pi/T$. Problem (42) is a nonlinear perturbation of the linear problem

$$\begin{gathered}\ddot{q}(t) - \lambda q(t) = 0 \\ q(0) - q(T) = \dot{q}(0) - \dot{q}(T) = 0\end{gathered} \tag{48}$$

at the zero eigenvalue and problem (47) a nonlinear perturbation of (48) at an arbitrary eigenvalue $-m^2\omega^2$.

Assume for a moment that for a.e. $t \in [0,T]$, $F(t,.)$ is strictly convex and such that

$$\frac{F(t,q)}{|q|} \to +\infty \quad \text{as} \quad |q| \to \infty, \tag{49}$$

so that $F^*(t,v)$ exists and is of class C^1 in v. Let us recall the elementary result that

$$\begin{gathered}\ddot{q}(t) + m^2\omega^2 q(t) + v(t) = 0 \\ q(0) - q(T) = \dot{q}(0) - \dot{q}(T) = 0\end{gathered} \tag{50}$$

with $v \in L^2(0,T;\mathbf{R}^N)$ has a solution if and only if

$$v \in V = \left\{ w \in L^2(0,T,\mathbf{R}^N) : \int_0^T w(t) \cos m\omega t \, dt \right.$$
$$\left. = \int_0^T w(t) \sin m\omega t \, dt = 0 \right\} \tag{51}$$

in which case (50) has the family of solutions given by

$$q(t) = a \cos m\omega t + b \sin m\omega t - \int_0^T \sin m\omega(t-s)v(s)\, ds, \quad (a,b) \in \mathbf{R}^N).$$

Then we immediately check that when $v \in V$, (50) has a unique solution belonging to V, which we shall denote by Kv. We define in this way a linear operator K in V, and the general solution of (50) can be written

$$q(t) = \overline{q} + (Kv)(t)$$

where

$$\overline{q} \in W = V^{\perp} = \{w \in L^2(0,T;\mathbf{R}^N) :$$
$$w(t) = a \cos m\omega t + b \sin m\omega t, \quad a,b \in \mathbf{R}^N\}.$$

Also, using Fourier series, it is easy to show that if

$$v(t) \sim \sum_{\substack{k \in \mathbf{Z} \\ |k| \neq m}} v_k e^{ik\omega t},$$

then

$$Kv(t) \sim \sum_{\substack{k \in \mathbf{Z} \\ |k| \neq m}} \frac{v_k}{(k^2 - m^2)\omega^2} e^{ik\omega t}, \tag{52}$$

and hence,

$$\int_0^T (Kv(t), v(t))\, dt \leq \frac{1}{(2m+1)\omega} \|v\|_{L^2}^2, \tag{53}$$

for all $v \in V$.

Let us introduce in (47) the change of unknown given by

$$v(t) = -\ddot{q} - m^2\omega^2 q(t) \tag{54}$$

with $v \in V$, or equivalently

$$q(t) = \overline{q}(t) + (Kv)(t) \tag{55}$$

where $\overline{q} \in W$. If q satisfies (47), then

$$v(t) = \nabla F^*(t, v(t)) \quad \text{a.e. on } [0,T] \tag{56}$$

which, together with (54), shows that v satisfies the equation

$$-(Kv)(t) + \nabla F^*(t, v(t)) = \overline{q}(t) \quad \text{a.e. on } [0,T] \tag{57}$$

or

$$-(Kv)(t) + \nabla F^*(t, v(t)) \in W \tag{58}$$

for a.e. $t \in [0,T]$. Conversely, if $v \in V$ satisfies (57) or (58), then defining q by (54), we see that the elimination of v implies (47). Now, (58) is the Euler equation for the critical points in V of the functional χ defined by

$$\chi(v) = \int_0^T [-(1/2)(Kv(t), v(t)) + F^*(t, v(t))]\, dt, \tag{59}$$

as it follows from Remark 1.1. As K is now a bounded linear operator on V, χ often has better properties than the direct action associated to (47). It is a variant of the dual action introduced for Hamiltonian systems in Section 2.4. To eliminate the unpleasant assumption of strict convexity and condition (49) on F, we shall use, like in Theorem 3.1, a perturbation argument.

Theorem 3.7. *Let $m \in \mathbf{N}^*$ and $F : [0,T] \times \mathbf{R}^N \to \mathbf{R}$, $(t,q) \to F(t,q)$ be such that $F(.,q)$ is measurable for each $q \in \mathbf{R}^N$ and $F(t,.)$ is convex and continuously differentiable for a.e. $t \in [0,T]$. Assume that the following conditions are satisfied.*

B_1. *There exists $l \in L^4(0,T;\mathbf{R}^N)$ such that, for all $q \in \mathbf{R}^N$ and a.e. $t \in [0,T]$, one has*

$$F(t,q) \geq (l(t), q).$$

B$_2$. *There exists* $\alpha \in]0,(2m+1)\omega[$ *and* $\gamma \in L^2(0,T;\mathbf{R}^+)$ *such that, for every* $q \in \mathbf{R}^N$ *and a.e.* $t \in [0,T]$, *one has*

$$F(t,q) \le (\alpha/2)|q|^2 + \gamma(t).$$

B$_3$. $\int_0^T F(t, a\cos m\omega t + b\sin m\omega t)\,dt \to +\infty$ *as*

$$|a| + |b| \to \infty, \quad a, b \in \mathbf{R}^N.$$

Then problem (47) *has at least one solution* q *such that* $v = -\ddot{q} - m^2\omega^2 q$ *minimizes on* V *the dual action* χ.

Proof. a) Existence of a solution for a perturbed problem. Let $\epsilon_0 > 0$ be such that

$$\alpha + \epsilon_0 < (2m+1)\omega \tag{60}$$

and let

$$F_\epsilon : [0,T] \times \mathbf{R}^N \to \mathbf{R}, \quad (t,q) \to (\epsilon/2)|q|^2 + F(t,q)$$

where $0 < \epsilon \le \epsilon_0$. Proceeding as in part (a) of the proof of Theorem 3.1, we see that $F_\epsilon^* : [0,T] \times \mathbf{R}^N \to \mathbf{R}$ is well defined and that the function

$$\varphi_\epsilon : V \to \mathbf{R}, \quad v \to \int_0^T F_\epsilon^*(t, v(t))\,dt$$

is well defined and continuously differentiable on V. The same is obviously true for the function

$$\psi : V \to \mathbf{R}, \quad v \to \int_0^T (1/2)(Kv(t), v(t))\,dt$$

and hence for the perturbed dual action

$$\chi^\epsilon : V \to \mathbf{R}, \quad v \to \int_0^T [-(1/2)(Kv(t), v(t)) + F_\epsilon^*(t, v(t))]\,dt.$$

Moreover, as in Theorem 3.1,

$$F_\epsilon^*(t,v) \ge \frac{1}{\alpha + \epsilon_0}\frac{|v|^2}{2} - \gamma(t)$$

and hence, by (52),

$$\chi^\epsilon(v) \ge \left[\frac{1}{\alpha+\epsilon_0} - \frac{1}{(2m+1)\omega}\right]\|v\|_{L^2}^2 - \int_0^T \gamma(t)\,dt = \delta_0\|v\|_{L^2}^2 - \gamma_0 \tag{61}$$

with $\delta_0 > 0$. Thus, every minimizing sequence for χ^ϵ is bounded. Now, φ_ϵ is weakly lower semi-continuous on V by Theorem 1.2 and $K(V) \subset V \cap W_T^{1,2}$.

By Proposition 1.2, ψ is weakly continuous on V and hence χ^ϵ is w.l.s.c. Thus by Theorem 1.1, χ^ϵ has a minimum at some $v_\epsilon \in V$, for which

$$\langle(\chi^\epsilon)'(v_\epsilon), w\rangle = 0$$

for all $w \in V$. It is easy to check that

$$\langle(\chi^\epsilon)'(v_\epsilon), w\rangle = \int_0^T [-(Kv_\epsilon(t), w(t)) + (\nabla F_\epsilon^*(t, v_\epsilon(t)), w(t))]\, dt$$

(see the proof of Theorem 1.4) and hence

$$-Kv_\epsilon + \nabla F_\epsilon^*(., v_\epsilon(.)) \in W.$$

The reasoning above shows then that if $\overline{q}^\epsilon = -Kv_\epsilon + \nabla F_\epsilon^*(., v_\epsilon(.))$, then $q_\epsilon = \overline{q}^\epsilon + Kv_\epsilon$ is a solution of

$$\begin{gathered}\ddot{q}(t) + m^2\omega^2 q(t) + \nabla F_\epsilon(t, q(t)) = 0 \\ q(0) - q(T) = \dot{q}(0) - \dot{q}(T) = 0.\end{gathered} \tag{62}$$

b) a posteriori estimates on q_ϵ. It follows from (B_1), (B_2), and Proposition 2.2 that

$$|\nabla F(t,q)| \leq 2\alpha[(1 + |l(t)|)\, |q| + \gamma(t)] + 1.$$

It is then easy to verify that the function

$$\overline{F} : \mathbf{R}^{2N} \to \mathbf{R}, \quad [a, b] \to \int_0^T F(t, a\cos m\omega t + b \sin m\omega t)\, dt$$

is continuously differentiable. By assumption (B_3), $\overline{F}$ has a minimum at some $[\overline{a}, \overline{b}] \in \mathbf{R}^{2N}$ for which

$$\begin{aligned} &\int_0^T \nabla F(t, \overline{a}\cos m\omega t + \overline{b}\sin m\omega t) \cos m\omega t\, dt \\ = &\int_0^T \nabla F(t, \overline{a}\cos m\omega t + \overline{b}\sin m\omega t) \sin m\omega t\, dt = 0. \end{aligned}$$

But then $\nabla F(., \overline{a}\cos m\omega(.) + \overline{b}\sin m\omega(.)) \in V_m$ and letting

$$w(t) = \nabla F(t, \overline{a}\cos m\omega t + \overline{b}\sin m\omega t),$$

we have, by duality,

$$\begin{aligned} F^*(t, w(t)) &= (w(t), \overline{a}\cos m\omega t + \overline{b}\sin m\omega t) \\ &\quad - F(t, \overline{a}\cos m\omega t + \overline{b}\sin m\omega t) \end{aligned}$$

for a.e. $t \in [0, T]$, so that $F^*(., w(.)) \in L^1(0, T; \mathbf{R})$. Consequently, using (61), we get

$$\delta_0 \|v_\epsilon\|_{L^2}^2 - \gamma_0 \leq \chi^\epsilon(v_\epsilon) \leq \chi^\epsilon(w) \leq \chi(w) = c_1 < \infty$$

and hence

$$\|v_\epsilon\|_{L^2} \le c_2$$

with c_2 independent of ϵ. Consequently, if $\tilde{q}^\epsilon = q_\epsilon - \overline{q}^\epsilon$,

$$\|\tilde{q}^\epsilon\|_{L^2} = \|Kv_\epsilon\|_{L^2} \le c_3$$

and

$$\|\ddot{q}_\epsilon + m^2\omega^2 q_\epsilon\|_{L^2} = \|v_\epsilon\|_{L^2} \le c_2.$$

By the convexity of $F(t,.)$ and (62), we obtain

$$\begin{aligned} F\left(t, \frac{\overline{q}^\epsilon(t)}{q}\right) &\le \frac{1}{2}F(t, q_\epsilon(t)) + \frac{1}{2}F(t, -\tilde{q}^\epsilon(t)) \\ &\le \frac{1}{2}(\nabla F(t, q_\epsilon(t)), q_\epsilon(t)) + \frac{1}{2}F(t,0) + \frac{\alpha}{4}|\tilde{q}^\epsilon(t)|^2 + \frac{\gamma(t)}{2} \\ &\le \frac{1}{2}(-\ddot{q}_\epsilon(t) - m^2\omega^2 q_\epsilon(t), q_\epsilon(t)) + \gamma(t) + \frac{\alpha}{4}|\tilde{q}^\epsilon(t)|^2. \end{aligned}$$

Hence,

$$\begin{aligned} \int_0^T F\left(t, \frac{\overline{q}^\epsilon(t)}{2}\right) &\le \frac{1}{2}\int_0^T (-\ddot{q}(t) - m^2\omega^2 q(t), \tilde{q}^\epsilon(t))\, dt \\ &\quad + \int_0^T \gamma(t)\, dt + \frac{\alpha}{4}\|\tilde{q}^\epsilon\|_{L^2}^2 \\ &\le \frac{1}{2}\|\ddot{q}_\epsilon + m^2\omega^2 q_\epsilon\|_{L^2}\|\tilde{q}^\epsilon\|_{L^2} + \gamma_0 + (\alpha/4)\|\tilde{q}^\epsilon\|_{L^2}^2 \le c_4, \end{aligned}$$

so that, by (B_3), $\|\overline{q}^\epsilon\|_{L^2} \le c_4$, all norms being equivalent in the finite-dimensional subspace W. Consequently,

$$\|q_\epsilon\|_{L^2} \le c_3 + c_4 = c_5$$

and

$$\|\ddot{q}_\epsilon\|_{L^2} \le c_2 + m^2\omega^2(c_3 + c_4) = c_6.$$

c) Existence of a solution for the original problem. By the above inequalities, there is a sequence (ϵ_n) in $]0, \epsilon_0]$ tending to zero and some T-periodic $q \in C^1([0,T], \mathbf{R}^N)$ such that $q_{\epsilon_n} \to q$ in $C^1([0,T], \mathbf{R}^N)$. From (62) in integrated form

$$\dot{q}_{\epsilon_n}(t) - \dot{q}_{\epsilon_n}(0) + \int_0^T [m^2\omega^2 q_{\epsilon_n}(s) + \nabla F_{\epsilon_n}(s, q_{\epsilon_n}(s))]\, ds = 0,$$

we deduce

$$\dot{q}(t) - \dot{q}(0) + \int_0^T [m^2\omega^2 q(s) + \nabla F(s, q(s))]\, ds = 0$$

and q is a solution of (47). The fact that the weak limit v of (v_{ϵ_n}) minimizes χ on V is proved as in Theorem 3.1. □

Remark 3.4. A similar result holds for the problem

$$-\ddot{q}(t) - m^2\omega^2 q(t) + \nabla F(t, q(t)) = 0 \quad (m \geq 1)$$
$$q(0) - q(T) = \dot{q}(0) - \dot{q}(T) = 0$$

if, in condition (B_2), $\alpha \in]0, (2m-1)\omega[$. The only difference consists in defining Kv as the unique solution in V of

$$-\ddot{q}(t) - m^2\omega^2 q(t) = v(t)$$
$$q(0) - q(T) = \dot{q}(0) - \dot{q}(T) = 0.$$

The same approach can also be used to study Hamiltonian problems of the form

$$\pm(J\dot{u}(t) + m\omega u(t)) + \nabla H(t, u(t)) = 0$$
$$u(0) = u(T)$$

where $m \in \mathbf{Z}$ and $H(t, .)$ is convex.

Corollary 3.3. *Assume that the conditions* B_1 *to* B_3 *of Theorem* 3.7 *are replaced by the existence of numbers*

$$0 < \beta \leq \alpha < (2m+1)\omega$$

such that, for every $q \in \mathbf{R}^N$ *and a.e.* $t \in [0, T]$, *one has*

$$(\beta/2)|q|^2 - \gamma(t) \leq F(t, q) \leq (\alpha/2)|q|^2 + \gamma(t)$$

where $\gamma \in L^2(0, T; \mathbf{R}^+)$. *Then problem* (47) *has at least one solution* q *such that* $v = -\ddot{q} - m^2\omega^2 q$ *minimizes on* V *the dual action* χ.

Proof. It is easy to show that the assumptions B_1 to B_3 of Theorem 3.7 hold. □

3.8 The Range of Some Second Order Nonlinear Operators with Periodic Boundary Conditions

We shall study the following special case of (47) with $N = 1$.

$$\begin{aligned} \ddot{q}(t) + m^2\omega^2 q(t) + g(t, q(t)) = h(t) \\ q(0) - q(T) = \dot{q}(0) - \dot{q}(T) = 0 \end{aligned} \tag{63}$$

where $h \in C(0, T)$, $m \in \mathbf{N} \setminus \{0\}$, $\omega = 2\pi/T$, $g : [0, T] \times \mathbf{R} \to \mathbf{R}$ is continuous and $g(t, .)$ is non-decreasing for each $t \in [0, T]$. Letting

$$F(t, q) = \int_0^T g(t, s)\, ds - h(t)q, \tag{64}$$

we see that (63) is a special case of (42). We write, as usual, $u^+ = \max(u, 0)$, $u^- = \max(-u, 0)$, and use the notations of Section 3.7.

Proposition 3.5. *Assume that F defined in* (64) *satisfies the assumptions* B_1 *and* B_2 *of Theorem* 3.7 (*with* $N = 1$). *Assume, moreover, that there exist* $q_1 \in L^\infty(0,T)$, $q_2 \in L^\infty(0,T)$, *such that*

$$\int_0^T h(t) \sin(m\omega t + \varphi)\, dt < \int_0^T g(t, q_1(t)) \sin^+(m\omega t + \varphi)\, dt$$

$$- \int_0^T g(t, q_2(t)) \sin^-(m\omega t + \varphi)\, dt$$

for all $\varphi \in \mathbf{R}$. *Then problem* (63) *has at least one solution.*

Proof. It suffices to prove that condition (B_3) with $N = 1$ holds in Theorem 3.7, or equivalently that

$$\int_0^T F(t, A \sin(m\omega t + \varphi))\, dt \to +\infty \tag{65}$$

as $A \to +\infty$ uniformly in $\varphi \in \mathbf{R}$. By convexity, we have

$$\int_0^T F(t, A\sin(m\omega t + \varphi))\, dt = \int_{\sin(m\omega t+\varphi)>0} G(t, A\sin(m\omega t + \varphi))\, dt$$

$$+ \int_{\sin(m\omega t+\varphi)<0} G(t, A\sin(m\omega t + \varphi))\, dt - A \int_0^T h(t) \sin(m\omega t + \varphi)\, dt$$

$$\geq \int_{\sin(m\omega t+\varphi)>0} G(t, q_1(t)) + g(t, q_1(t))(A\sin(m\omega t + \varphi) - q_1(t))\, dt$$

$$+ \int_{\sin(m\omega t+\varphi)<0} G(t, q_2(t)) + g(t, q_2(t))(A\sin(m\omega t + \varphi) - q_2(t))\, dt$$

$$- A \int_0^T h(t) \sin(m\omega t + \varphi)\, dt \geq A \int_0^T g(t, q_1(t) \sin^+(m\omega t + \varphi)\, dt$$

$$- A \int_0^T g(t, q_2(t)) \sin^-(m\omega t + \varphi)\, dt + c_1 + c_2 - A \int_0^T h(t) \sin(m\omega t + \varphi)\, dt,$$

and the result follows from (65), because, by continuity and periodicity, the difference between the right and the left members has a positive lower bound independent of φ. □

Let now $g_+ : [0,T] \to]-\infty, +\infty]$ and $g_- : [0,T] \to]-\infty, +\infty[$ be defined on $[0,T]$ by

$$g_+(t) = \lim_{q \to +\infty} g(t, q), \quad g_-(t) = \lim_{q \to -\infty} g(t, q), \tag{66}$$

so that $g_-(t) \le g(t,0) \le g_+(t)$ on $[0,T]$ and hence g_- and g_+ are Lebesgue-integrable on $[0,T]$, with possibly

$$\int_0^T g_-(t)\,dt = -\infty \quad \text{or} \quad \int_0^T g_+(t)\,dt = +\infty.$$

Theorem 3.8. *Assume that F defined in* (64) *satisfies the assumptions* B_1 *and* B_2 *of Theorem* 3.7 (*with* $N=1$). *Assume, moreover, that $h \in C(0,T)$ is such that*

$$\int_0^T h(t)\sin(m\omega t+\varphi)\,dt < \int_0^T [g_+(t)\sin^+(m\omega t+\varphi) - g_-(t)\sin^-(m\omega t+\varphi)]\,dt \tag{67}$$

for all $\varphi \in \mathbf{R}$. Then problem (63) *has at least one solution.*

Proof. We show that (65) holds for some $q_1 \in \mathbf{R}$ and $q_2 \in \mathbf{R}$. If it is not the case we can find a sequence φ_k in $[0,T]$ converging to some φ_0 such that

$$\int_0^T [g(t,k)\sin^+(m\omega t+\varphi_k) - g(t,-k)\sin^-(m\omega t+\varphi_k) - h(t)\sin(m\omega t+\varphi_k)]\,dt \le 0,$$

and hence, by Fatou's lemma,

$$0 \ge \int_0^T [g_+(t)\sin^+(m\omega t+\varphi_0) - g_-(t)\sin^-(m\omega t+\varphi_0) - h(t)\sin(m\omega t+\varphi_0)]\,dt,$$

a contradiction with (67). □

Remark 3.5. As

$$g_-(t) \le g(t,q) \le g_+(t)$$

for a.e. $t \in [0,T]$ and all $q \in \mathbf{R}$, we see that if (63) has a solution q, then necessarily for all $\varphi \in \mathbf{R}$,

$$\begin{aligned}\int_0^T h(t)\sin(m\omega t+\varphi)\,dt &= \int_0^T g(t,q(t))\sin(m\omega t+\varphi)\,dt \\ &= \int_{\sin(m\omega t+\varphi)>0} g(t,q(t))\sin(m\omega t+\varphi)\,dt \\ &\quad + \int_{\sin(m\omega t+\varphi)<0} g(t,q(t))\sin(m\omega t+\varphi)\,dt \\ &\le \int_0^T g_+(t)\sin^+(m\omega t+\varphi)\,dt - \int_0^T g_-(t)\sin^-(m\omega t+\varphi)\,dt\end{aligned}$$

which shows that condition (65) with $\leq$ instead of $<$ is necessary for the solvability of (63). Such an "almost" necessary and sufficient condition for the existence of a solution to (63) when B_1 and B_2 hold is called a *Landesman–Lazer condition.*

Remark 3.6. When g is independent of t, g_- and g_+ are constant and (67) takes the simpler form

$$\left[\left(\frac{1}{T}\int_0^T h(t)\sin m\omega t\, dt\right)^2 + \left(\frac{1}{T}\int_0^T h(t)\cos m\omega t\, dt\right)^2\right]^{1/2} < \frac{1}{\pi}(g_+ - g_-).$$

Of course, similar results hold for the problem

$$-\ddot{q}(t) - m^2\omega^2 q(t) + g(t, q(t)) = h(t)$$

$$q(0) - q(T) = \dot{q}(0) - \dot{q}(T) = 0.$$

Historical and Bibliographical Notes

Theorem 3.1 is a special case of a more general existence theorem due to Mawhin–Willem [MaW_1] and the same is true for Corollary 3.1. Those results were motivated by earlier ones of Clarke [Cl_1], who introduced the duality technique, and of Brezis–Coron [BrC_1], who introduced the perturbation argument. The more general conditions of Theorem 3.1 require new arguments in the obtention of the corresponding a posteriori estimates. Proposition 3.3 is due to Willem [Wil_2]. Theorem 3.2, which can be found in [Wil_2], extends some results of Rabinowitz [Rab_1] and Clarke–Ekeland [ClE_2]. Theorem 3.3 slightly improves a result of Brezis–Coron [BrC_1], motivated by earlier contributions of Clarke–Ekeland [ClE_1] (the first paper on solutions with prescribed minimal period). The example in Remark 3.2 is due to Rabinowitz [Rab_2] and Proposition 3.4 is due to Yorke [Yor_1], with a short proof of Lasry–Willem. See Busenberg–Martelli [BuM_1] for another approach and generalizations.

Lemmas 3.1 and 3.2 are due to Rabinowitz [Rab_3] and Theorem 3.4 to Weinstein [Wei_1], the present proof being modeled on Clarke–Ekeland [ClE_1]. An extension to some starshaped energy surfaces has been devised by Rabinowitz [$\mathrm{Rab}_{3,4}$], using a Galerkin and a minimax argument. The results of the Sections 3.6 and 3.7 are special cases of more general existence theorems of Mawhin–Willem [MaW_1], Propositions 3.5 and Theorem 3.8 being close to recent results of Caristi [Car_1]. The Landeman–Lazer condition [LaLa_1] in Theorem 3.8 was first introduced for periodic problems by Lazer–Leach [LaLe_1] (using Schauder's fixed point theorem). See [GaM_1] and [Fuc_1] for various approaches to Landesman–Lazer type problems.

Corollary 3.3 is a special case of a result of [Wil_6] and generalizes [Dol_1], [LaS_1], [$Maw_{9,10}$].

Abstract versions of Theorems 3.1, 3.5, 3.7, and 3.8 can be found also in [Bre_2], [Car_1], [Dim_1], [EkL_2], [$Maw_{3,4}$], [Wil_6]. Other results on periodic solutions of Hamiltonian systems based upon the dual least action principle are given in [$Amb_{4,5}$], [$Bere_1$], [Blo_1], [Cla_6], [Eke_{11}], [$Manc_1$], [$Wil_{8,9}$].

For applications of the dual least action principle to other boundary value problems, the reader can consult [AmS_1], [$Maw_{5,6,7}$], [$MWW_{1,2}$]. Results in this line based upon a combination of Lyapunov–Schmidt arguments and variational methods are given in [$Bat_{1,2}$], [BaC_1], [$Cas_{3,4,5}$], [Cal_2], [$Fac_{2,3,4}$], [Laz_1], [LLM_1], [LMc_1], [Ter_1], [The_1].

The importance of getting a posteriori estimates for the solutions of variational problems was first emphasized by Dolph ([Dol_1]) in a different context.

Some of the problems considered in this chapter can also be treated, under supplementary conditions, by the method of natural isoperimetric conditions (or natural constraints) initiated by Poincaré [Poi_2] and developed in [$BiH_{1,2}$], [HeS_1], [$Ber_{3,4}$], [BeB_1], [VGr_{1-4}].

Periodic solutions of conservative mechanical systems with fixed energy was first considered by a geometrical approach in [Sei_1]. For further results on periodic solutions of Hamiltonian systems, see also [Cro_1], [Den_1], [The_2], [Wei_3], [Gia_1], [Rab_{17}].

Exercises

1. Show that Theorem 3.1 also holds, under the same assumptions A_1 and A_2 upon H, for the problem

$$J\dot{u}(t) + \nabla H(t, u(t)) = e(t),$$
$$u(0) = u(T),$$

where $e \in L^2(0, T; \mathbf{R}^{2N})$ is such that

$$\int_0^T H(t,u)\,dt - \left(u, \int_0^T e(t)\,dt\right) \to +\infty$$

as $|u| \to \infty$ in $\mathbf{R}^{2N}$.

Hint: Writing $e = \overline{e} + \tilde{e}$, with $\overline{e} = T^{-1} \int_0^T e(t)\,dt$, make the change of variable $u = w + \tilde{E}$, where $\tilde{E}$ is the unique solution of

$$J\dot{v} = \tilde{e}(t), \quad v(0) = v(T), \quad \overline{v} = 0,$$

and apply Theorem 3.1 to the equivalent problem

$$J\dot{w} + \nabla H(t, w(t) + \tilde{E}(t)) = 0$$
$$w(0) = w(T).$$

2. Show that if $0 < \alpha < 2$ and $c \in C(\mathbf{R}, \mathbf{R})$ is such that $c(t) \geq \gamma > 0$ for all $t \in \mathbf{R}$ and is T-periodic, the problem
$$J\dot{u}(t) + \alpha c(t)\, |u(t)|^{\alpha-2} u(t) = 0$$
has, for each $k \in \mathbf{N} \backslash \{0\}$, a kT-periodic solution u_k such that $\|u_k\| \to \infty$ and its minimal period $T_k \to +\infty$ when $k \to \infty$.

3. Consider the problem
$$J\dot{u} + c\alpha\, |u|^{\alpha-2} u = 0$$
where $c > 0$ and $\alpha > 1$ and show by direct computation and using the energy integral that its general solution is given by
$$u(t) = \cos(\omega t)\xi + \sin(\omega t)\, J\xi$$
where $\xi \in \mathbf{R}^{2N}$ and $\omega = \alpha c^{2/\alpha} h^{\alpha - 2/\alpha}$, $h = c\,|\xi|^\alpha$. For $\alpha < 2$, compare this result with that of Theorem 3.3.

4. Let $H \in C^1(\mathbf{R}^{2N}, \mathbf{R})$. If $S = H^{-1}(c)$ is a sphere and if, for each $u \in S$, $\nabla H(u) \neq 0$, then all the solutions of $J\dot{u} + \nabla H(u) = 0$ on S are periodic.

5. Let $H(u) = \sum_{n=1}^{N} \frac{\omega_n}{2}(u_n^2 + u_{N+n}^2)$ where the ω_n are positive real numbers which are rationally independent. Show that, for each $c > 0$, $H^{-1}(c)$ contains exactly N orbits of periodic solutions of the system
$$J\dot{u} + \nabla H(u) = 0.$$

6. Show that under the assumptions A_1 and A_2 of Theorem 3.5, the conclusion of this theorem holds for
$$\ddot{q}(t) + \nabla F(t, q(t)) = e(t),$$
$$q(0) - q(T) = \dot{q}(0) - \dot{q}(T) = 0,$$
where $e \in L^2(0, T; \mathbf{R}^N)$ is such that
$$\int_0^T F(t,q)\, dt - \left(q, \int_0^T e(t)\, dt \right) \to +\infty$$
as $|q| \to \infty$ in $\mathbf{R}^N$.

Hint. Adapt the argument of Exercise 4.

7. Formulate and prove the analog of Theorem 3.7 for problems of the form
$$\pm(J\dot{u}(t) + m\omega u(t)) + \nabla H(t, u(t)) = 0,$$
$$u(0) = u(T),$$
where $\omega = 2\pi/T$, $m \in \mathbf{Z} \setminus \{0\}$ and $H(t, .)$ is convex.

4

Minimax Theorems for Indefinite Functionals

Introduction

The dual least action principle has provided sharp existence theorems for the periodic solutions of Hamiltonian systems when the Hamiltonian is convex in u. When it is not the case, the existence of critical points of saddle point type can be proved by using some *minimax arguments.* To motivate them, we can consider the following intuitive situation. If $\varphi \in C^1(\mathbf{R}^2, \mathbf{R})$, we can view $\varphi(x, y)$ as the altitude of the point of the graph of φ having (x, y) as projection on $\mathbf{R}^2$. Assume that there exists points $u_0 \in \mathbf{R}^2$, $u_1 \in \mathbf{R}^2$ and a bounded open neighborhood Ω of u_0 such that $u_1 \in \mathbf{R}^2 \setminus \Omega$ and $\varphi(u) > \max(\varphi(u_0), \varphi(u_1))$ whenever $u \in \partial\Omega$ (that is the case for example if u_0 and u_1 are two isolated local minimums of φ).

Looking at the graph of φ in a topographical way, we can thus consider the point $[u_0, \varphi(u_0)]$ as located in a valley surrounded by a ring of mountains pictured by the set $\{[u, \varphi(u)] : u \in \partial\Omega\}$, the point $[u_1, \varphi(u_1)]$ being located outside of the ring. To go from $[u_0, \varphi(u_0)]$ to $[u_1, \varphi(u_1)]$ in a way which minimizes the highest altitude on the path, we must cross the mountain ring through the lowest *mountain pass.* The projection on $\mathbf{R}^2$ of the top of this mountain pass will provide a critical point of φ with critical value

$$c = \inf_{g \in \Gamma} \max_{s \in [0,1]} \varphi(g(s)),$$

where Γ denotes the set of paths joining u_0 to u_1 (i.e. the set of continuous mappings $g : [0, 1] \to \mathbf{R}^2$ with $g(0) = u_0$, $g(1) = u_1$). The validity of this result will be insured only when some compactness condition is satisfied by φ and variants of the result will be obtained by modifying the class Γ.

Those minimax theorems will be deduced from *Ekeland's variational principle* for a semi-continuous real function f which is bounded below on a complete metric space M (another approach, based upon deformations along the paths of steepest descent of φ can be used and is developed in Chapter 6). If a real function achieves its minimum on M at a, its graph will lie entirely in the "half-space" $\{[u, s] \in M \times \mathbf{R} : s \geq f(a)\}$. Ekeland's variational principle insures that, for each $\epsilon > 0$, there is some $a_\epsilon \in M$ such that

$$f(a_\epsilon) \leq \inf_M f + \epsilon$$

and the graph of f lies entirely in the "cone" $\{[u,s] \in M \times \mathbf{R} : s \geq f(a_\epsilon) - \epsilon d(a_\epsilon, u)\}$. This theorem, given in Section 4.1, has many applications (to optimization, optimal control, fixed points, dynamical systems, differential geometry, ...) and variants which will not be developed here. To apply it to the mountain pass situation described above, one will take in Section 4.5 $M = \Gamma$ and $f(g) = \max_{s\in[0,1]} \varphi(g(s))$.

To go from the existence of an "almost critical point" in Ekeland's principle to that of a critical point, a *compactness condition* of the type introduced by Palais and Smale in their extensions of Lusternik–Schnirelmann and Morse theories to infinite-dimensional spaces is required. Such conditions are analyzed in Section 4.2 where an example shows how those conditions are related to the obtention of suitable a priori bounds. This example concerns functionals φ of the form

$$\varphi(u) = \int_0^T L(t, u(t), \dot{u}(t))\, dt$$

where $L(t,x,y) = (1/2)(M(t,x)y,y) - V(t,x) + (f(t),x)$, $M(t,.)$ and $V(t,.)$ are T_i-periodic in each variable x_i and f has mean value zero, so that they generalize the ones considered in Section 1.6.

Another minimax result, *Rabinowitz saddle point theorem,* is used in Section 4.3 to provide existence results for the periodic solutions of equations like

$$\ddot{u} + g(u) = h(t),$$

when g, not necessarily monotone, is bounded, and the corresponding φ is indefinite. Again, the existence conditions are related to Landesman–Lazer conditions. The same saddle point theorem is applied to Section 4.4 to the periodic solutions of systems describing a Josephson multipoint junction.

Each method to obtain critical points can, of course, be combined to the dual least action principle and Section 4.5 provides an interesting application to the periodic solutions with fixed point of an autonomous Hamiltonian system

$$J\dot{u}(t) + \nabla H(u(t)) = 0$$

with H convex, superquadratic and such that $\nabla H(0) = 0$. The corresponding dual action χ is unbounded from above and from below, but the existence of a nontrivial critical point can be deduced from the mountain pass theorem.

In Section 4.6, we introduce the concept of Lusternik–Schnirelman category of a subset A of a topological space Y (namely the smallest integer such that A can be covered by k closed sets contractible in Y) to prove a multiplicity result for the critical points of functionals $\varphi : X \to \mathbf{R}$ which are bounded from below and invariant under the action of a discrete subgroup G of the Banach space X. This is done as an application of Ekeland's variational principle. Let $\pi : X \to X/G$ be the canonical surjection and

$1 \leq j \leq \dim \operatorname{span} G + 1$. The critical levels are obtained by a minimax characterization of the type

$$\inf_{A \in \mathcal{A}_j} \max_A \varphi$$

where $\mathcal{A}_j$ denotes the compact subsets $\pi(A)$ or $\pi(X)$ such that the category in $\pi(X)$ of $\pi(A)$ is not less than j.

A natural application to the systems described in Section 4.2 provide the existence of at least $N + 1$ geometrically distinct critical points. This gives in particular two T-periodic solutions at least for the forced pendulum equation under the assumption of Section 1.6. This is specially interesting if we notice that, in the unforced case, the periodic solution which minimizes the action corresponds to the unstable equilibrium.

4.1 Ekeland's Variational Principle and the Existence of Almost Critical Points

Theorem 4.1. *Let M be a complete metric space and let $\Phi : M \to]-\infty, +\infty]$ be a l.s.c. function, bounded from below and not identical to $+\infty$. Let $\epsilon > 0$ be given and $u \in M$ be such that*

$$\Phi(u) \leq \inf_M \Phi + \epsilon.$$

Then there exists $v \in M$ such that

$$\begin{aligned} \Phi(v) &\leq \Phi(u) \\ d(u,v) &\leq 1 \end{aligned} \tag{1}$$

and, for each $w \neq v$ in M,

$$\Phi(w) > \Phi(v) - \epsilon d(v,w). \tag{2}$$

Proof. The relation

$$w \leq v \Leftrightarrow \Phi(w) + \epsilon d(v,w) \leq \Phi(v)$$

defines an ordering on M, as checked immediately. Let us construct inductively a sequence (u_n) as follows, starting with $u_0 = u$. If we suppose that u_n is known, let

$$S_n = \{w \in M : w \leq u_n\}$$

and let us choose $u_{n+1} \in S_n$ such that

$$\Phi(u_{n+1}) \leq \inf_{S_n} \Phi + \frac{1}{n+1}.$$

Clearly, $S_{n+1} \subset S_n$, as $u_{n+1} \le u_n$, and, since Φ is l.s.c., S_n is closed. Now, if $w \in S_{n+1}$, $w \le u_{n+1} \le u_n$ and hence

$$\epsilon d(w, u_{n+1}) \le \Phi(u_{n+1}) - \Phi(w) \le \inf_{S_n} \Phi + \frac{1}{n+1} - \inf_{S_n} \Phi = \frac{1}{n+1}$$

so that

$$\operatorname{diam} S_{n+1} \le \frac{2}{\epsilon(n+1)},$$

i.e. diam $S_n \to 0$ as $n \to \infty$. M being complete, this implies that

$$\bigcap_{n \in \mathbf{N}} S_n = \{v\} \tag{3}$$

for some $v \in M$. In particular, $v \in S_0$, i.e.

$$v \le u_0 = u$$

so that

$$\Phi(v) \le \Phi(u) - \epsilon d(u, v) \le \Phi(u)$$

and

$$d(u, v) \le \epsilon^{-1}(\Phi(u) - \Phi(v)) \le \epsilon^{-1}\left(\inf_M \Phi + \epsilon - \inf_M \Phi\right) = 1.$$

To obtain (2), it suffices to prove that $w \le v$ implies $w = v$. If $w \le v$, then, for each $n \in \mathbf{N}$,

$$w \le u_n$$

so that $w \in \cap_{n \in \mathbf{N}} S_n$ and, by (3), $w = v$. □

Remark 4.1. By using the equivalent distance λd with $\lambda > 0$, the conclusions (1) and (2) can be respectively replaced by

$$d(u, v) \le 1/\lambda$$

and

$$\Phi(w) > \Phi(v) - \epsilon \lambda d(v, w). \tag{4}$$

The choice $\lambda = \epsilon^{-1/2}$ is then particularly interesting. We first prove a result for functions bounded from below on a Banach space.

Theorem 4.2. *Let X be a Banach space, $\varphi : X \to \mathbf{R}$ be a function bounded from below, and differentiable on X. Then, for each $\epsilon > 0$ and for each $u \in X$ such that*

$$\varphi(u) \le \inf_X \varphi + \epsilon, \tag{5}$$

there exists $v \in X$ such that

$$\varphi(v) \le \varphi(u) \tag{6}$$

$$|u - v| \le \epsilon^{1/2} \tag{7}$$

$$|\varphi'(v)| \le \epsilon^{1/2}. \tag{8}$$

Proof. Let us take $M = X$, $\Phi = \varphi$ and, for $\epsilon > 0$ given, let us choose $\lambda = \epsilon^{-1/2}$ like in Remark 4.1 of Theorem 4.1. Then, if u satisfies (5), there exists $v \in X$ such that (6), (7) hold and

$$\varphi(w) > \varphi(v) - \epsilon^{1/2}|v - w| \tag{9}$$

for all $w \neq v$ in X. Therefore, taking $w = v + th$ with $t > 0$, $h \in X$, $|h| = 1$ in (9) we get

$$\varphi(v + th) - \varphi(v) > -\epsilon^{1/2}t.$$

Dividing both members by t and letting $t \to 0$, we obtain

$$-\epsilon^{1/2} \le \langle \varphi'(v), h \rangle$$

for all $h \in X$ with $|h| = 1$, and hence (8). □

Corollary 4.1. *Let X be a Banach space, $\varphi : X \to \mathbf{R}$ be a function bounded from below and differentiable on X. Then, for each minimizing sequence (u_k) of φ, there exists a minimizing sequence (v_k) of φ such that*

$$\varphi(v_k) \le \varphi(u_k)$$

$$|u_k - v_k| \to 0 \ \text{ if } \ k \to \infty$$

$$|\varphi'(v_k)| \to 0 \ \text{ if } \ k \to \infty.$$

Proof. If (u_k) is a minimizing sequence for φ, take

$$\begin{aligned} \epsilon_k &= \varphi(u_k) - \inf_X \varphi && \text{if } \varphi(u_k) - \inf_X \varphi > 0 \\ &= 1/k && \text{if } \varphi(u_k) - \inf_X \varphi = 0 \end{aligned}$$

and then take v_k associated to u_k and ϵ_k in Theorem 4.2. □

For an indefinite function, we shall state and prove a minimax theorem modeled on the intuitive situation described in the introduction.

Theorem 4.3. *Let K be a compact metric space, $K_0 \subset K$ a closed set, X a Banach space, $\chi \in C(K_0, X)$ and let us define the complete metric space M by*

$$M = \{g \in C(K, X) : g(s) = \chi(s) \text{ if } s \in K_0\}$$

with the usual distance d. Let $\varphi \in C^1(X, \mathbf{R})$ and let us define

$$c = \inf_{g \in M} \max_{s \in K} \varphi(g(s)), \quad c_1 = \max_{\chi(K_0)} \varphi.$$

If

$$c > c_1, \tag{10}$$

then for each $\epsilon > 0$ *and each* $f \in M$ *such that*

$$\max_{s\in K} \varphi(f(s)) \leq c + \epsilon, \tag{11}$$

there exists $v \in X$ *such that*

$$c - \epsilon \leq \varphi(v) \leq \max_{s\in K} \varphi(f(s)),$$

$$\text{dist}\,(v, f(K)) \leq \epsilon^{1/2},$$
$$|\varphi'(v)| \leq \epsilon^{1/2}.$$

Proof. Without loss of generality, we can assume that

$$0 < \epsilon < c - c_1. \tag{12}$$

Let $f \in M$ satisfying the condition (11). We define the function $\Phi : M \to \mathbf{R}$ by

$$\Phi(g) = \max_{s\in K} \varphi(g(s)),$$

so that $c = \inf_M \Phi > c_1$. To show that Φ is continuous, one uses the uniform continuity of φ on $g(K)$.

Now Theorem 4.1 implies the existence of $h \in M$ such that

$$\Phi(h) \leq \Phi(f) \leq c + \epsilon,$$

$$\max_{s\in K} |h(s) - f(s)| \leq \epsilon^{1/2}$$

and

$$\Phi(g) > \Phi(h) - \epsilon^{1/2} d(h, g) \tag{13}$$

whenever $g \in M$ and $g \neq h$. Thus our theorem will be proved if we show the existence of some $s \in K$ such that

$$c - \epsilon \leq \varphi(h(s))$$

and

$$|\varphi'(h(s))| \leq \epsilon^{1/2},$$

i.e.

$$\langle \varphi'(h(s)), v\rangle \geq -\epsilon^{1/2}$$

whenever $v \in X$ and $|v| = 1$. If it is not the case, then, for each $s \in S$, where

$$S = \{s \in K : c - \epsilon \leq \varphi(h(s))\},$$

there exist $\delta_s > 0$, $v_s \in X$ with $|v_s| = 1$ and an open ball B_s in K containing s such that for $t \in B_s$ and $u \in X$ with $|u| \leq \delta_s$, we have

$$\langle \varphi'(h(t)) + u, v_s\rangle < -\epsilon^{1/2}. \tag{14}$$

(We have used the continuity of φ'.) Since S is compact, there exists a finite subcovering $B_{s_1}, \ldots, B_{s_k}$ of S and we define $\psi_j : K \to [0,1]$ by

$$\psi_j(t) = \frac{\operatorname{dist}(t, \mathbf{C}B_{s_j})}{\sum_{i=1}^k \operatorname{dist}(t, \mathbf{C}B_{s_i})}, \quad \text{if } t \in \bigcup_{i=1}^k B_{s_i}$$

$$\psi_j(t) = 0, \quad \text{if } t \in K \setminus \bigcup_{i=1}^k B_{s_i}.$$

Finally, let $\delta = \min(\delta_{s_1}, \ldots, \delta_{s_k})$, let $\psi : K \to [0,1]$ be a continuous function such that

$$\begin{aligned} \psi(t) \quad &= 1 \quad \text{if } c \leq \varphi(h(t)) \\ &= 0 \quad \text{if } \varphi(h(t)) \leq c - \epsilon, \end{aligned}$$

and let $g \in C(K, X)$ be defined by

$$g(t) = h(t) + \delta\psi(t) \sum_{j=1}^k \psi_j(t) v_{s_j}.$$

It follows from (12) that, for $t \in K_0$,

$$\varphi(h(t)) = \varphi(\chi(t)) \leq c_1 < c - \epsilon$$

and hence $\psi(t) = 0$. Thus, for $t \in K_0$,

$$g(t) = h(t) = \chi(t),$$

i.e. $g \in M$. Let us now estimate $\Phi(g)$ from above. The mean value theorem and (14) imply that, for each $t \in S$, there is some $0 < \tau < 1$ for which

$$\begin{aligned} \varphi(g(t)) - \varphi(h(t)) &= \langle \varphi' \left(h(t) + \tau\delta\psi(t) \sum_{j=1}^k \psi_j(t) v_{s_j} \right), \delta\psi(t) \sum_{j=1}^k \psi_j(t) v_{s_j} \rangle \\ &= \delta\psi(t) \sum_{j=1}^k \psi_j(t) \langle \varphi' \left(h(t) + \tau\delta\psi(t) \sum_{j=1}^k \psi_j(t) v_{s_j} \right), v_{s_j} \rangle \\ &\leq -\epsilon^{1/2} \delta\psi(t). \end{aligned} \tag{15}$$

If $t \notin S$, $\psi(t) = 0$ and $\varphi(g(t)) = \varphi(h(t))$. Now, if $\bar{t}$ is such that $\varphi(g(\bar{t})) = \Phi(g)$, we obtain

$$\varphi(h(\bar{t})) \geq \varphi(g(\bar{t})) \geq c$$

so that $\bar{t} \in S$ and $\psi(\bar{t}) = 1$. By (15), we get

$$\varphi(g(\bar{t})) - \varphi(h(\bar{t})) \leq -\epsilon^{1/2} \delta$$

and in particular

$$\Phi(g) + \epsilon^{1/2}\delta \leq \varphi(h(\bar{t})) \leq \Phi(h)$$

so that $g \neq h$. But, by the definition of g, we have

$$d(g, h) \leq \delta$$

and hence

$$\Phi(g) + \epsilon^{1/2} d(g, h) \leq \Phi(h),$$

which contradicts (13) and completes the proof. □

The following result gives sufficient conditions insuring that (10) is satisfied.

Corollary 4.2. *Let K, K_0, X, χ, M, φ, c, and c_1 be defined as in Theorem* 4.3. *Assume that there exists $S \subset X$ such that*

$$g(K) \cap S \neq \phi \quad \text{for all} \quad g \in M, \tag{16}$$

and let

$$c_0 = \inf_S \varphi.$$

Then, if

$$c_1 < c_0, \tag{17}$$

the condition (10) *of Theorem* 4.3 *holds and hence also its conclusion.*

Proof. By (16), we have

$$c = \inf_{g \in M} \max_{s \in K} \varphi(g(s)) \geq c_0$$

and then (10) follows from (17). □

Corollary 4.3. *Under the conditions of Theorem* 4.3, *for each sequence (f_k) in M such that*

$$\max_K \varphi(f_k) \to c,$$

there exists a sequence (v_k) in X such that

$$\varphi(v_k) \to c$$

$$\operatorname{dist}(v_k, f_k(K)) \to 0$$

$$|\varphi'(v_k)| \to 0$$

when $k \to \infty$.

Proof. We define $\epsilon_k = \max_K \varphi(f_k) - c$ if $\max_K \varphi(f_k) - c > 0$ and $\epsilon_k = 1/k$ in the other case and we apply, for each $k \in \mathbf{N}^*$, Theorem 4.3 to ϵ_k and f_k. □

4.2 A Closedness Condition and the Existence of Critical Points

We have seen in the previous section how to obtain "almost" critical points of functions of class $\mathbf{C}^1$ on a Banach space X. Some auxiliary closedness condition is required to obtain the existence of critical points. We shall require that if $(C, 0) \subset \mathbf{R} \times X^*$ is in the closure of the range of $\varphi \times \varphi'$, then it must be in the range of $\varphi \times \varphi'$.

Definition 4.1. *Let $\varphi : X \to \mathbf{R}$ differentiable and $c \in \mathbf{R}$. We say that φ satisfies the $(\mathrm{PS})_c$-condition if the existence of a sequence (u_k) in X such that*

$$\varphi(u_k) \to c, \quad \varphi'(u_k) \to 0$$

as $k \to \infty$, implies that c is a critical value of φ.

Remark 4.2. We shall use later a compactness condition called the *Palais–Smale condition* (PS-*condition*) and which requires that every sequence (u_j) in X such that $(\varphi(u_j))$ is bounded and $\varphi'(u_j) \to 0$ as $j \to \infty$ contains a convergent subsequence. It is clear that the PS-condition implies the $(\mathrm{PS})_c$-condition for each $c \in \mathbf{R}$. Example 2 below shows that the converse is not true.

Examples

1.Let $X = \mathbf{R}$ and $\varphi(u) = \exp u$. As $\varphi'(u) = \exp u$, only sequences (u_k) such that $u_k \to -\infty$ are such that $\varphi'(u_k) \to 0$; for such a sequence, $\varphi(u_k) \to 0$ but 0 is not a critical value for φ. Thus, exp does not satisfy the $(\mathrm{PS})_0$-condition. It trivially satisfies the $(\mathrm{PS})_c$-condition when $c \neq 0$.

2. Let $X = \mathbf{R}$ and $\varphi(u) = \sin u$; if (u_k) is such that

$$\sin u_k \to c, \quad \cos u_k \to 0$$

as $k \to \infty$, we can write

$$u_k = 2m_k\pi + v_k \quad \text{with} \quad v_k \in [0, 2\pi]$$

so that

$$\sin v_k \to c, \quad \cos v_k \to 0$$

as $k \to \infty$. Now (v_k) has a convergent subsequence with limit v such that

$$\sin v = c, \quad \cos v = 0,$$

and hence c is a critical for sin. Thus sin satisfies the $(\mathrm{PS})_c$-condition for every $c \in \mathbf{R}$.

3. Let $X = \mathbf{R}$ and $\varphi(u) = u$; then $\varphi'(u) = 1$ and the $(\mathrm{PS})_c$-condition holds for each $c \in \mathbf{R}$.

4. Let $X = \mathbf{R}$ and $\varphi(u) = u^2/2$; then $\varphi'(u) = u$ and $\varphi'(u_k) \to 0$ if and only if $u_k \to 0$, in which case $\varphi(u_k) \to 0$. Thus $(\mathrm{PS})_0$ holds and, trivially, $(\mathrm{PS})_c$ holds for all $c \neq 0$.

The examples above show that functions satisfying the $(\mathrm{PS})_c$-condition for each $c \in \mathbf{R}$ may have an infinity, a finite number of critical points, or no ones. Corollary 4.1 can be used to provide a sufficient condition for the existence of a critical point.

Theorem 4.4. *Let X be a Banach space, $\varphi : X \to \mathbf{R}$ a function bounded from below and differentiable on X. If φ satisfies the $(\mathrm{PS})_c$-condition with $c = \inf_X \varphi$, then φ has a minimum on X.*

Proof. By Corollary 4.1, there exists a minimizing sequence (v_k) such that $\varphi'(v_k) \to 0$ as $k \to \infty$. By the $(\mathrm{PS})_c$-condition with $c = \inf \varphi$, c is a critical value and the proof is complete. □

As an example of application to differential equations, let us consider the following problem

$$\begin{gathered}\frac{d}{dt} D_y L(t, u(t), \dot{u}(t)) \\ u(0) - u(T) = \dot{u}(0) - \dot{u}(T) = 0\end{gathered} \tag{18}$$

where

$$L(t,x,y) = (1/2)(M(t,x)y,y) - V(t,x) + (f(t),x)$$

is such that the following conditions hold:

(L_1) $M(t,x)$ is a symmetric matrix of order N continuously differentiable on $[0,T] \times \mathbf{R}^N$ and such that

$$(M(t,x)y,y) \geq \alpha|y|^2$$

for some $\alpha > 0$ and every $(t,x,y) \in [0,T] \times \mathbf{R}^N \times \mathbf{R}^N$.

(L_2) $V(t,x)$ is measurable in t for every $x \in \mathbf{R}^N$ and continuously differentiable in x for almost every $t \in [0,T]$, and there exists $h \in L^1(0,T;\mathbf{R})$ such that

$$|V(t,x)| + |D_x V(t,x)| \leq h(t)$$

for every $x \in \mathbf{R}^N$ and almost every $t \in [0,T]$.

(L_3) $M(t,x)$ and $V(t,x)$ are T_i-periodic in x_i, $1 \leq i \leq N$.

(L_4) $f \in L^1(0,T;\mathbf{R}^N)$ and $\int_0^T f(t)\,dt = 0$. If $u \in L^1(0,T;\mathbf{R}^N)$, we shall write $u = \overline{u} + \tilde{u}$, where

$$\overline{u} = T^{-1} \int_0^T u(t)\,dt.$$

Proposition 4.1. *Under the assumptions* (L_1) *to* (L_4), *the function* φ *defined on* H_T^1 *by*

$$\varphi(u) = \int_0^T L(t, u(t), \dot{u}(t))\, dt$$

is continuously differentiable and bounded from below. Moreover, every sequence (u_k) *such that* $\varphi'(u_k) \to 0$, $\varphi(u_k)$ *is bounded and* $(\overline{u}_k)$ *is bounded contains a convergent subsequence.*

Proof. Using assumptions (L_1), (L_2), (L_4), and Sobolev inequality, we obtain

$$\varphi(u) \geq \alpha\|\dot{u}\|_{L^2}^2 - \int_0^T h(t)\, dt - \|f\|_{L^1}\|\tilde{u}\|_\infty \geq \alpha\|\dot{\tilde{u}}\|_{L^2}^2 - c_1 - c_2\|\dot{\tilde{u}}\|_{L^2}, \quad (19)$$

so that φ is bounded from below. Theorem 1.4 implies that φ is continuously differentiable on H_T^1 and that

$$\begin{aligned}\langle\varphi'(u), v\rangle = \int_0^T & [(M(t, u(t))\dot{u}(t), \dot{v}(t)) \\ & + (1/2)\sum_{i=1}^N (D_{x_i}M(t, u(t))\dot{u}(t), \dot{u}(t))v_i(t) \\ & - (D_x V(t, u(t))v(t)) + (f(t), v(t))]\, dt. \end{aligned} \quad (20)$$

Now let (u_k) be a sequence in H_T^1 such that $\varphi'(u_k) \to 0$, $\varphi(u_k)$ is bounded and $(\overline{u}_k)$ is bounded. It follows from (19) that $(\dot{\tilde{u}}_k)$ is bounded in L^2. Consequently, (u_k) is bounded in H_T^1 and, going if necessary to a subsequence, we can assume that $u_k \rightharpoonup u$ in H_T^1 and $u_k \to u$ in $C([0,T], \mathbf{R}^N)$. But then

$$\langle\varphi'(u_k) - \varphi'(u), u_k - u\rangle \to 0 \text{ as } k \to \infty. \quad (21)$$

Using (20) and assumption (L_1), we obtain

$$\begin{aligned}\langle\varphi'(u_k) - \varphi'(u), u_k - u\rangle &\geq \alpha\|\dot{u}_k - \dot{u}\|_{L^2}^2 \\ &+ \int_0^T ([M(t, u_k(t)) - M(t, u(t))]\dot{u}(t), \dot{u}_k(t) - \dot{u}(t))\, dt \\ &+ (1/2)\int_0^T \sum_{i=1}^N [D_{x_i}M(t, u_k(t))\dot{u}_k(t), \dot{u}_k(t)) - (D_{x_i}M(t, u(t))\dot{u}(t), \dot{u}(t))] \\ &[u_{k,i}(t) - u_i(t)]\, dt - \int_0^T (D_x V(t, u_k(t)) - D_x V(t, u(t)), u_k(t) - u(t))\, dt \\ &+ \int_0^T (f(t), u_k(t) - u(t))\, dt.\end{aligned}$$

It is then easy to verify, using (21), that $\|\dot{u}_k - \dot{u}\|_{L^2} \to 0$ as $k \to \infty$, and hence that $u_k \to u$ in H_T^1. □

Theorem 4.5. *Under assumptions* (L_1) *to* (L_4), *problem* (18) *has a weak solution.*

Proof. By Proposition 1, $c = \inf_{H_T^1} \varphi$ is finite. Let us verify the $(PS)_c$-condition. Let (u_k) be a sequence in H_T^1 such that $\varphi'(u_k) \to 0$ and $\varphi(u_k) \to c$. Because of (L_3) and (L_4), we have

$$\varphi(u + T_i e_i) = \varphi(u), \quad (1 \le i \le N) \tag{22}$$

and hence we can assume, without loss of generality, that

$$0 \le (\overline{u}_k, e_i) \le T_i, \quad 1 \le i \le N.$$

But then $(\overline{u}_k)$ is bounded and, by Proposition 4.1, we can assume, going if necessary to a subsequence, that $u_k \to u$ in H_T^1. Hence $\varphi'(u) = 0$ and $\varphi(u) = c$, i.e. c is a critical value of φ. Now Theorem 4.4 implies the existence of a minimum for φ at some point $u \in H_T^1$ at which, necessarily, $\varphi'(u) = 0$. Since

$$\langle \varphi'(u), v \rangle = 0$$

for all $v \in C_T^\infty$, relation (20) implies that $\dot{u}$ has a weak derivative. But then $\dot{u}(0) = \dot{u}(T)$, $u(0) = u(T)$ and u is a weak solution of (18). □

Remark 4.3. We shall prove in Section 4.6 that problem (18) has at least $N + 1$ "geometrically distinct" weak solutions.

We can now deduce the existence of critical points of functions $\varphi \in C^1(X, \mathbf{R})$ by adding a $(PS)_c$ condition to the assumptions of Theorem 4.3.

Theorem 4.6. *Under the assumptions of Theorem* 4.3, *if* φ *satisfies the* $(PS)_c$*-condition, then* c *is a critical value for* φ.

Proof. From Corollary 4.3, we obtain a sequence (v_k) such that

$$\varphi(v_k) \to c, \quad \varphi'(v_k) \to 0$$

as $k \to \infty$. By condition $(PS)_c$, c is a critical value and the proof is complete. □

Particular choices of K, K_0, and χ will provide interesting existence theorems for critical points of φ. Two important examples will be described in Sections 4.3 and 4.5.

4.3 The Saddle Point Theorem and Periodic Solutions of Second Order Systems with Bounded Nonlinearity

To motivate a first important special case of Theorem 4.6 which is due to Rabinowitz, let us return to the problem of Section 1.8

$$\begin{aligned} \ddot{u}(t) &= \nabla F(t,u(t)) \quad \text{(a.e. on } [0,T]) \\ u(0)-u(T) &= \dot{u}(0)-\dot{u}(T)=0 \end{aligned} \tag{23}$$

with the same regularity assumptions under F and with the condition

$$|\nabla F(t,x)| \le h(t) \tag{24}$$

for some $h \in L^1(0,T)$, a.e. $t \in [0,T]$ and all $x \in \mathbf{R}^N$. It was proved in Chapter 1 that (23) has a solution when

$$\int_0^T F(t,x)\,dt \to +\infty \quad \text{as } |x| \to \infty. \tag{25}$$

It is natural to raise the question of the existence of a solution when (25) is replaced by

$$\int_0^T F(t,x)\,dt \to -\infty \quad \text{as } |x| \to \infty. \tag{26}$$

In this case, the function φ is neither bounded from below, now from above. Indeed, if w is a constant function,

$$\varphi(w) = \int_0^T F(t,w)\,dt \to -\infty \quad \text{as } |w| \to \infty$$

and, if $v \in H_T^1$ has mean zero,

$$\begin{aligned} \varphi(v) &= \int_0^T (|\dot{v}(t)|^2/2)\,dt + \int_0^T F(t,0)\,dt + \int_0^T [F(t,v(t)) - F(t,0)]\,dt \\ &= \int_0^T (|\dot{v}(t)|^2/2)\,dt + \int_0^T F(t,0)\,dt + \int_0^T \int_0^1 (\nabla F(t,sv(t)), v(t))\,ds\,dt \\ &\ge \int_0^T (|\dot{v}(t)|^2/2)\,dt - c_1 - \left(\int_0^T h(t)\,dt\right) \|v\|_\infty \\ &\ge \int_0^T (|\dot{v}(t)|^2/2)\,dt - c_1 - c_2 \left(\int_0^T |\dot{v}(t)|^2 dt\right)^{1/2}, \end{aligned}$$

where c_1 and c_2 are positive constants, so that φ is not bounded from above. In particular, if X^+ denotes the subspace of functions with mean value zero in H_T^1 and X^- the subspace of constant functions in H_T^1, we see that

$$\inf_{X^+} \varphi > -\infty$$

and that there exists $R > 0$ such that

$$\sup_{S_R^-} \varphi < \inf_{X^+} \varphi,$$

where $S_R^- = \{u \in X^- : |u| = R\}$. We shall obtain a solution of (23) from the following abstract result.

Theorem 4.7. *Let X be a Banach space and let $\varphi \in C^1(X, \mathbf{R})$. Assume that X splits into a direct sum of closed subspaces $X = X^- \oplus X^+$ with*

$$\dim X^- < \infty$$

and

$$\sup_{S_R^-} \varphi < \inf_{X^+} \varphi, \tag{27}$$

where $S_R^- = \{u \in X^- : |u| = R\}$. Let

$$B_R^- = \{u \in X^- : |u| \leq R\}$$

$$M = \{g \in C(B_R^-, X) : g(s) = s \ \text{ if } s \in S_R^-\}$$

and

$$c = \inf_{g \in M} \max_{s \in B_R^-} \varphi(g(s)). \tag{28}$$

Then, if φ satisfies the (PS)$_c$*-condition, c is a critical value of φ.*

Proof. We shall apply Theorem 4.6 with $c_1 = \sup_{S_R^-} \varphi$, $K = B_R^-$, $K_0 = S_R^-$ and $\chi(s) = s$, $s \in S_R^-$. We have only to prove that $c > c_1$. By Corollary 4.2 with $c_0 = \inf_{X^+} \varphi$ and (27), it suffices to prove that, for each $g \in M$, there exists some $s^* \in B_R^-$ such that $g(s^*) \in X^+$. If P denotes the projector into X^- with null-space X^+, this is equivalent to findings $s^* \in B_R^-$ such that $Pg(s^*) = 0$. Now, $Pg \in C(B_R^-, X^-)$ is such that $Pg = Id$ on $\partial B_R^- = S_R^-$. Then, by Corollary 2 of Section 5.3 on topological degree, $d(Pg, B_R^-) = d(Id, B_R^-) = 1$ and Pg has a zero in B_R^- by the existence property of degree (in the easy case where dim $X^- = 1$, the intermediate value theorem suffices). □

Theorem 4.8. *Let F satisfy the regularity assumptions of Section* 1.8 *as well as the assumptions* (24) *and* (26). *Then problem* (23) *has at least one solution u for which $\varphi(u) = c$, where c is given by* (28).

Proof. By Theorem 4.7, it is sufficient to prove that φ satisfies the (PS)$_c$-condition, and we shall show that (PS)$_c$ indeed holds for each $c \in \mathbf{R}$. Let (u_k) be a sequence in H_T^1 such that

$$\varphi(u_k) \to c, \quad \varphi'(u_k) \to 0 \ \text{ as } k \to \infty.$$

Writing $u_k = \overline{u}_k + \tilde{u}_k$ with $\overline{u}_k = (1/T)\int_0^T u_k(t)\,dt$, and using the fact that there is some k_0 such that

$$|\langle \varphi'(u_k), h\rangle| \le \|h\|$$

for all $k \ge k_0$ and $h \in H_T^1$, we obtain, for $k \ge k_0$

$$|\langle \varphi'(u_k), \tilde{u}_k\rangle| = \left| \int_0^T [|\dot{u}_k(t)|^2 + (\nabla F(t, u_k(t)), \tilde{u}_k(t))]dt \right| \le \|\tilde{u}_k\|$$

and hence

$$\|\tilde{u}_k\| \le C_1, \quad k \ge k_0 \tag{29}$$

because of (24), the Sobolev inequality and the equivalence of the L^2-norm for $\dot{u}$ and the H_T^1-norm on X^+. Now, $\varphi(u_k)$ is bounded and hence

$$\int_0^T (|\dot{u}_k(t)|^2/2)\,dt + \int_0^T F(t, \overline{u}_k)\,dt + \int_0^T [F(t, u_k(t)) - F(t, \overline{u}_k)]dt \ge C_2,$$

$$k \in \mathbf{N}$$

so that, using (29) and (24), we obtain

$$\int_0^T F(t, \overline{u}_k)\,dt \ge C_3, \quad k \in \mathbf{N}$$

and then

$$|\overline{u}_k| \le C_4, \quad k \in \mathbf{N}$$

by condition (26). Thus (u_k) is bounded in H_T^1 and hence contains a subsequence, relabeled (u_k) which converges to some $u \in H_T^1$, weakly in H_T^1 and strongly in $C([0,T], \mathbf{R}^N)$. Arguing then as in Proposition 4.1, we conclude that $(\mathrm{PS})_c$ is satisfied. □

4.4 Periodic Solutions of Josephson-Type Systems

Systems of the form

$$\ddot{u}(t) + N^2 D u(t) + f(u(t)) = h(t) \tag{30}$$

where

$$D = \begin{pmatrix} 1 & -1 & & & & \\ -1 & 2 & -1 & & 0 & \\ & \cdot & \cdot & \cdot & & \\ & & \cdot & \cdot & \cdot & \\ & 0 & & -1 & 2 & -1 \\ & & & & -1 & 1 \end{pmatrix}$$

and $f(u) = (a_1 \sin u_1, \ldots, a_N \sin u_N)$ arise in the theory of Josephson multipoint junctions in solid physics as well as in the space discretization of some boundary value problems for the sine-Gordon equation. Similar systems also describe the motion of forced linearly coupled pendulums.

More generally, let us consider the periodic problem

$$\begin{aligned} \ddot{u}(t) + Au(t) - \nabla F(t, u(t)) &= h(t) \quad \text{(a.e. on } [0,T]) \\ u(0) - u(T) = \dot{u}(0) - \dot{u}(T) &= 0, \end{aligned} \tag{31}$$

where A is a $(N \times N)$-symmetric matrix, $h \in L^1(0,T;\mathbf{R}^N)$, $F(t,.)$ is continuously differentiable for a.e. $t \in [0,T]$ and $F(., u)$ is measurable on $[0,T]$ for each $u \in \mathbf{R}^N$. We shall use the saddle point theorem to obtain the existence of solutions of (31) when $F(t,.)$ satisfies some periodicity conditions depending upon A and which reduce to those considered in Section 4.3 when $A = 0$.

Theorem 4.9. *Assume that the following conditions are satisfied.*

1. *There exists $g \in L^1(0,T)$ such that*

$$|F(t,u)| \leq g(t), \quad |\nabla F(t,u)| \leq g(t)$$

for all $u \in \mathbf{R}^N$ and a.e. $t \in [0,T]$.

2. dim $N(A) = m \geq 1$ *and A has no eigenvalue of the form $k^2\omega^2$*

$$(k \in \mathbf{N} \setminus \{0\}), \quad \textit{where } \omega = 2\pi/T.$$

3. *There exist $\alpha_j \in \mathbf{R}^N$ and $T_j > 0$ $(1 \leq j \leq m)$ such that*

$$N(A) = \operatorname{span}(\alpha_1, \ldots, \alpha_m)$$

and

$$F(t, u + T_j\alpha_j) = F(t,u) \quad (1 \leq j \leq m)$$

for a.e. $t \in [0,T]$ and all $u \in \mathbf{R}^N$.

4. $$\int_0^T (h(t), \alpha_j)\, dt = 0 \quad (1 \leq j \leq m).$$

Then (31) has at least one solution.

The proof of Theorem 4.9 requires several preliminary results. We know that the solutions of (31) correspond to the critical points of the function φ defined on H_T^1 by

$$\varphi(u) = \int_0^T \{(1/2)[|\dot{u}(t)|^2 - (Au(t), u(t))] + F(t, u(t)) + (h(t), u(t))\}\, dt.$$

Letting

$$q(u) = \int_0^T (1/2)[|\dot{u}(t)|^2 - (Au(t), u(t))]\, dt,$$

we see that

$$q(u) = (1/2)\|u\|^2 - (1/2)\int_0^T ((A+I)u(t), u(t))\, dt$$

$$= (1/2)(((I-K)u, u))$$

where $K : H_T^1 \to H_T^1$ is the linear self-adjoint operator defined, using Riesz representation theorem, by

$$\int_0^T ((A+I)u(t), v(t))\, dt = ((Ku, v))$$

$(u, v \in H_T^1)$. The compact imbedding of H_T^1 into $C([0,T], \mathbf{R}^N)$ implies that K is compact. By classical spectral theory, we can decompose H_T^1 into the orthogonal sum of invariant subspaces for $I - K$

$$H_T^1 = H^- \oplus H^0 \oplus H^+$$

where $H^0 = N(I-K)$ and H^- and H^+ are such that, for some $\delta > 0$,

$$q(u) \le -(\delta/2)\|u\|^2 \quad \text{if } u \in H^- \tag{32}$$

$$q(u) \ge (\delta/2)\|u\|^2 \quad \text{if } u \in H^+. \tag{33}$$

Notice that H^- is finite dimensional (as K has only finitely many eigenvalues λ_i with $\lambda_i > 1$) and H^0 corresponds to the critical points of $q(u)$, i.e. to the solution of

$$\ddot{u}(t) + Au(t) = 0$$

$$u(0) - u(T) = \dot{u}(0) - \dot{u}(T) = 0.$$

Consequently, assumption 2 and elementary theory of linear differential systems imply that

$$H^0 = N(A).$$

If $u \in H_T^1$, we shall write $u = u^- + u^0 + u^+$ where $u^- \in H^-$, $u^0 \in H^0$, and $u^+ \in H^+$.

In the following proposition, we assume that the assumptions of Theorem 4.9 hold.

Proposition 4.2. $\dim H^- = 0$ *if and only if* A *is semi-negative definite.*

Proof. If $\dim H^- = 0$, then $q(u) \ge 0$ for each $u \in H_T^1$ and, in particular, for each constant function c; thus,

$$0 \le q(c) = -(T/2)(Ac, c)$$

for all $c \in \mathbf{R}^N$. Conversely, if A is semi-negative definite, $q(u) \geq 0$ for all $u \in H_T^1$ and $\dim H^- = 0$. □

Proposition 4.3. *Each sequence* (u_k) *in* H_T^1 *such that* (u_k^0) *is bounded and*

$$\nabla\varphi(u_k) \to 0$$

contains a convergent subsequence.

Proof. Let (u_k), be such a sequence; if C_1 is such that

$$\|\nabla\varphi(u_k)\| \leq C_1$$

for all $k \in \mathbf{N}$, then, using (32), (33), and Sobolev inequality, we get

$$C_1(\|u_k^+\|^2 + \|u_k^-\|^2)^{1/2} = C_1\|u_k^+ - u_k^-\| \geq ((\nabla\varphi(u_k), u_k^+ - u_k^-))$$

$$= (((I-K)u_k, u_k^+ - u_k^-)) + \int_0^T (\nabla F(t, u_k(t)) + h(t), u_k^+(t) - u_k^-(t))\, dt$$

$$\geq \delta(\|u_k^+\|^2 + \|u_k^-\|^2) - C_2(\|u_k^+\|^2 + \|u_k^-\|^2)^{1/2},$$

where C_2 depends only on h and g. Thus,

$$(\|u_k^+\|^2 + \|u_k^-\|^2)^{1/2} \leq C_3$$

and (u_k) is bounded. Going, if necessary, to a subsequence, we can assume that $u_k \rightharpoonup u$ in H_T^1 and $u_k \to u$ in $C([0,T], \mathbf{R}^N)$. Now, the equality

$$((\nabla\varphi(u_k) - \nabla\varphi(u), u_k - u)) = \|u_k - u\|^2 - ((K(u_k - u), u_k - u))$$

$$+ \int_0^T (\nabla F(t, u_k(t)) - \nabla F(t, u(t)), u_k(t) - u(t))\, dt$$

implies that $\|u_k - u\| \to 0$ as $k \to \infty$ and the proof is complete. □

Proposition 4.4. *For each* $c \in \mathbf{R}$, φ *satisfies the* $(\mathrm{PS})_c$*-condition.*

Proof. Let $c \in \mathbf{R}$ and (u_k) be such that

$$\varphi(u_k) \to c \quad \text{and} \quad \nabla\varphi(u_k) \to 0 \tag{34}$$

as $k \to \infty$. If we write

$$u_k^0 = \sum_{j=1}^m c_j \alpha_j,$$

then $c_j = \hat{c}_j + k_j T_j$ $(1 \leq j \leq m)$ for some $k_j \in \mathbf{Z}$ and $\hat{c}_j \in [0, T_j[$ $(1 \leq j \leq m)$. Set $\hat{u}_k = u_k^- + \sum_{j=1}^m \hat{c}_j \alpha_j + u_k^+$, so that $\hat{u}_k^- = u_k^-$, $\hat{u}_k^+ = u_k^+$ and $(\hat{u}_k^0) = (\sum_{j=1}^m \hat{c}_j \alpha_j)$ is bounded. Moreover, $\hat{u}_k - u_k \in N(I-K)$, so that

$$q(\hat{u}_k) = q(u_k), \quad \nabla q(\hat{u}_k) = \nabla q(u_k).$$

On the other hand, by assumptions 3 and 4 of Theorem 4.10, we have

$$\int_0^T [F(t,u_k(t)) - (h(t),u_k(t))]\,dt = \int_0^T [F(t,\hat{u}_k(t) + \sum_{j=1}^m k_j T_j \alpha_j)$$

$$- (h(t), u_k^+(t) + u_k^-(t))]\,dt = \int_0^T [F(t,\hat{u}_k(t)) - (h(t),\hat{u}_k(t))]\,dt$$

so that

$$\varphi(u_k) = \varphi(\hat{u}_k), \quad k \in \mathbf{N}.$$

Similarly,

$$\nabla\varphi(u_k) = \nabla\varphi(\hat{u}_k), \quad k \in \mathbf{N}$$

so that $\nabla\varphi(\hat{u}_k) \to 0$ as $k \to \infty$. Proposition 4.3 implies that $(\hat{u}_k)$ contains a converging subsequence $(\hat{u}_{j_k})$ with limit $\hat{u}$. Hence, by (34)

$$c = \lim_{k\to\infty} \varphi(u_{j_k}) = \lim_{k\to\infty} \varphi(\hat{u}_{j_k}) = \varphi(\hat{u})$$

$$0 = \lim_{k\to\infty} \nabla\varphi(u_{j_k}) = \lim_{k\to\infty} \nabla\varphi(\hat{u}_{j_k}) = \nabla\varphi(\hat{u})$$

and condition $(PS)_c$ holds. □

Proof of Theorem 4.9. Let us first notice that if $u = u^0 + u^+ \in H^0 \oplus H^+$, then, using Sobolev inequality,

$$\varphi(u) = (1/2)(((I-K)u^+,u^+)) + \int_0^T [F(t,u(t)) + (h(t),u^+(t))]\,dt$$

$$\geq (\delta/2)\|u^+\|^2 - \|g\|_{L^1} - C\|h\|_{L^1}\|u^+\|$$

and hence φ is bounded below on $H^0 \oplus H^+$. Hence, if $\dim H^- = 0$, φ is bounded below on H and has a minimum by Proposition 4.4 and Theorem 4.4. If $\dim H^- > 0$ and $u = u^- \in H^-$, then

$$\varphi(u) = (1/2)(((I-K)u,u)) + \int_0^T [F(t,u(t)) + (h(t),u(t))]\,dt$$

$$\leq -(\delta/2)\|u\|^2 + \|g\|_{L^1} + C\|h\|_{L^1}\|u\|$$

and

$$\varphi(u) \to -\infty \text{ if } \|u\| \to \infty \text{ in } H^-.$$

Taking $X = H_T^1$, $X^- = H^-$, $X^+ = H^0 \oplus H^+$ in Theorem 4.7, we see that $\dim X^- < \infty$ and that there exists $R > 0$ such that

$$\sup_{S_R^-} \varphi < \inf_{X^+} \varphi$$

where $S_R^- = \{u \in X^- : \|u\| = R\}$. The existence of a critical point for φ follows then from Theorem 4.7 and Proposition 4.4. □

In the example given at the beginning of the section, we have

$$(Dv, v) = \sum_{j=1}^{N-1} (v_j - v_{j+1})^2,$$

so that dim $N(A) = 1$ and $A = \text{span}\{(1, 1, \ldots, 1)\}$. On the other hand,

$$F(t, u) = \sum_{j=1}^{N} a_j \cos u_j$$

satisfies assumption 3 with $T_j = 2\pi$. Therefore, the conditions for the existence of a periodic solution for (30) reduce to

$$\det(N^2 D - k^2\omega^2) \neq 0, \quad k \in \mathbf{N} \setminus \{0\}$$

and

$$\sum_{j=1}^{N} \int_0^T h_j(t)\, dt = 0.$$

4.5 The Mountain Pass Theorem and Periodic Solutions of Superlinear Convex Autonomous Hamiltonian Systems

Let us first state and prove the *mountain pass theorem* which was described vaguely in the introduction.

Theorem 4.10. *Let X be a Banach space and $\varphi \in C^1(X, \mathbf{R})$. Assume that there exist $u_0 \in X$, $u_1 \in X$, and a bounded open neighborhood Ω of u_0 such that $u_1 \in X \setminus \Omega$ and*

$$\inf_{\partial\Omega} \varphi > \max(\varphi(u_0), \varphi(u_1)).$$

Let

$$\Gamma = \{g \in C([0,1], X) : g(0) = u_0, g(1) = u_1\}$$

and

$$c = \inf_{g \in \Gamma} \max_{s \in [0,1]} \varphi(g(s)).$$

If φ satisfies the $(\text{PS})_c$-condition, then c is a critical value of φ and $c >$ $\max(\varphi(u_0), \varphi(u_1))$.

Proof. Let us take $K = [0,1]$, $K_0 = \{0,1\}$, $\chi(0) = u_0$, $\chi(1) = u_1$, $M = \Gamma$, $S = \partial\Omega$, $c_0 = \inf_{\partial\Omega}\varphi$, $c_1 = \max(\varphi(u_0), \varphi(u_1))$ in Theorem 4.3 and Corollary 4.2. It remains only to show that

$$g([0,1]) \cap \partial\Omega \neq 0$$

for all $g \in \Gamma$, and this follows at once from the fact that $g(0) = u_0 \in \Omega$, $g(1) = u_1 \in X \setminus \overline{\Omega}$ and from a classical connectedness result. □

We shall combine the dual least action principle with the mountain pass theorem to prove the existence of non-trivial periodic solutions for the autonomous Hamiltonian system

$$J\dot{u}(t) + \nabla H(u(t)) = 0 \tag{35}$$

where $H \in C^1(\mathbf{R}^{2N}, \mathbf{R})$ is strictly convex, $H(0) = 0$, $\nabla H(0) = 0$ but, in contrast to the results of Section 3.4, H is superquadratic instead of being subquadratic.

Theorem 4.11. *If there exists $q > 2$, $\alpha > 0$, such that, for every $u \in \mathbf{R}^{2N}$, one has*

$$qH(u) \leq (\nabla H(u), u) \tag{36}$$

and

$$H(u) \leq \alpha |u|^q, \tag{37}$$

then, for each $T > 0$, (35) has a non-trivial T-periodic solution.

The proof of Theorem 4.11 will be a combination of the dual least action principle and the mountain pass lemma, and requires several preliminary results.

Lemma 4.1. *If*

$$M = \max_{|u|=1} H(u), \quad m = \min_{|u|=1} H(u),$$

then

$$|u| \leq 1 \Rightarrow H(u) \leq M|u|^q, \quad |u| \geq 1 \Rightarrow H(u) \geq m|u|^q. \tag{38}$$

Proof. If $f(s) = H(sv)$ for some fixed v, assumption (36) implies that $sf'(s) \geq qf(s)$. Thus, if $s \geq 1$, $f(s) \geq s^q f(1)$, i.e. $H(sv) \geq s^q H(v)$. If $|u| \leq 1$, this implies

$$H(u/|u|) \geq |u|^{-q} H(u)$$

and if $|u| \geq 1$, this implies

$$H(u) = H(|u|(u/|u|)) \geq |u|^q H(u/|u|). \quad □$$

Lemma 4.2. *The function H^* is continuously differentiable on $\mathbf{R}^{2N}$ and, if*

$$\frac{1}{p} + \frac{1}{q} = 1, \quad m^* = \min_{|v|=1} H^*(v), \quad M^* = \max_{|v|=1} H^*(v), \quad \alpha^* = (\alpha q)^{-p/q}/p,$$

we have $m^* > 0$ *and*

$$pH^*(v) \geq (\nabla H^*(v), v) \tag{39}$$

$$|v| \leq 1 \Rightarrow H^*(v) \geq m^*|v|^p \tag{40}$$

$$|v| \geq 1 \Rightarrow H^*(v) \leq M^*|v|^p \tag{41}$$

$$H^*(v) \geq \alpha^*|v|^p \tag{42}$$

for all $v \in \mathbf{R}^{2N}$.

Proof. Since H is strictly convex and, by (38), such that $H(u)/|u| \to +\infty$ as $|u| \to \infty$, Proposition 2.4 implies that $H^* \in C^1(\mathbf{R}^{2N}, \mathbf{R})$. Now it follows from Theorem 2.2 and Proposition 2.3 that

$$v = \nabla H(u) \Leftrightarrow u = \nabla H^*(v) \Leftrightarrow H^*(v) = (v, u) - H(u).$$

Hence, assumption (36) implies that

$$H^*(v) = (v, u) - H(u) \geq \left(1 - \frac{1}{q}\right)(v, u) = \frac{1}{p}(v, \nabla H^*(v)).$$

Like in the proof of Lemma 4.1, (39) implies (40) and (41). Finally, (42) follows from (37) by relation 2.4. □

Remark 4.4. By Lemma 4.1, assumption (36) implies for H a superquadratic growth at infinity and a subquadratic growth near the origin. In particular, the results of Chapter 3 are not applicable.

Remark 4.5. Since, by (41), $H^*(v) \leq M^*(1 + |v|^p)$ for all $v \in \mathbf{R}^{2N}$, Proposition 2.2 and Theorem 1.4 imply that the dual action φ defined by

$$\varphi(v) = \int_0^T [(1/2)(J\dot{v}(t), v(t)) + H^*(\dot{v}(t))]\, dt$$

is continuously differentiable on $X = \tilde{W}_T^{1,p}$.

We shall apply the mountain pass theorem to φ.

Remark 4.6. It will be convenient, in this section, to use the norm $\|v\| = \|\dot{v}\|_{L^p}$ on X. By Proposition 1.1, this norm is equivalent to the $W_T^{1,p}$-norm on X.

Lemma 4.3. *There exists* $c > 0$ *such that, for each* $v \in X$, *one has*

$$\int_0^T (J\dot{v}(t), v(t))\, dt \geq -C\|v\|^2.$$

Proof. Hölder's inequality and Proposition 1.1 imply that

$$\int_0^T (J\dot{v}(t), v(t)) \geq -\|J\dot{v}\|_{L^p}\|v\|_{L^q} \geq -T^{1/q}\|\dot{v}\|_{L^p}\|v\|_{\infty}$$

$$\geq -cT^{1/q}\|\dot{v}\|_{L^p}^2. \qquad \square$$

Lemma 4.4. *For every $\ell \in X^*$, there exists $f \in L^q(0,T;\mathbf{R}^{2N})$ such that, for all $v \in X$,*

$$\langle \ell, v\rangle = \int_0^T (f(t), \dot{v}(t))\,dt,$$

$$\|\ell\|_{X^*} = \|f\|_{L^q}.$$

Proof. The mapping $L : X \to L^p(0,T;\mathbf{R}^{2N})$, $v \to \dot{v}$ is an isometry. Let $Y = L(X)$ and $M = L^{-1} : Y \to X$. If $\ell \in X^*$, the function

$$\Phi : Y \to \mathbf{R}, \quad u \to \langle \ell, Mu\rangle$$

is linear and continuous. By the Hahn–Banach theorem, Φ has an extension $\Psi \in (L^p)^*$ such that $\|\Phi\|_{(L^p)^*} = \|\Psi\|_{(L^p)^*}$. The Riesz representation theorem implies the existence of $f \in L^q(0,T;\mathbf{R}^{2N})$ such that $\langle \Psi, u\rangle = \int_0^T (f(t), u(t))dt$ for every $u \in L^p(0,T;\mathbf{R}^{2N})$. Thus, for each $v \in X$,

$$\begin{aligned}\langle \ell, v\rangle &= \langle \ell, MLv\rangle = \langle \Psi, Lv\rangle = \int_0^T (f(t), Lv(t))\,dt \\ &= \int_0^T (f(t), \dot{v}(t))\,dt.\end{aligned}$$

Moreover,

$$\|\ell\|_{X^*} = \|\Phi\|_{(L^p)^*} = \|\Psi\|_{(L^p)^*} = \|f\|_{L^q}. \qquad \square$$

Lemma 4.5. *Every sequence (v_j) in X such that $(\varphi(v_j))$ is bounded and $\varphi'(v_j) \to 0$ contains a convergent subsequence.*

Proof. Theorem 1.4 and Lemma 4.4 imply the existence of a sequence (f_j) in $L^q(0,T;\mathbf{R}^{2N})$ such that $\|f_j\|_{L^q} \to 0$ as $j \to \infty$ and

$$\int_0^T (-Jv_j(t) + \nabla H^*(\dot{v}_j(t)) - f_j(t), \dot{w}(t))\,dt = 0$$

for all $w \in X$. By a standard Fourier series argument, we obtain

$$-Jv_j(t) + \nabla H^*(\dot{v}_j(t)) + c_j = f_j(t) \tag{43}$$

a.e. on $[0,T]$, for some $c_j \in \mathbf{R}^{2N}$. This implies that

$$\int_0^T (J\dot{v}_j(t), v_j(t))dt + \int_0^T (\nabla H^*(\dot{v}_j(H), \dot{v}_j(t))dt = \int_0^T (f_j(t), \dot{v}_j(t))\,dt. \tag{44}$$

Using (39), (42), and (44), we obtain

$$\begin{aligned}\varphi(v_j) &= (1/2)\int_0^T (f_j(t), \dot v_j(t))\,dt - (1/2)\int_0^T (\nabla H^*(\dot v_j(t)), \dot v_j(t))\,dt \\ &+ \int_0^T H^*(\dot v_j(t))\,dt \geq \left(1-\frac{p}{2}\right)\int_0^T H^*(\dot v_j(t))\,dt - (1/2)\|f_j\|_{L^q}\|\dot v_j\|_{L^p} \\ &\geq \left(1-\frac{p}{2}\right)\alpha^*\|\dot v_j\|_{L^p}^p - c'\|\dot v_j\|_{L^p} = \left(1-\frac{p}{2}\right)\alpha^*\|v_j\|^p - c'\|v_j\|.\end{aligned}$$

Since $1 < p < 2$ and $(\varphi(v_j))$ is bounded, (v_j) is bounded in X. Going if necessary to a subsequence, we can assume that $v_j \rightharpoonup v$ in X, $v_j \to v$ uniformly on $[0,T]$ and that

$$c_j = (1/T)\int_0^T (f_j(t) - \nabla H^*(v_j(t)))\,dt \to c$$

in $\mathbf{R}^{2N}$. By duality, (43) implies that

$$\dot v_j(t) = \nabla H(f_j(t) + Jv_j(t) - c_j) \tag{45}$$

for a.e. $t \in [0,T]$. Now, assumption (37) and Proposition 2.2 imply that ∇H maps continuously L^q into L^p, so that, letting $j \to \infty$ in (45), we get

$$\dot v_j \to \nabla H(Jv - c)$$

in $L^p(0,T;\mathbf{R}^{2N})$, and hence $v_j \to v$ in X. □

Proof of Theorem 4.11.

1) We shall apply Theorem 4.10 to φ. By Lemma 4.5, φ satisfies the $(\mathrm{PS})_c$-condition for every $c \in \mathbf{R}$.

2) Since

$$\varphi(v) \geq -(C/2)\|v\|^2 + \alpha^*\|v\|^p$$

there exists $\rho > 0$ such that

$$\begin{aligned}\varphi(v) &> 0 = \varphi(0) \quad \text{if } 0 < \|v\| < \rho \\ &> \delta > 0 \quad \text{if } \|v\| = \rho.\end{aligned}$$

3) Let

$$v_1(t) = \left(\cos\frac{2\pi t}{T}\right)e + \left(\sin\frac{2\pi t}{T}\right)Je$$

where $|e| \geq 1$. Then,

$$\int_0^T (J\dot v_1(t), v_1(t))\,dt = -(T^2/2\pi)\,|e|^2$$

and Lemma 4.2 implies that

$$\varphi(v_1) \leq -(T^2/2\pi)\,|e|^2 + TM^*\,|e|^p.$$

Since $p < 2$, we can choose $|e|$ such that $\varphi(v_1) < 0$.

4) Theorem 4.10, with $u_0 = 0$, $u_1 = v_1$ and $\Omega = B(0,\rho)$, implies the existence of a critical point v of φ such that $\varphi(v) > \varphi(0)$. By Theorem 2.3,

$$u(t) = \nabla F^*(\dot{v}(t))$$

is a nontrivial T-periodic solution of (35). □

4.6 Multiple Critical Points of Periodic Functionals

Let G be a discrete subgroup of a Banach space X and let $\pi : X \to X/G$ be the canonical surjection. A subset A of X is *G-invariant* if $\pi^{-1}(\pi(A)) = A$. A function f defined on X is *G-invariant* if $f(u+g) = f(u)$ for every $u \in X$ and every $g \in G$. If a differentiable functional $\varphi : X \to \mathbf{R}$ is G-invariant, then φ' is also G-invariant. Consequently, if u is a critical point of such a φ, then $\pi^{-1}(\pi(u))$ is a set of critical points of φ, and is called a *critical orbit* of φ.

We shall use the following compactness condition.

Definition 4.2. *A G-invariant differentiable functional $\varphi : X \to \mathbf{R}$ satisfies the* (PS)$_G$*-condition if, for every sequence (u_k) in X such that $\varphi(u_k)$ is bounded and $\varphi'(u_k) \to 0$, the sequence $(\pi(u_k))$ contains a convergent subsequence.*

If we consider for example the function φ associated to (18) in Section 4.2 and the discrete subgroup

$$G = \left\{ \sum_{i=1}^{N} k_i T_i e_i \ : \ k_i \in \mathbf{Z},\ 1 \le i \le N \right\}$$

of H_T^1, where the T_i are the positive real numbers of Assumption (L_3), we see from (22), that φ is G-invariant. Moreover, if $u \in H_T^1$, with $u = \overline{u} + \tilde{u}$, $\overline{u} = (1/T)\int_0^T u(t)\,dt$, there exists unique $k_i \in \mathbf{Z}$ $(1 \le i \le N)$ such that $(\overline{u}, e_i) - k_i T_i \in [0, T_i[$ $(1 \le i \le N)$, and we set $\overline{u}^0 = ((\overline{u}, e_1) - k_1 T_1, \ldots, (\overline{u}, e_N) - k_N T_N)$. If (u_k) is a sequence in X with $(\varphi(u_k))$ bounded and $\varphi'(u_k) \to 0$, then $v_k = \overline{u}_k^0 + \tilde{u}_k$ is a representative of $[u_k] \in H_T^1/G$ which, by Proposition 4.1, will have a convergent subsequence. Therefore, $(\pi(u_k)) = (\pi(v_k))$ has the same property and φ satisfies the (PS)$_G$-condition.

This G-invariance of the functional will provide a substantial improvement of the conclusion of Theorem 4.4.

Theorem 4.12. *Let $\varphi \in C^1(X, \mathbf{R})$ be a G-invariant functional satisfying the* (PS)$_G$*-condition. If φ is bounded from below and if the dimension N of the space generated by G is finite, then φ has at least $N+1$ critical orbits.*

The proof of Theorem 4.12 depends on the notion of *Lusternik–Schnirelman category*. Let us first recall that a subset C of a topological space Y is *contractible* in Y if there exists $h \in C([0,1] \times C, Y)$ and $y \in Y$ such that

$$h(0,u) = u, \quad h(1,u) = y$$

for all $u \in C$.

Definition 4.3. *A subset A of a topological space Y has category k in Y if k is the least integer such that A can be covered by k closed sets contractible in Y. The category of A in Y is denoted by $\mathrm{Cat}_Y(A)$.*

Lemma 4.6. *Let Y, Z be topological spaces and let $A, B \subset Y$.*

i) *If $A \subset B$, then $\mathrm{cat}_Y(A) \leq \mathrm{cat}_Y(B)$.*

ii) $\mathrm{cat}_Y(A \cup B) \leq \mathrm{cat}_Y(A) + \mathrm{cat}_Y B$.

iii) *If A is closed and $B = \eta(1, A)$, where $\eta \in C([0,1] \times A, Y)$ is such that $\eta(0,u) = u$ for every $u \in A$, then $\mathrm{cat}_Y(A) \leq \mathrm{cat}_Y(B)$.*

iv) *If $z \in Z$, $\mathrm{cat}_{Y \times Z}(A \times \{z\}) = \mathrm{cat}_Y(A)$.*

Proof. Properties i), ii), and iv) are obvious. If $\mathrm{cat}_Y(B) = +\infty$, property iii) is also obvious. Let us assume that $k = \mathrm{cat}_Y(B)$ is finite, let $(B_j)_{1 \leq j \leq k}$ be the corresponding covering of B by closed sets which are contractible in Y, and let $(h_j)_{1 \leq j \leq k}$ be the corresponding deformations. Define

$$A_j = (\eta_1)^{-1}(B_j) \quad (1 \leq j \leq k)$$

where $\eta_1 = \eta(1,.)$. The sets A_j are closed and the mapping $h_j \star \eta : [0,1] \times A_j \to Y$ defined by

$$\begin{aligned} (hj \star \eta)(t,u) &= \eta(2t,u), & 0 \leq t \leq 1/2 \\ &= h_j(2t-1, \eta(1,u)), & 1/2 \leq t \leq 1 \end{aligned}$$

is continuous and such that $(h_j \star \eta)(0,u) = u$ and $(h_j \star \eta)(1,u) = a_j$ for each $u \in A_j$ $(1 \leq j \leq k)$. Since $(A_j)_{1 \leq j \leq k}$ is a covering of A, $\mathrm{cat}_Y(A) \leq k$. □

Definition 4.4. A metric space Y is an *absolute neighborhood extensor*, shortly an ANE, if, for every metric space E, every closed subset F of E and every $f \in C(F, Y)$, there exists a continuous extension of f defined on a neighborhood of F in E.

Examples

a) *A finite product of* ANE's *is an* ANE.

b) *A convex subset of a normed space is an* ANE.

c) *A circle is an* ANE.

Proof. Property a) is obvious. Property b) follows from Dugundji's extension theorem: if F is a closed subspace of a metric space E and if $f \in C(F, X)$ with X a normed space, then there exists an extension $g \in C(E, X)$ of f such that $g(E)$ is contained in the convex hull of $f(F)$. To prove property c), let F be a closed subspace of a metric space E, and let $f \in C(F, S^1)$. By Tietze's extension theorem, there exists an extension $g \in C(E, \mathbf{R}^2)$ of f. On the other hand, there exists a neighborhood U of S^1 in $\mathbf{R}^2$ and a retraction $r : U \to S^1$. The map $r \circ g$ defined on $g^{-1}(U)$ is the desired extension of f. □

Lemma 4.7. *If A is a closed subset of an* ANE Y, *then there exists a closed neighborhood U of A such that*

$$\text{cat}_Y(A) = \text{cat}_Y(U).$$

Proof. 1) If $\text{cat}_Y(A) = +\infty$, the result is clear. Let us assume that $k = \text{cat}_Y(A)$ is finite. Let $(A_j)_{1 \le j \le k}$ be the corresponding covering of A by closed sets contractible in Y. It suffices to prove that each A_j has a closed neighborhood U_j contractible in Y, since then

$$U = \bigcup_{j=1}^{k} U_j$$

is a closed neighborhood of A and

$$k = \text{cat}_Y(A) \le \text{cat}_Y(U) \le k.$$

2) There exists $h_j \in C([0,1] \times A_j, Y)$ and $a_j \in Y$ such that

$$h_j(0, u) = u, \quad h_j(1, u) = a_j$$

for each $u \in A_j$. The set

$$F = ([0,1] \times A_j) \cup (\{0\} \times Y) \cup (\{1\} \times Y)$$

is closed in $E = [0,1] \times Y$. The function $f : E \to Y$ defined by

$$\begin{aligned} f(t, u) &= h_j(t, u), & t \in [0,1], \quad & u \in A_j \\ f(0, u) &= u, & & u \in Y \\ f(1, u) &= a_j, & & u \in Y \end{aligned}$$

is continuous. Since Y is an ANE, f has a continuous extension η defined on a neighborhood $\mathcal{N}$ of E. We can assume that $\mathcal{N}$ is closed since Y is normal. Using the compactness of $[0,1]$, it is easy to verify the existence of a closed neighborhood U_j of A_j since that $[0,1] \times U_j \subset \mathcal{N}$. Since

$$\eta(0, u) = u, \quad \eta(1, u) = a_j$$

for each $u \in U_j$, U_j is contractible in Y and the proof is complete. □

Let G be a discrete subgroup of a Banach space X and assume that the dimension N of the space V generated by G is finite. Then X is isomorphic to $\mathbf{R}^N \times Z$, where Z is a complement of V, G is isomorphic to $\mathbf{Z}^N$ and $\pi(X)$ is isomorphic to $T^N \times Z$ where T^N is the N-torus. Property iv) of Lemma 4.6 implies that if $A = [0,1]^N \times \{0\}$,

$$\text{cat}_{\pi(X)}(\pi(A)) = \text{cat}_{T^N \times Z}(T^N \times \{0\}) = \text{cat}_{T^N}(T^N).$$

By a standard result in algebraic topology,

$$\text{cat}_{T^N}(T^N) = N + 1.$$

Thus, we have

$$\text{cat}_{\pi(X)}(\pi(A)) = N + 1$$

and, for $1 \leq j \leq N+1$, the set

$$\mathcal{A}_j = \{A \subset X \,:\, A \text{ is compact and } \text{cat}_{\pi(X)}(\pi(A)) \geq j\}$$

is nonempty.

In order to apply Ekeland's variational principle, we need the following lemmas.

Lemma 4.8. *For $1 \leq j \leq N+1$, the space $\mathcal{A}_j$ with the Hausdorff distance*

$$\delta(A,B) = \max\{\sup_{a\in A} \text{dist}(a,B), \sup_{b\in B} \text{dist}(b,A)\}$$

is a complete metric space.

Proof. 1) It is easy to verify that the space

$$\mathcal{M} = \{A \subset X \,:\, A \text{ is closed, bounded and nonempty}\}$$

with the Hausdorff distance is a metric space. Let us prove that $(\mathcal{M}, \delta)$ is complete. Let (A_n) be a Cauchy sequence in $(\mathcal{M}, \delta)$ and define the closed bounded set

$$A_\infty = \bigcap_{n=1}^{\infty} \overline{\bigcup_{m=n}^{\infty} A_m}.$$

For every $\epsilon > 0$, there exists $n(\epsilon)$ such that

$$p, q \geq n(\epsilon) \Rightarrow \delta(A_p, A_q) \leq \epsilon. \tag{46}$$

It suffices to prove that $\delta(A_\infty, A_{n(\epsilon)}) \leq \epsilon$ since then $q \geq n(\epsilon)$ implies $\delta(A_\infty, A_q) \leq 2\epsilon$. It follows from (46) that

$$A_\infty \subset \overline{\bigcup_{m=n(\epsilon)}^{\infty} A_m} \subset (A_{n(\epsilon)})_\epsilon.$$

Hence, we obtain

$$\sup_{a \in A_\infty} \operatorname{dist}(a, A_{n(\epsilon)}) \leq \epsilon. \tag{47}$$

Let $b = b_{n_0} \in A_{n(\epsilon)}$. There exists an increasing sequence (n_k) and a sequence (b_{n_k}) in X such that $b_{n_k} \in A_{n_k}$ and $\|b_{n_k} - b_{n_{k-1}}\| \leq 2^{-k}\epsilon$. Since (b_{n_k}) is a Cauchy sequence, there exists $a \in X$ such that $b_{n_k} \to a$ as $k \to \infty$. Clearly $a \in A_\infty$. For every k, $\|b_{n_k} - b_{n_0}\| \leq \epsilon$, so that $\|a - b_{n_0}\| \leq \epsilon$ and we have proved that

$$\sup_{b \in A_{n(\epsilon)}} \operatorname{dist}(b, A_\infty) \leq \epsilon. \tag{48}$$

Inequalities (47) and (48) imply that $\delta(A_\infty, A_{n(\epsilon)}) \leq \epsilon$.

2) Let us prove that

$$\mathcal{K} = \{A \in \mathcal{M} \,:\, A \text{ is compact}\}$$

is closed in $(\mathcal{M}, \delta)$. Let (A_n) be a sequence in $\mathcal{K}$ and $A \in \mathcal{M}$ be such that $\delta(A_n, A) \to 0$ in $n \to \infty$. Since A is closed in the complete metric space X, it suffices to prove that, for every $\epsilon > 0$, there exists a finite covering of A by open balls of radius ϵ. For $\epsilon > 0$, there exists n such that $\delta(A_n, A) < \epsilon/2$. Since A_n is compact, there exists a finite covering of A_n by open balls with radius $\epsilon/2$. Hence there exists a finite covering of A by open balls of radius ϵ.

3) It remains only to prove that

$$\mathcal{A}_j = \{A \in \mathcal{K} \,:\, \operatorname{cat}_{\pi(X)}(\pi(A)) \geq j\}$$

is closed in $(\mathcal{K}, \delta)$. Let (A_n) be a sequence in $\mathcal{A}_j$ and let $A \in \mathcal{K}$ be such that $\delta(A_n, A) \to 0$ as $n \to \infty$. By Examples (a) - (b) - (c),

$$\pi(X) \cong T^N \times Z \cong (S^1)^N \times Z$$

is an ANE. Lemma 4.7 implies the existence of a closed neighborhood U of $\pi(A)$ such that

$$\operatorname{cat}_{\pi(X)}(\pi(A)) = \operatorname{cat}_{\pi(X)}(U).$$

Since $\pi^{-1}(U)$ is a closed neighborhood of the compact A, there exists n such that $A_n \subset U$. Hence

$$\operatorname{cat}_{\pi(X)}(\pi(A)) = \operatorname{cat}_{\pi(X)}(U) \geq \operatorname{cat}_{\pi(X)}(\pi(A_n)) \geq j,$$

and $A \in \mathcal{A}_j$. □

Lemma 4.9. *If* $1 \leq j \leq N+1$ *and* $\varphi \in C(X, \mathbf{R})$, *then the function* $\Phi : \mathcal{A}_j \to \mathbf{R}$ *defined by*

$$\Phi(A) = \max_{u \in A} \varphi(u)$$

is lower semi-continuous.

Proof. Let (A_n) be a sequence in $\mathcal{A}_j$ and $A \in \mathcal{A}_j$ be such that $\delta(A_n, A) \to 0$ in $n \to \infty$. For each $u \in A$, there exists a sequence (u_n) in X such that $u_n \in A_n$ and $u_n \to u$. Therefore,

$$\varphi(u) = \lim_{n\to\infty} \varphi(u_n) \leq \underline{\lim}_{n\to\infty} \Phi(A_n),$$

and, since $u \in A$ is arbitrary,

$$\Phi(A) \leq \underline{\lim}_{n\to\infty} \Phi(A_n). \quad \square$$

Proof of Theorem 4.12. 1) Let us define, for $1 \leq j \leq N+1$,

$$c_j = \inf_{A \in \mathcal{A}_j} \max_A \varphi.$$

Since $\mathcal{A}_{j+1} \subset \mathcal{A}_j$, one has clearly

$$-\infty < \inf \varphi \leq c_1 \leq c_2 \leq \ldots \leq c_{N+1} < +\infty.$$

It suffices now to prove that, if $c_k = c_j$ for some $1 \leq j \leq k \leq N+1$, then

$$K_{c_j} = \{u \in X : \varphi'(u) = 0 \text{ and } \varphi(u) = c_j\}$$

contains $k - j + 1$ critical orbits.

2) Assume that $c_k = c_j = c$ for some $1 \leq j \leq k \leq N+1$ and that K_c contains $n \leq k - j$ distinct critical orbits $\pi(u_1), \ldots, \pi(u_n)$. Let $\rho > 0$ be such that π restricted to $B(u_m, 2\rho)$ is one-to-one, $1 \leq m \leq n$, and define

$$\mathcal{N}_\rho = \bigcup_{m=1}^{n} \bigcup_{g \in G} B(u_m + g, \rho).$$

By the $(\mathrm{PS})_G$-condition, there exists $\epsilon \in]0, \rho^2[$ such that, if $u \in \varphi^{-1}([c - \epsilon, c + \epsilon]) \setminus \mathcal{N}_\rho$, then

$$\|\varphi'(u)\| > \epsilon^{1/2}. \tag{49}$$

Indeed, if it is not the case, there is a sequence (u_i) such that

$$u_i \in X \setminus \mathcal{N}_\rho, \quad c - \frac{1}{i} \leq \varphi(u_i) \leq c + \frac{1}{i}, \quad \|\varphi'(u_i)\| \leq \sqrt{\frac{1}{i}},$$

and, by $(\mathrm{PS})_G$, we can assume, going if necessary to a subsequence, that $\pi(u_i) \to \pi(u)$, $i \to +\infty$, for some $u \in X$. We can also assume that $u_i \in [0,1]^N \times Z$, since φ and φ' are G-invariant. But then $u_i \to u$, $u \in X \setminus \mathcal{N}_\rho$, $\varphi(u) = c$, $\varphi'(u) = 0$, a contradiction since $\mathcal{N}_\rho$ is a neighborhood of K_c. By the definition of c_k, there exists $A \in \mathcal{A}_k$ such that

$$\Phi(A) = \max_A \varphi \leq c + \epsilon.$$

Let $B = A \setminus \mathcal{N}_{2\rho}$. Lemma 4.6 implies that

$$\begin{aligned} k &\leq \operatorname{cat}_{\pi(X)}(\pi(A)) \leq \operatorname{cat}_{\pi(X)}(\pi(B) \cup \pi(\mathcal{N}_{2\rho})) \\ &\leq \operatorname{cat}_{\pi(X)}(\pi(B)) + \operatorname{cat}_{\pi(X)}(\mathcal{N}_{2\rho}). \end{aligned}$$

It follows from the definition of category that $\operatorname{cat}_{\pi(X)}(\mathcal{N}_{2\rho}) \leq n$. But then

$$k \leq \operatorname{cat}_{\pi(X)}(\pi(B)) + n \leq \operatorname{cat}_{\pi(X)}(\pi(B)) + k - j,$$

i.e. $B \in \mathcal{A}_j$.

3) Lemmas 4.8 and 4.9 and Ekeland's variational principle imply the existence of $C \in \mathcal{A}_j$ such that

$$\Phi(C) \leq \Phi(B) \leq \Phi(A) \leq c + \epsilon,$$

$$\delta(B, C) \leq \epsilon^{1/2}$$

$$\Phi(D) > \Phi(C) - \epsilon^{1/2}\delta(C, D) \tag{50}$$

whenever $D \in \mathcal{A}_j$ and $D \neq C$. Since $B \cap \mathcal{N}_{2\rho} = \phi$ and $\delta(B, C) \leq \epsilon \leq \rho$, $C \cap \mathcal{N}_\rho = \phi$. In particular, the set

$$S = \{u \in C : c - \epsilon \leq \varphi(u)\}$$

is contained in $\varphi^{-1}([c-\epsilon, c+\epsilon]) \setminus \mathcal{N}_\rho$. By (49) and the continuity of φ', we can find, for each $u \in S$, $\delta_u > 0$ and $v_u \in X$ with $|v_u| = 1$, such that, for every $g \in G$, one has

$$|h| \leq \delta_u \Rightarrow \langle \varphi'(u + g + h), v_u \rangle < -\epsilon^{1/2}. \tag{51}$$

By the compactness of S, there exists a finite covering of S of the form $\{B_1, \dots, B_\ell\}$ with

$$B_i = B(u_i, \delta_{u_i}), \quad 1 \leq i \leq \ell.$$

Let us define $\psi_i : X \to [0,1]$ by

$$\psi_i(u) = \frac{\sum_{g \in G} \operatorname{dist}(u + g, CB_i)}{\sum_{j=1}^{\ell} \sum_{g \in G} \operatorname{dist}(u + g, CB_j)}, \quad \text{if } u \in \bigcup_{j=1}^{\ell} B_j$$

$$\psi_i(u) = 0, \quad \text{if } u \in X \setminus \bigcup_{j=1}^{\ell} B_j.$$

Finally, let $\delta = \min\{\delta_{u_1}, \dots, \delta_{u_\ell}\}$, let $\psi : X \to [0,1]$ be a continuous invariant function such that

$$\begin{aligned} \psi(u) &= 1 \quad \text{if } c \leq \varphi(u) \\ &= 0 \quad \text{if } \varphi(u) \leq c - \epsilon, \end{aligned}$$

and let $\eta \in C([0,1] \times X, X)$ be defined by

$$\eta(t,u) = u + t\delta\psi(u)\sum_{i=1}^{\ell} \psi_i(u) v_{u_i}.$$

For each $t \in [0,1]$, the mapping $\eta(t,.)$ satisfies the equivariance property

$$\eta(t, u+g) = \eta(t,u) + g$$

for all $u \in X$ and all $g \in G$. It follows from Lemma 4.6 that, if $D = \eta(1, C)$,

$$\mathrm{cat}_{\pi(X)}(\pi(D)) \geq \mathrm{cat}_{\pi(X)}(\pi(C)) \geq j,$$

and hence, D being compact, $D \in \mathcal{A}_j$.

The mean value theorem and (51) imply that, for each $u \in S$, there is some $\tau \in \ [0,1[$ for which

$$\begin{aligned}\varphi(\eta(1,u)) - \varphi(u) &= \langle \varphi'(\eta(\tau,u)), \delta\psi(u)\sum_{i=1}^{\ell}\psi_i(u)v_{u_i}\rangle \\ &= \delta\psi(u)\sum_{i=1}^{\ell}\psi_i(u)\langle\varphi'(u + \tau\delta\psi(u)\sum_{i=1}^{\ell}\psi_i(u)v_{u_i}), v_{u_i}\rangle \\ &\leq -\epsilon^{1/2}\delta\psi(u).\end{aligned}$$

If $u \notin S$, $\psi(u) = 0$, and $\varphi(\eta(1,u)) = \varphi(u)$. If $\overline{u} \in C$ is such that $\varphi(\eta(1,\overline{u})) = \Phi(D)$, we obtain

$$c \leq \varphi(\eta(1,\overline{u})) \leq \varphi(\overline{u})$$

so that $\overline{u} \in S$ and $\psi(\overline{u}) = 1$. By (52), we get

$$\varphi(\eta(1,\overline{u})) - \varphi(\overline{u}) \leq -\epsilon^{1/2}\delta$$

and, in particular,

$$\Phi(D) + \epsilon^{1/2}\delta \leq \varphi(\overline{u}) \leq \Phi(C).$$

But, by the definition of D, we have

$$\delta(C,D) \leq \delta$$

and hence

$$\Phi(D) + \epsilon^{1/2}\delta(C,D) \leq \Phi(C),$$

which contradicts (50) and completes the proof. □

Assume now that $L : [0,T] \times \mathbf{R}^N \times \mathbf{R}^N \to \mathbf{R}$ satisfies the assumptions (L_1) to (L_4) of Section 4.2. We shall say that two weak solutions u and v of (18) are *geometrically distinct* if

$$u \not\equiv v \ (\mathrm{mod}\, T_i e_i, \ 1 \leq i \leq N).$$

Theorem 4.13. *Under assumptions* (L_1) *to* (L_4) *of Section* 4.2, *problem* (18) *has at least* $N+1$ *geometrically distinct weak solutions.*

Proof. We shall apply Theorem 4.12 with $X = H_T^1$ and

$$G = \left\{ \sum_{i=1}^{N} k_i T_i e_i \ :\ k_i \in \mathbf{Z},\ 1 \le i \le N \right\},$$

so that the function φ defined on H_T^1 by

$$\varphi(u) = \int_0^T L(t, u(t), \dot{u}(t))\, dt$$

is G-invariant, continuously differentiable, bounded from below, and satisfies the $(PS)_G$-condition, as shown in the beginning of this section. By Theorem 4.12, φ has at least $N+1$ critical orbits, since N is the dimension of the space generated by G. It is then easy to obtain, as in Theorem 4.6, $N+1$ geometrically distinct solutions of (18). □

Remark 4.7. We shall prove in Section 9.4 that, under a nondegeneracy assumption, problem (18) has at least 2^N geometrically distinct weak solutions.

Examples

1) For each $m > 0$, $g > 0$, $\ell \in \mathbf{R}$, and $f \in L^1(0,T;\mathbf{R})$ such that

$$\int_0^T f(t)\, dt = 0,$$

the equation of the forced simple pendulum

$$m\ell^2 \ddot{u}(t) + g\ell \sin u(t) = f(t)$$

has at least two geometrically distinct weak T-periodic solutions. It suffices to take $G = \{2k\pi \ :\ k \in \mathbf{Z}\}$.

2) For each $m_i > 0$, $\ell_i \in \mathbf{R}$, $g > 0$ and $f_i \in L^1(0,T;\mathbf{R})$ such that

$$\int_0^T f_i(t)\, dt = 0$$

($i = 1, 2$), the equations of the forced double pendulum, which correspond to

$$\begin{aligned} L(t, x_1, x_2, y_1, y_2) &= (1/2)(m_1 + m_2)\ell_1^2 y_1^2 + (1/2) m_2 \ell_2^2 y_2^2 \\ &+ m_2 \ell_1 \ell_2 y_1 y_2 \cos(x_1 - x_2) + (m_1 + m_2) g \ell_1 \cos x_1 \\ &+ m_2 g \ell_2 \cos x_2 + f_1(t) x_1 + f_2(t) x_2, \end{aligned}$$

have at least three geometrically distinct weak T-periodic solutions, as L satisfies conditions (L_1) to (L_4) of Section 4.2 and is G-invariant for $G = \{2\pi k_1 e_1 + 2\pi k_2 e_2 : k_1, k_2 \in \mathbf{Z}\}$.

3) For each $m_i > 0$, $\ell_i > 0$, $0 < a \leq \ell_1$, $0 < b \leq \ell_2$, $g > 0$, $k > 0$, and $h_i \in L^1(0,T;\mathbf{R})$, $(i = 1,2)$, the equations of a double pendulum coupled by a linear spring with spring constant k attached at distances a and b of the respective fixed points of the pendulums and driven by horizontal forces $-2adk + h_1(t)$ and $2bdk + h_2(t)$ where d is the distance between the fixed points, correspond to

$$L(t, x_1, x_2, y_1, y_2) = (1/2)[m_1\ell_1^2 y_1^2 + m_2\ell_2^2 y_2^2]$$
$$- [a^2 + b^2 - 2ab\cos(x_1 - x_2) - 2d(b\sin x_2 - a\sin x_1)]$$
$$+ m_1 g\ell_1 \cos x_1 + m_2 g\ell_2 \cos x_2 + h_1(t)\sin x_1 + h_2(t)\sin x_2.$$

Again, those equations have at least three geometrically distinct weak T-periodic solutions.

Historical and Bibliographical Notes

Theorems 4.1 and 4.2 are due to Ekeland [Eke$_4$] and the given proof to Lasry and Crandall (see [Eke$_{5,6}$] for more details and applications and [Aub$_1$], [Aub$_2$], [Wil$_3$] for other proofs). Theorem 4.3 and Corollary 4.2 are "approximate" versions of a minimax theorem given in Shi [Shi$_1$] to unify the mountain pass lemma of Ambrosetti–Rabinowitz [AmR$_1$] (Theorem 4.10) and the saddle point theorem of Rabinowitz [Rab$_5$] (Theorem 4.7), which have been widely applied and generalized (see e.g. [Rab$_6$] for a clear survey and references). Their original proof used a deformation technique of the type described in Chapter 6, proofs via Ekeland's principle being given independently in [AuE$_1$] and [Shi$_1$] at various levels of generality. Compactness conditions like in Definition 4.1 were first introduced by Palais and Smale [PaS$_1$, Pal$_1$, Sma$_1$] and are generally called PS-conditions. The specific one given in Definition 4.1 seems to be due to Penot [Pen$_1$] and Brézis–Coron–Nirenberg [BCN$_1$]. Although they appear very naturally in the frame of Ekeland's variational principle, the Palais–Smale conditions were first introduced in the context of deformation techniques. Theorem 4.4 is essentially due to Palais–Smale [PaS$_1$] and Proposition 4.1 and Theorem 4.5 goes back in spirit to Mawhin–Willem [MaW$_{2,3}$]. See also [CFS]. Theorem 4.8 is the version, for periodic solutions of second order differential equations, of Rabinowitz approach [Rab$_5$] to a result first proved by Ahmad–Lazer–Paul [ALP$_1$] for Dirichlet problems. Theorem 4.9 comes from [Maw$_{12}$]. Theorem 4.11 was first proved by Rabinowitz [Rab$_{3,7}$] by a direct minimax approach and without the convexity assumption. The proof given here is due to Ekeland [Eke$_1$] and has motivated subsequent work by Ambrosetti–Mancini [AmM$_1$] and Ekeland–Hofer [EkH$_1$].

Theorem 4.12 follows also from the Lusternik–Schnirelman theory on a Finsler manifold (see [LuS] for the finite-dimensional case and [Pal_3] for the general situation). The simple minimax characterization of the critical values is due to Rabinowitz ([Rab_{23}]. The use of the Hausdorff distance in connection with category and Ekeland's variational principle was introduced in [Szu_8] where Lemmas 4.8 and 4.9 are proved.

Theorem 4.13 is due to Mawhin–Willem [MaW_2] in the case of a simple pendulum, to Fournier–Willem [FoW_1] in the case of a double pendulum and to Mawhin [Maw_{12}] and Rabinowitz [Rab_{23}] in the general case. Example 4.3 was considered, for odd h_i, by Marlin [Mar_1].

The geometry of a function near a critical point of mountain pass type has been discussed by Pucci-Serrin [$\text{PuS}_{1,2,3}$] and Hofer [Hof_1]. For the saddle point theorem, see [LaSo_1].

For other applications of Ekeland's variational principle to critical point theory, see e.g. [BaE_1], [Eke_{12}], [Ele_1], [McL_1], [Pen_2], [ShC_1], [$\text{Szu}_{1,2}$], [Wil_7]. That this principle characterizes complete metric spaces is shown in [Sul_1].

Variants of the mountain pass and of the saddle point theorems as well as various applications can be found in [AhL_1], [BaM_1], [BBF_1], [Ben_3], [BeR_1], [Cas_2], [$\text{Cha}_{3,4,5}$], [DFS_1], [GrS_1], [Hof_4], [LiP_1], [LuS_1], [Maw_8], [Nir_2], [$\text{Rab}_{9,14}$], [RuS_1], [$\text{Sol}_{1,2}$], [$\text{Tia}_{1,2}$], [War_1], [Wa_1].

Further discussions of Palais–Smale type conditions can be found in [Gor_2], [$\text{Str}_{2,3}$].

See [AmC_1], [ASt_1], and [Cor_2] for other applications of the mountain pass theorem combined with the dual least action principle.

The periodic solution of nonlinear wave equations (an infinite-dimensional Hamiltonian problem) were first attacked through critical point theory by Rabinowitz ([$\text{Rab}_{5,15,16}$]). See also [ChH_1], [CLD_1], [Snc_1], [Wil_9], [Wu_2], [WuL_1], [Gia_1], [Wei_3].

Exercises

1. Let M be a complete metric space, $\varphi : M \to \mathbf{R}$ a l.s.c. nonnegative function and $T : M \to M$ a mapping such that

$$d(u, Tu) \leq \varphi(u) - \varphi(Tu)$$

for each $u \in M$. Show that T has a fixed point (Caristi).

Hint. Apply Ekeland's principle with $\epsilon = \frac{1}{2}$ to get $v \in M$ such that

$$(1/2)d(v, Tv) \geq \varphi(v) - \varphi(Tv).$$

2. Show that if $T : M \to M$ satisfies

$$d(Tu, Tv) \leq kd(u, v)$$

for some $k \in [0,1[$ and all $u, v \in M$, then a function φ exists having the properties of Exercise 4.1.

Hint. Try φ of the form $\varphi(u) = cd(u, Tu)$ and determine c.

3. Given an elementary proof of Ekeland's principle when $M = \mathbf{R}^N$.

   Hint. Apply the results of Chapter 1 to the coercive function

$$\psi(w) = \Phi(w) + \epsilon\|u - w\| \text{ where } \Phi(u) \leq \inf \Phi + \epsilon$$

   and show that if ψ achieves its minimum at v, then v satisfies the conditions of Ekeland's principle ([HiU$_1$]).

4. Let H be a Hilbert space, $C \subset H$ a closed convex set and let $\Phi \in C^1(H, \mathbf{R})$ be bounded from below and such that $(I - \nabla\Phi)(C) \subset C$. Show that for each $\epsilon > 0$ and each $u \in C$ with $\Phi(u) \leq \inf_C \Phi + \epsilon$ there exists $v \in C$ such that

   i) $\Phi(v) \leq \Phi(u)$
   ii) $\|v - u\| \leq \epsilon^{1/2}$
   iii) $\|\nabla\Phi(u)\| \leq \epsilon^{1/2}$.

   Hint. Apply Ekeland's principle with $M = C$ and $w = (1-t)v + t(v - \nabla\Phi(v))$, $t \in [0,1]$. ([Ho$_5$], [DFS$_1$]).

5. Let X be a Banach space and $\varphi \in C^1(X, \mathbf{R})$ be such that φ achieves its minimum on X at u_0 and such that, for some $R > 0$,

$$\{u \in X \,:\, \varphi'(u) = 0\} \subset B(u_0, R).$$

   If φ satisfies the (PS)-condition, on $\partial B(u_0, R)$, then

$$\inf_{\partial B(u_0,R)} f > f(u_0).$$

   Hint. Proceed by contradiction and apply Ekeland's variational principle.

6. Let X be a Banach space and let $\varphi : X \to \mathbf{R}$ be a differentiable function satisfying the Palais–Smale condition.

   a) If φ is bounded from below then φ is coercive.
   b) If $\varphi^{-1}(c)$ is bounded for some $c \in \mathbf{R}$ then $|\varphi|$ is coercive.

   Hint. Apply Ekeland's principle and a contradiction argument.

7. Let X be a Banach space and let $\varphi \in C^1(X, \mathbf{R})$ satisfying the (PS)-condition. If φ is not bounded from below and has a local minimum, show that φ has at least two critical points. If φ has $n \geq 2$ local minima, show that it has at least $n + 1$ critical points.

8. Show that Theorem 4.9 still holds for the problem

$$\ddot{u}(t) = \nabla F(t, u(t))$$
$$u(0) - u(T) = \dot{u}(0) - \dot{u}(T) = 0$$

if $F(t,u) = G(u) + H(t,u)$ with

$$(G(u) - G(v), u - v) \geq -\gamma |u - v|^2$$

for some $\gamma < \omega^2$ and all $u, v \in \mathbf{R}^N$ and

$$|\nabla H(t,u)| \leq h(t)$$

for some $h \in L^1(0,T)$, all $u \in \mathbf{R}^N$ and a.e. $t \in [0,T]$. ([AhL1]).

Hint. Use Wirtinger inequality.

9. Show that, when F satisfies the regularity conditions of Section 1.8 and

$$|\nabla F(t,x)| \leq h(t)$$

for some $h \in L^1(0,T)$, a.e. $t \in [0,T]$ and each $x \in \mathbf{R}^N$, the problem

$$\ddot{u}(t) + m^2\omega^2 u(t) = \nabla F(t, u(t))$$
$$u(0) - u(T) = \dot{u}(0) - \dot{u}(T) = 0$$

with $m \in \mathbf{N} \setminus \{0\}$ and $\omega = 2\pi/T$ has at least one solution if either

$$\int_0^T F(t, a\cos m\omega t + b \sin m\omega t)\, dt \to +\infty$$

or

$$\int_0^T F(t, a\cos m\omega t + b \sin m\omega t)\, dt \to -\infty$$

when $|(a,b)| \to \infty$ in $\mathbf{R}^{2N}$.

Hint. Use Rabinowitz saddle point theorem with, according to the case, $X^- = H^-$, $X^+ = H^0 \oplus H^+$ or $X^- = H^- \oplus H^0$, $X^+ = H^+$ where

$$H^- = \left\{ \sum_{j=0}^{m-1} a_j \cos j\omega t + b_j \sin j\omega t \ : \ a_j \in \mathbf{R}^N, \right.$$
$$\left. b_j \in \mathbf{R}^N, 0 \leq j \leq m-1 \right\}$$

$$H^0 = \{a \cos m\omega t + b \sin m\omega t \ : \ a \in \mathbf{R}^N, b \in \mathbf{R}^N\}$$

$$H^+ = \left\{ u \in H_T^1 \ : \ \int_0^T u(t) \cos j\omega t\, dt = \int_0^T u(t) \sin j\omega t\, dt = 0, \right.$$

$0 \leq j \leq m\}$ and, to verify the PS$_c$-condition, estimate

$$|\langle \varphi'(u_k), u_k^+ - u_k^- \rangle|$$

if $u = u^- + u^0 + u^+$ with $u^- \in H^-$, $u^0 \in H^0$, $u^+ \in H^+$.

10. Compare the results of Exercise 3.3 with the conclusions of Theorem 4.11.

11. Let Y be a closed subspace of a normed space. The following properties are equivalent:

    i) Y is an ANE;

    ii) there is a neighborhood U of Y of which Y is a retract.

12. Use the mountain pass lemma to show that if $a \in \mathbf{R}$ and $f \in L^1(0,T)$ is such that $\int_0^T f(t)\,dt = 0$, then the problem

$$u'' + a \sin u = f(t)$$

has a weak solution geometrically distinct from the one which minimizes the action functional ([MaW$_2$]).

5

A Borsuk–Ulam Theorem and Index Theories

Introduction

The (classical) *Borsuk–Ulam theorem* is a result which ensures that if $\Omega \subset \mathbf{R}^n$ is an open bounded symmetric neighborhood of the origin and if $f : \partial\Omega \to \mathbf{R}^{n-1}$ is continuous and odd, then $0 \in f(\partial\Omega)$. This result can be proved using degree theory, a way of making an algebraic count of the zeros, in the closure $\overline{D}$ of an open bounded set $D \subset \mathbf{R}^n$, of continuous mappings $g : \overline{D} \subset \mathbf{R}^n$ having no zeros on ∂D. A short account of degree theory is given in Section 5.3.

The Borsuk–Ulam theorem is fundamental in measuring the "size" of some subsets of a Banach space X which are symmetric with respect to the origin. If A is such a set, its "size" is measured by its *index,* namely the smallest integer k such that there exists an odd mapping $f \in C(A, \mathbf{R}^k \backslash \{0\})$. For $\partial\Omega$ like above, taking $f = Id$ shows that ind $\partial\Omega \leq n$ and Borsuk–Ulam theorem implies that ind $\partial\Omega = n$.

Symmetry with respect to the origin and oddness of a mapping are nothing but invariance properties with respect to a representation of the group $\mathbf{Z}_2$. To cover more general situations, we sketch in Section 5.1 the elements of the theory of *representation of compact topological groups* in Banach spaces (or how to represent groups by linear operators). The case of the group S^1 will be particularly important for the study of periodic solutions of autonomous Hamiltonian systems. Indeed, if $u(t)$ is a solution of such a system, the same is true for $u(t+\omega)$ for each $\omega \in \mathbf{R}$ and hence, if $u(t)$ is T-periodic, we see the appearance of the representation $T(\omega)$ of the group $\mathbf{R}/T\mathbf{Z} \sim S^1$ given by

$$(T(\omega)u)(t) = u(t+\omega).$$

The computation of the index associated to the group S^1 requires a Borsuk–Ulam theorem in this setting. This is done in Section 5.3 using the *parametrized Sard theorem* introduced in Section 5.2 and *degree theory.*

One then disposes of the elements necessary to develop in Section 5.4 a general index theory containing the previously mentioned ones as special cases.

5.1 Group Representations

Let G be a topological group. A *representation* of G over a Banach space X is a family $\{T(g)\}_{g\in G}$ of linear operators $T(g) : X \to X$ such that

$$T(0) = Id,$$

$$T(g_1 + g_2) = T(g_1)T(g_2)$$

$$(g, u) \to T(g)u \text{ is continuous.}$$

A subset A of X is *invariant* (under the representation) if $T(g)A = A$ for all $g \in G$. A representation $\{T(g)\}_{g\in G}$ of G over X is *isometric* if $||T(g)u|| = ||u||$ for all $g \in G$ and all $u \in X$.

Lemma 5.1. *Let $\{T(g)\}_{g\in G}$ be a representation of a compact group over a Banach space X. Let Y be a closed invariant vector subspace of X which admits a topological complement. Then Y has an invariant topological complement.*

Proof. Let P be a (continuous) projector onto Y. Then the linear continuous operator

$$Q = \int_G T(-g)PT(g)\,dg,$$

where dg is the normalized Haar measure on G, clearly maps X into Y and, if $y \in Y$,

$$Qy = \int_G T(-g)PT(g)y\,dg = \int_G T(-g)T(g)y\,dg = y.$$

Thus Q is a continuous projector onto Y and is easily checked to be equivariant. Consequently, the range of $Id - Q$ is an invariant topological complement of Y. □

Lemma 5.2. *Let $\{T(g)\}_{g\in G}$ be a representation of a compact topological group over $\mathbf{R}^N$. Then there exists an invertible matrix $L : \mathbf{R}^N \to \mathbf{R}^N$ such that $LT(g)L^{-1}$ is an isometric representation of G over $\mathbf{R}^N$ (with the Euclidean norm).*

Proof. Since the quadratic form

$$Q(u) = \int_G |T(g)u|^2 dg$$

is positive definite, there exists a nonsingular matrix $L : \mathbf{R}^N \to \mathbf{R}^N$ such that $Qu = |Lu|^2$. It is then easy to verify that $LT(g)L^{-1}$ is an isometric representation of G over $\mathbf{R}^N$. □

Theorem 5.1. *Let $\{T(\theta)\}_{\theta\in S^1}$ be an isometric representation of S^1 over $\mathbf{R}^N$ (with the Euclidean norm). Then $T(\theta)$ has the matrix representation*

$$\operatorname{diag}\{M_1, \ldots, M_k\}, \tag{1}$$

where M_j is either of order 1 and

$$M_j = 1, \tag{2}$$

or is of order 2 and for some $n \in \mathbf{N} \setminus \{0\}$ one has

$$M_j = \begin{pmatrix} \cos n\theta & -\sin n\theta \\ \sin n\theta & \cos n\theta \end{pmatrix}. \tag{3}$$

Proof. Since $S^1 = \mathbf{R}/2\pi\mathbf{Z}$, we can consider $T(\theta)$ as a continuous 2π-periodic mapping over $\mathbf{R}$. We first prove that $T(\theta)$ is differentiable. Since

$$\lim_{\alpha \to 0} \left[\alpha^{-1} \int_0^\alpha T(\omega) d\omega \right] = T(0) = Id$$

we see that $\int_0^\alpha T(\theta) d\theta$ is invertible for all $\alpha \neq 0$ sufficiently small. On the other hand, for any $\alpha \in \mathbf{R}$ and $\beta \in \mathbf{R}$, we have

$$T(\beta) \int_0^\alpha T(\theta) d\theta = \int_0^\alpha T(\beta + \theta) d\theta = \int_\beta^{\beta+\alpha} T(\theta)\, d\theta,$$

so that, for $\alpha \neq 0$ sufficiently small, we get

$$T(\beta) = \left(\int_\beta^{\beta+\alpha} T(\theta) d\theta \right) \left[\int_0^\alpha T(\theta) d\theta \right]^{-1}.$$

This implies clearly the differentiability of $T(\beta)$ at any $\beta \in \mathbf{R}$. Differentiating both sides of the identity

$$T(\theta + \beta) = T(\theta) T(\beta)$$

with respect to θ and setting $\theta = 0$, we obtain

$$T'(\beta) = T'(0) T(\beta),$$

so that

$$T(\theta) = \exp(\theta L)$$

where $L = T'(0)$. Since the representation is isometric, $T(.)u$ is bounded over $\mathbf{R}$ for every $u \in \mathbf{R}^N$. By the theory of linear differential equations, $T(\theta)$ has a matrix representation (1), where M_j is either of order 1 and $M_j = 1$, or is of order 2 and

$$M_j = \begin{pmatrix} \cos \beta_j \theta & -\sin \beta_j \theta \\ \sin \beta_j \theta & \cos \beta_j \theta \end{pmatrix}$$

for some $\beta_j > 0$. Since $T(2\pi) = Id$, we necessarily have $\beta_j \in \mathbf{N} \setminus \{0\}$. □

Remark 5.1. The matrices M_j are the *irreductible representations* of S^1.

Remark 5.2. Let

$$\mathrm{Fix}(S^1) = \{u \in \mathbf{R}^N : T(\theta)u = u \text{ for all } \theta \in S^1\}.$$

Theorem 5.1 implies that if $\mathrm{Fix}(S^1) = \{0\}$ and if $T(\theta)$ is isometric, then N must be even.

5.2 The Parametrized Sard Theorem

Let $U \subset \mathbf{R}^p$ be open. A point $v \in \mathbf{R}^q$ is a *regular value* of $f \in C^m(U, \mathbf{R}^q)$ $(m \geq 1)$ if $f'(u)$ is onto for each $u \in f^{-1}(v)$.

We shall use the following important result.

Sard Theorem. *Let $U \subset \mathbf{R}^p$ be open and let $f \in C^m(U, \mathbf{R}^q)$. If $m >$ $\max(0, p-q)$, then almost every point of $\mathbf{R}^q$ is a regular value of f.*

A subset Z of $\mathbf{R}^p$ is a *C^m-submanifold of dimension d* if, for every $z \in Z$, there exists an open neighborhood A of 0 in $\mathbf{R}^p$, an open neighborhood B of z in $\mathbf{R}^p$ and a C^m-diffeomorphism $\Phi : A \to B$ such that

$$\Phi(A \cap \mathbf{R}^d) = B \cap Z.$$

It is easy to construct submanifolds by using regular values.

Preimage Theorem. *Let $U \subset \mathbf{R}^p$ be open and let $f \in C^m(U, \mathbf{R}^q)$, $(m \geq 1)$. If v is a regular value of f and if $f^{-1}(v) \neq \phi$, then $f^{-1}(v)$ is a C^m-submanifold of dimension $p-q$.*

Proof. Let $z \in Z = f^{-1}(v)$. Without loss of generality, we can assume that $v = 0$ and $z = 0$. Since 0 is a regular value of f, the kernel X of $f'(0)$ has dimension $p - q$. Define now $\psi : X \times X^{\perp} \to \mathbf{R}^p$ by

$$\psi(x, y) = (x, f(x, y)).$$

It is easy to verify that $\psi'(0)$ is invertible. By the inverse function theorem, ψ is a C^m-diffeomorphism from a open neighborhood B of 0 in $\mathbf{R}^p$ to an open neighborhood A of 0 in $\mathbf{R}^p$. Moreover,

$$\psi(B \cap Z) = A \cap \mathbf{R}^{p-q}.$$

Setting $\Phi = \psi^{-1}$ completes the proof. □

We now state and prove the *parametrized Sard theorem*.

Theorem 5.2. *Let $U \subset \mathbf{R}^p$, $\Lambda \subset \mathbf{R}^n$ be open and let $f \in C^m(U \times \Lambda, \mathbf{R}^q)$. If $m > \max(0, p-q)$ and $0 \in \mathbf{R}^q$ is a regular value of f, then, for almost every $\lambda \in \Lambda$, 0 is a regular value of $f(., \lambda)$.*

Proof. We can assume that $Z = f^{-1}(0) \neq \phi$. By the preimage theorem, Z is a C^m-submanifold of dimension $d = p + n - q$. Since Z is Lindelöf, there is a sequence (Φ_j, A_j, B_j) satisfying the definition of a submanifold such that

$$Z \subset \bigcup_{j \in \mathbf{N}} B_j.$$

Let us denote by x_j the restriction of Φ_j to $A_j \cap \mathbf{R}^d = D(x_j)$ and let $x_j(v) = (y_j(v), z_j(v)) \in U \times \Lambda$.

By Sard's theorem, for each $j \in \mathbf{N}$, almost every $\lambda \in \Lambda$ is a regular value of z_j and then almost every $\lambda \in \Lambda$ will be a regular value of every z_j.

Assume that $f(u,\lambda) = 0$ and that λ is a regular value for every z_j. Then there is some $j \in \mathbf{N}$ and $v \in D(x_j)$ such that

$$u = y_j(v), \quad \lambda = z_j(v).$$

Now 0 is a regular value of f and hence, for each $c \in \mathbf{R}^q$ there will be $a \in \mathbf{R}^p$ and $b \in \mathbf{R}^n$ such that

$$f'_u(u,\lambda)a + f'_\lambda(u,\lambda)b = c. \tag{4}$$

From

$$f(y_j(s), z_j(s)) = 0, \quad s \in D(x_j),$$

we deduce

$$f'_u(u,\lambda)y'_j(v) + f'_\lambda(u,\lambda)z'_j(v) = 0. \tag{5}$$

Moreover, λ being a regular value of z_j, there will exist $w \in \mathbf{R}^{p+n-q}$ such that

$$z'_j(v)w = b. \tag{6}$$

From (4) - (5) - (6) we deduce that

$$c = f'_u(u,\lambda)[a - y'_j(v)w].$$

Hence 0 is a regular value of $f(.,\lambda)$ and the proof is complete. □

Let $\{T(\theta)\}_{\theta \in S^1}$ be an isometric representation of S^1 over $\mathbf{R}^{2k}$ (with the Euclidean norm) such that $\mathrm{Fix}(S^1) = \{0\}$. By Theorem 5.1 and Remark 5.2, if we identify $\mathbf{R}^{2k}$ with $\mathbf{C}^k$, $T(\theta)$ will have the representation

$$T(\theta)u = \left(e^{in_1\theta}u_1, \ldots, e^{in_k\theta}u_k\right), \tag{7}$$

where $n_1, \ldots, n_k \in \mathbf{N} \setminus \{0\}$.

Let D be an open invariant neighborhood of 0 and let $\Phi \in C(\overline{D}, \mathbf{R}^{2k})$ and $n \in \mathbf{N} \setminus \{0\}$ be such that

$$\Phi \in C^1(D, \mathbf{R}^{2k}), \tag{8}$$

$$\Phi(T(\theta)u) = e^{in\theta}\Phi(u), \quad \theta \in S^1, \quad u \in \partial D, \tag{9}$$

$$0 \notin \Phi(\partial D). \tag{10}$$

By (7), (9), and (10), $m_j = n/n_j \in \mathbf{N} \setminus \{0\}$, because if $(0, \ldots, u_j, \ldots, 0) \in \partial D$, then, for $\theta = 2\pi/n_j$,

$$\Phi(T(\theta)u) = \Phi(0, \ldots, u_j, \ldots, 0) = e^{2i\pi n/n_j}\Phi(0, \ldots, u_j, \ldots, 0).$$

Moreover, for all $\lambda = (\lambda_1, \ldots, \lambda_k) \in \mathbf{C}^{k \times k}$, the mapping Φ_λ defined by

$$\Phi_\lambda(u) = \Phi(u) + \lambda_1 u_1^{m_1} + \ldots + \lambda_k u_k^{m_k}$$

satisfies the equivariance property (9).

Corollary 5.1. *Under the above assumptions* $0 \in \mathbf{C}^k$ *is a regular value of the restriction of* Φ_λ *to* $D \setminus \{0\}$ *for almost every* $\lambda \in \mathbf{C}^{k\times k}$.

Proof. Set $\Lambda = \mathbf{C}^{k\times k}$ and define $f : U \times \Lambda \to \mathbf{C}^k$ by $f(u,\lambda) = \Phi_\lambda(u)$. In order to apply Theorem 5.2 with $p = q = 2k$, $m = 1$ and $n = k^2$, it suffices to check that $f'(u,\lambda)$ is onto for each $(u,\lambda) \in U \times \Lambda$, i.e. to be able to solve the equation in $(a,b) \in \mathbf{C}^k \times \mathbf{C}^{k^2}$

$$f'u(u,\lambda)a + f'\lambda(u,\lambda)b = c$$

for each $c \in \mathbf{C}^k$. Choosing $a = 0$, it remains to solve

$$b_1 u_1^{m_1} + \ldots + b_k u_k^{m_k} = c$$

which is always possible as $u \neq 0$. □

5.3 Topological Degree

Let $D \subset \mathbf{R}^N$ be bounded and open and let $f \in C(\overline{D}, \mathbf{R}^N)$. If

$$\begin{gathered} 0 \notin f(\partial D), \\ f \in C^2(D, \mathbf{R}^N), \qquad (11) \\ 0 \text{ is a regular value of } f|_D, \end{gathered}$$

the implicit function theorem implies that $f^{-1}\{0\}$ is finite. The *topological degree of* f *in* D (with respect to 0) is then defined by

$$d(f,D) = \sum_{u \in f^{-1}\{0\}} \text{sign}\det f'(u)$$

and represents, therefore in the "generic" case (11), the "algebraic" number of zeros of f in D.

The following lemma is basic to extend the concept of degree in more general situations and prove its basic properties.

Lemma 5.3. *Let* $\mathcal{F} \in C(\overline{D} \times [0,1], \mathbf{R}^N)$ *be such that*

$$\begin{gathered} 0 \notin \mathcal{F}(\partial D \times [0,1]), \\ \mathcal{F} \in C^2(D \times [0,1]), \qquad (12) \end{gathered}$$

0 *is a regular value of* $\mathcal{F}(\cdot, j)|_D$ $(j = 0,1)$. *Then one has* $d(\mathcal{F}(.,0),D) = d(\mathcal{F}(.,1),D)$.

Let $g \in C(\overline{D}, \mathbf{R}^N)$ be such that $0 \notin g(\partial D)$. The Weierstrass approximation theorem and the Sard theorem imply the existence of $f \in C(\overline{D}, \mathbf{R}^N)$ satisfying (11) and the inequality

$$|f(u) - g(u)| < \operatorname{dist}(0, g(\partial D)) \tag{13}$$

for all $u \in \partial D$. We then define the *topological degree of g in D* by

$$d(g, D) = d(f, D).$$

Such a definition is meaningful if we prove that for $\tilde{f} \in C(\overline{D}, \mathbf{R}^N)$ satisfying (11) and (13) we have

$$d(\tilde{f}, D) = d(f, D).$$

If we set

$$\mathcal{F}(u,t) = (1-t)f(u) + t\tilde{f}(u),$$

we obtain

$$|\mathcal{F}(u,t) - g(u)| < \operatorname{dist}(0, g(\partial D))$$

for all $(u,t) \in \partial D \times [0,1]$. Consequently,

$$0 \notin \mathcal{F}(\partial D \times [0,1])$$

and we conclude by applying Lemma 5.3.

By reduction to the C^2-regular case (11), it is easy to prove the following properties.

(i) If $0 \notin g(D)$ then $d(g, D) = 0$, which immediately gives the *existence property:* if $d(g, D) \neq 0$, then $0 \in g(D)$.

(ii) (*Excision*). If $U \subset D$ is open and $0 \notin g(\overline{D} \backslash U)$, then $d(g, D) = d(g, U)$.

(iii) (*Additivity*). If $D = D_1 \cup D_2$ where D_1 and D_2 are open disjoint sets and $0 \notin g(\partial D_1 \cup \partial D_2)$, then

$$d(g, D) = d(g, D_1) + d(g, D_2).$$

(iv) (*Cartesian product*). Let $D_1 \subset \mathbf{R}^p$ and $D_2 \subset \mathbf{R}^q$ be open and bounded and let $f_1 \in C(\overline{D}_1, \mathbf{R}^p)$, $f_2 \in C(\overline{D}_2, \mathbf{R}^q)$ be such that $0 \notin f_j(\partial D_j)$ $(j = 1, 2)$. Then

$$d((f_1, f_2), D_1 \times D_2) = d(f_1, D_1) d(f_2, D_2).$$

(In the C^2-regular case, (iv) follows from the relation $\det(f_1, f_2)'(u_1, u_2) = \det f_1'(u_1) \det f_2'(u_2)$.)

Example. Let $\rho > 0$, $D = \{u \in \mathbf{R}^{2k} : |u| < \rho\}$, $m_j \in \mathbf{N} \setminus \{0\}$ $(1 \leq j \leq k)$ and identify $\mathbf{R}^{2k}$ with $\mathbf{C}^k$. If we define $f : \overline{D} \to \mathbf{C}^k$ by

$$f(u_1, \ldots u_k) = (u_1^{m_1}, \ldots, u_k^{m_k}),$$

then

$$d(f, D) = m_1 \dots m_k. \tag{14}$$

Proof. By properties (ii) and (iv), it suffices to consider the case $k = 1$. Define $f : \mathbf{C} \to \mathbf{C}$ by $f(u) = u^m$, i.e. $f(x + iy) = (x + iy)^m$, where $m \in \mathbf{N} \setminus \{0\}$. For each $z \in \mathbf{C} \setminus \{0\}$, the equation $f(u) = z$ has exactly m solutions and, by the Cauchy–Riemann equations, $\det f'(u) > 0$ at each solution. It follows from the definition of the degree that $d(f, D) = m$. □

Theorem 5.3 (*Homotopy invariance*). *Let $\mathcal{G} \in C(\overline{D} \times [0,1], \mathbf{R}^N)$ be such that $0 \notin \mathcal{G}(\partial D \times [0,1])$. Then*

$$d(\mathcal{G}(.,0), D) = d(\mathcal{G}(.,1), D).$$

Proof. By Weierstrass approximation theorem and Sard theorem, there is a $\mathcal{F} \in C(\overline{D} \times [0,1], \mathbf{R}^N)$ satisfying (12) and such that

$$|\mathcal{F}(u,t) - \mathcal{G}(u,t)| < \operatorname{dist}(0, \mathcal{G}(\partial D \times [0,1]))$$

for all $(u,t) \in \partial D \times [0,1]$. One concludes by using Lemma 5.3 and the definitions of the degree. □

Corollary 5.2. (*Continuity property*). *Let $f \in C(\overline{D}, \mathbf{R}^N)$ and $g \in C(\overline{D}, \mathbf{R}^N)$ be such that $0 \notin f(\partial D)$ and*

$$|f(u) - g(u)| < \operatorname{dist}(0, f(\partial D))$$

for all $u \in \partial D$. Then $d(f, D) = d(g, D)$.

Proof. Use the homotopy

$$\mathcal{G}(u,t) = (1-t)f(u) + tg(u)$$

and Theorem 5.3. □

Theorem 5.4. *Let $\{T(\theta)\}_{\theta \in S^1}$ be an isometric representation of S^1 over $\mathbf{R}^{2k}$ (with the Euclidean norm) such that $\operatorname{Fix}(S^1) = \{0\}$ and let D be an open bounded invariant neighborhood of 0. If $\Phi \in C(\overline{D}, \mathbf{R}^{2k})$ and $n \in \mathbf{N} \setminus \{0\}$ verify (9) and (10), then $d(\Phi, D) \neq 0$.*

Proof. First step. Identifying $\mathbf{R}^{2k}$ and $\mathbf{C}^k$ and using Theorem 5.1, we see that $T(\theta)$ has the representation (7). Let $m_j = n/n_j$ $(1 \le j \le k)$ where the n_j are given in (7) and let $\rho > 0$ be such that when $|u| \le \rho$ then $u \in D$. Define Ψ by

$$\begin{aligned} \Psi(u) &= (u_1^{m_1}, \dots, u_k^{m_k}) && \text{if } |u| \le \rho, \\ \Psi(u) &= \Phi(u) && \text{if } u \in \partial D, \end{aligned}$$

and use Tietze theorem to obtain a continuous extension Φ_1 of Ψ over $\overline{D}$. By the continuity property of the degree we immediately obtain

$$d(\Phi, D) = d(\Phi_1, D).$$

Second step. Let $\epsilon > 0$. By Weierstrass approximation theorem, there exists a polynomial $\Phi_2^\epsilon : \mathbf{R}^{2k} \to \mathbf{R}^{2k}$ such that

$$|\Phi_2^\epsilon(u) - \Phi_1(u)| < \epsilon \tag{15}$$

for all $u \in \overline{D}$.

Third step. Define Φ_3^ϵ on $\overline{D}$ by

$$\Phi_3^\epsilon(u) = \frac{1}{2\pi} \int_0^{2\pi} e^{-in\theta} \Phi_2^\epsilon(T(\theta)u)\, d\theta$$

so that

$$\Phi_3^\epsilon(T(\theta)u) = e^{in\theta}\Phi_3^\epsilon(u), \quad u \in \overline{D}, \quad \theta \in S^1.$$

If $|u| \le \rho$ or if $u \in \partial D$ we deduce from (15) that

$$|\Phi_3^\epsilon(u) - \Phi_1(u)| = \left| \frac{1}{2\pi} \int_0^{2\pi} e^{-in\theta} (\Phi_2^\epsilon(T(\theta)u) - \Phi_1(T(\theta)u))\, d\theta \right| < \epsilon. \tag{16}$$

By continuity of the degree, this implies that

$$d(\Phi_1, D) = d(\Phi_3^\epsilon, D)$$

for $\epsilon > 0$ small enough.

Fourth step. Let $\eta > 0$, since Φ_3^ϵ verifies (8), (9), and, for $\epsilon > 0$ small enough, (10), the Corollary 5.1 implies the existence of $\lambda \in \mathbf{C}^{k \times k}$ such that $|\lambda| < \eta$ and 0 is a regular value of $\Phi_4^\epsilon|_{D \setminus \{0\}}$, where

$$\Phi_4^\epsilon(u) = \Phi_3^\epsilon(u) + \lambda_1 u_1^{m_1} + \ldots + \lambda_k u_k^{m_k}. \tag{17}$$

Consequently, $(\Phi_4^\epsilon)^{-1}\{0\} \cap (D \setminus \{0\})$ is made of isolated points. But, since

$$\Phi_4^\epsilon(T(\theta)u) = e^{in\theta}\Phi_4^\epsilon(u), \quad u \in \overline{D}, \quad \theta \in S^1,$$

we see that $\Phi_4^\epsilon(T(\theta)u) = 0$, $\theta \in S^1$ whenever $\Phi_4^\epsilon(u) = 0$. Therefore, $0 \notin \Phi_4^\epsilon(D \setminus \{0\})$ and, letting $U = \{u \in D : |u| < \rho\}$ and using the continuity and excision properties of degree, we obtain, for η small enough,

$$d(\Phi_3^\epsilon, D) = d(\Phi_4^\epsilon, D).$$

Fifth step. For ϵ and η small enough, it follows from (16), (17), and (14) that

$$d(\Phi_4^\epsilon, U) = d(\Phi_1, U) = d(\Psi, U) = m_1 \ldots m_k \neq 0$$

and the proof is complete. □

Theorem 5.5. *Let $\{T(\theta)\}_{\theta \in S^1}$ be a representation of S^1 over $\mathbf{R}^{2k}$ such that* $\mathrm{Fix}(S^1) = \{0\}$ *and let D be an open bounded invariant neighborhood of 0. If $\Phi \in C(\partial D, \mathbf{C}^{k-1})$ and $n \in \mathbf{N} \setminus \{0\}$ are such that*

$$\Phi(T(\theta)u) = e^{in\theta}\Phi(u), \quad \theta \in S^1, \quad u \in \partial D,$$

then $0 \in \Phi(\partial D)$.

Proof. Replacing $T(\theta)$ by $LT(\theta)L^{-1}$ and Φ by $\Phi \circ L^{-1}$ with L given in Lemma 5.2 allows us to assume that $T(\theta)$ is an isometric representation of S^1 over $\mathbf{R}^{2k}$ with the Euclidean norm. Φ has a continuous extension $\psi : \overline{D} \to \mathbf{C}^{k-1} \subset \mathbf{C}^k$ by Tietze theorem and if $0 \notin \Phi(\partial D)$ Theorem 5.4 implies that $d(\psi, D) \neq 0$. But if $z \in \mathbf{C}^k$ is close enough to the origin and not in $\mathbf{C}^{k-1}$, property (i) of the degree and its continuity imply that

$$d(\psi, D) = d(\psi - z, D) = 0,$$

a contradiction. □

5.4 Index Theories

Let G be a compact topological group and let $\{T(g)\}_{g \in G}$ be an isometric representation of G over a Banach space X.

A mapping R between two invariant subsets of X (under the representation of G) is *equivariant* if

$$R \circ T(g) = T(g) \circ R$$

for all $g \in G$.

Definition 5.1. An *index* (for $\{T(g)\}_{g \in G}$) is a mapping from the closed invariant subsets of X into $\mathbf{N} \cup \{\infty\}$ such that:

(i) $\operatorname{ind} A = 0$ if and only if $A = \phi$;

(ii) if $R : A_1 \to A_2$ is equivariant and continuous, then

$$\operatorname{ind} A_1 \leq \operatorname{ind} A_2;$$

(iii) if A is compact invariant, there is a closed invariant neighborhood N of A such that

$$\operatorname{ind} N = \operatorname{ind} A;$$

(iv) $\operatorname{ind}(A_1 \cup A_2) \leq \operatorname{ind} A_1 + \operatorname{ind} A_2$ for all invariant subsets A_i $(i = 1, 2)$.

We shall give two important examples.

Example 1. Let $G = S^1$ and define the *S^1-index of a closed invariant subset A of X* as the smallest integer k such that there exists a $n \in \mathbf{N} \setminus \{0\}$ and a $\Phi \in C(A, \mathbf{C}^k \setminus \{0\})$ satisfying the following equivariance property

$$\Phi(T(\theta)u) = e^{in\theta}\Phi(u), \quad \theta \in S^1, \quad u \in A. \tag{18}$$

If such a mapping Φ does not exist, we define

$$\operatorname{ind} A = \infty.$$

Finally, we define ind $\phi = 0$.

We now have to show that this is an index in the sense of Definition 5.1. We first notice that if $\Phi \in C(A, \mathbf{C}^k \setminus \{0\})$ satisfies (18), there exists a continuous extension Ψ of Φ over X satisfying (18). It suffices to use first Tietze theorem to obtain a continuous extension $\tilde{\Phi}$ of Φ over X and then the mapping Ψ defined by

$$\Psi(u) = \frac{1}{2\pi} \int_0^{2\pi} e^{-in\theta} \tilde{\Phi}(T(\theta)u)\, d\theta$$

will satisfy the requirements.

We now check that the properties (i) to (iv) are satisfied.

(i) Follows directly from the definition.

(ii) If $\operatorname{ind} A_2 = \infty$, the result is trivial. If $\operatorname{ind} A_2 = k < \infty$, then there is a $\Phi \in C(A_2, \mathbf{C}^k \setminus \{0\})$ satisfying (18). Since R is equivariant and continuous, $\Psi = \Phi \circ R \in C(A_1, \mathbf{C}^k \setminus \{0\})$ and satisfies (18). Thus $\operatorname{ind} A_1 \leq k$.

(iii) Let V be any closed invariant neighborhood of the invariant subset A. Property (ii) implies that

$$\operatorname{ind} V \geq \operatorname{ind} A$$

since the inclusion map is equivariant. Thus, the result is trivial if $\operatorname{ind} A = \infty$. If $\operatorname{ind} A = k < \infty$, there is a $\Phi \in C(A, \mathbf{C}^k \setminus \{0\})$ satisfying (18) and hence, a continuous extension Ψ of Φ over X satisfying (18). Since $0 \notin \Psi(A) = \Phi(A)$ and $\Phi(A)$ is compact, there is, by uniform continuity of Ψ on A, a $\delta > 0$ such that $0 \notin \Psi(A_\delta)$. Thus, A_δ is a closed invariant neighborhood of A and $\operatorname{ind} A_\delta \leq k$. Consequently, $\operatorname{ind} A_\delta = \operatorname{ind} A$.

(iv) If $\operatorname{ind} A_1$ or $\operatorname{ind} A_2$ is infinite, the result is trivial. So, assume that $\operatorname{ind} A_j = k_j < \infty$ $(j = 1, 2)$. Then there exists $\Phi_j \in C(A_j, \mathbf{C}^k \setminus \{0\})$ satisfying (18) with $n = n_j$ and a continuous extension Ψ_j of Φ_j over X satisfying (18) with $n = n_j$ $(j = 1, 2)$. Define

$$\Psi : X \to \mathbf{C}^{k_1+k_2}$$

by

$$\Psi(u) = (\Psi_{1,1}^{n_2}(u), \ldots, \Psi_{1,k_1}^{n_2}(u), \Psi_{2,1}^{n_1}(u), \ldots, \Psi_{2,k_2}^{n_1}(u)).$$

Then one has

$$\Psi : A_1 \cup A_2 \to \mathbf{C}^{k_1+k_2} \setminus \{0\}$$

and

$$\Psi(T(\theta)u) = (e^{in_2 n_1\theta} \Psi_{1,1}^{n_2}(u), \ldots, e^{in_2 in_1\theta} \Psi_{1,k_1}^{n_2}(u),$$
$$e^{in_1 n_2\theta} \Psi_{2,1}^{n_1}(u), \ldots, e^{in_1 n_2\theta} \Psi_{2,k_2}^{n_1}(u)) = e^{in_1 n_2\theta} \Psi(u),$$

so that $\operatorname{ind}(A_1 \cup A_2) \leq k_1 + k_2$. □

Example 2. Let $G = \mathbf{Z}_2$ and define

$$T(0) = \mathrm{Id}, \quad T(1) = -\mathrm{Id}.$$

A subset A of X is invariant under this representation of $\mathbf{Z}_2$ if and only if it is symmetric with respect to the origin. The *$\mathbf{Z}_2$-index of a closed invariant subset* A of X is defined as the smallest integer k such that there exists an odd mapping $\Phi \in C(A, \mathbf{R}^k \setminus \{0\})$. If such a mapping does not exist, one defines $\operatorname{ind} A = \infty$ and, finally, one defines $\operatorname{ind} \phi = 0$. The properties of Definition 5.1 are checked as before.

We now give some results on the computation of the S^1-index. Recall that the *S^1-orbit* of $u \in X$ is the set $\mathcal{O}(u) = \{T(\theta)u : \theta \in S^1\}$.

Proposition 5.1. *The S^1-index of a finite union of S^1-orbits which do not meet* $\operatorname{Fix}(S^1)$ *is one.*

Proof. Assume that A is the union of disjoint S^1-orbits $\mathcal{O}(u_1), \ldots, \mathcal{O}(u_j)$ where $u_k \notin \operatorname{Fix}(S^1)$ $(1 \leq k \leq j)$. Let us look at $\theta \to T(\theta)u_k$ as a continuous mapping on $\mathbf{R}$ with minimal period $2\pi/n_k$ for some $n_k \in \mathbf{N} \setminus \{0\}$. The mapping $\Phi : A \to \mathbf{C} \setminus \{0\}$ defined by

$$\Phi(T(\theta)u_k) = e^{in\theta},$$

where $n = n_1 n_2 \ldots n_j$, satisfies (18). Thus $\operatorname{ind} A = 1$. □

Proposition 5.2. *Let Y be a closed invariant subspace of X of finite codimension and let A be a closed invariant subset of X. If* $\operatorname{Fix}(S^1) \subset Y$ *and $A \cap Y = \phi$, then*

$$\operatorname{ind} A \leq \frac{1}{2} \operatorname{codim} Y.$$

Proof. Lemma 5.1 implies the existence of an invariant topological complement Z of Y. As $\operatorname{Fix}(S^1) \cap Z = \{0\}$, it follows from Remark 5.2 that the dimension of Z is even and, identifying Z with $\mathbf{C}^N$, we deduce from Theorem 5.1 and Lemma 5.3 that the restriction of $T(\theta)$ to Z admits the representation

$$T(\theta) = L^{-1}\tilde{T}(\theta) \circ L$$

$$\tilde{T}(\theta)u = (e^{in_1\theta}u_1, \ldots, e^{in_N\theta}u_N), \quad u \in \mathbf{C}^N.$$

The mapping $\Phi : Z \setminus \{0\} \to \mathbf{C}^N \setminus \{0\}$ defined by

$$\Phi = \tilde{\Phi} \circ L,$$

$$\tilde{\Phi}(u) = (u_1^{n/n_1}, \ldots, u_N^{n/n_N})$$

with n the least common multiple of $n_1, \ldots n_N$ satisfies (18) as it is easily checked. Let P be the projector onto Z along Y. Since $A \cap Y = \phi$, $P \in C(A, Z \setminus \{0\})$ and, by the invariance of Y and Z, P is equivariant. Thus $\Phi \circ P \in C(A, \mathbf{C}^N \setminus \{0\})$ and satisfies (18). Consequently, $\operatorname{ind} A \leq N = \frac{1}{2}\operatorname{codim} Y$. □

Proposition 5.3. *Let Z be a finite-dimensional invariant subspace of X and let D be an open bounded invariant neighborhood of 0 in Z. If $Z \cap \text{Fix}(S^1) = \{0\}$, then*

$$\text{ind}\,\partial D = \frac{1}{2}\dim Z.$$

Proof. Again, $\dim Z$ is even and let $N = \frac{1}{2}\dim Z$. By Theorem 5.5, $\text{ind}\,\partial D \geq N$. Lemma 5.1 implies the existence of an invariant topological complement Y of Z. Clearly, $\text{Fix}(S^1) \subset Y$ and $\partial D \cap Y = \phi$, so that, by Proposition 5.2,

$$\text{ind}\,\partial D \leq \frac{1}{2}\text{codim}\,Y = N$$

and the proof is complete. □

Historical and Bibliographical Notes

For the element of group representations, the reader can consult [Brd$_1$] and Sard theorem is proved in [Mil$_1$]. The parametrized Sard theorem is a particular case of Thom transversality theorem [Tho$_1$] (see [AuE$_1$]).

The topological degree for C^1-mappings from regular subsets of $\mathbf{R}^N$ into $\mathbf{R}^N$ was initiated in 1869 by Kronecker and extended by Brouwer in 1912 to the continuous case using simplicial techniques. See [Mil$_1$], [Llo$_1$], [Zei$_1$], and [Rot$_4$] for recent expositions and discussions. The approach used here relies upon the so-called generic method already used in 1922 by Birkhoff–Kellogg [BiK$_1$] to prove the Brouwer fixed point theorem, and in 1952 by Nagumo [Nag$_1$] to define the Brouwer degree. This approach has been widely rediscovered and developed since.

Theorem 5.4 constitutes the S^1-version of the Borsuk–Ulam theorem: $d(f, D) = 1 \pmod 2$ for continuous and odd mappings $f : \overline{D} \subset \mathbf{R}^N \to \mathbf{R}^N$ such that $0 \notin f(\partial D)$ and D is a symmetric open bounded neighborhood of the origin. The original proof, due to Borsuk [Bor$_1$], gave a positive answer to a conjecture of Ulam. Generic proofs of this result were given independently by Borisovich–Zvyagin–Sapronov [BZS$_1$] and Alexander–Yorke [AlY$_1$]. A simpler case of Theorem 5.4 was first proved by Marzantowicz [Mar$_1$] and Geba–Granas [GeG$_1$] observed that the generic approach to the Borsuk–Ulam theorem could be extended to other group actions. This is realized in the given proof of Theorem 5.4, which is taken from Nirenberg [Nir$_1$].

Theorem 5.5 is the S^1-version of a result of Borsuk [Bor$_1$] for odd mappings. See [FHR$_1$] for another approach to an extended version of Theorem 5.5. Propositions 5.1, 5.2, 5.3 are due to Benci [Ben$_1$] for the case of a Hilbert space.

Historically, the first index theory was the genus of Krasnosel'skii [Kra$_{1,2}$] which corresponds to $G = \mathbf{Z}_2$ and odd maps. It has been introduced as an alternative to the Ljusternik–Schnirelmann category in the minimax approach of those authors to critical point theory [LJS$_1$]. An extension of the genus to more general group actions was first published by Svarc [Sva$_1$] and variants are due to Yang [Yan$_1$] and Conner–Floyd [CoF$_{1,2}$]. Other indices were developed by Yang [Yan$_1$], Conner–Floyd [CoF$_{1,2}$], Holm–Spanier [HoS$_1$], and Fadell–Rabinowitz [FaR$_{1,2}$]. Index theories when $G = \{S^1\}$ have been introduced by Fadell–Rabinowitz [FHR$_1$] and Fadell–Husseini [FaH$_1$].

Further information on index theories can be found in [Ano$_1$], [Bar$_1$], [Fad$_{1,2,3}$], [Neh$_1$], and [Ste$_1$].

Exercises

1. A representation $\{T(g)\}_{g\in G}$ of a topological group is *irreductible* if $T(g)$ has no non-trivial invariant subspace and is *completely reductible* if it is a direct sum of irreductible representations. A representation of a compact topological group over $\mathbf{R}^N$ is always completely irreductible.

   Hint. Use Lemma 5.1 and induction.

2. Let D be an open bounded symmetric neighborhood of 0 in $\mathbf{R}^N$ and let $\Phi \in C(\overline{D}, \mathbf{R}^N) \cap C^1(D, \mathbf{R}^N)$ be such that $\Phi(-u) = -\Phi(u)$, $u \in \partial D$ and $0 \notin \Phi(\partial D)$. For each real $(n \times n)$-matrix A, define $\Phi_A(u)$ by
$$\Phi_A(u) = \Phi(u) + Au.$$
   Then $0 \in \mathbf{R}^N$ is a regular value of the restriction of Φ_A to $D \setminus \{0\}$ for almost every $A \in \mathcal{L}(\mathbf{R}^N, \mathbf{R}^N)$.

   Hint. Following the line of Corollary 5.1 and use the density of the subset of non-singular $(n \times n)$-matrices. ([AlY$_1$], [BZS$_1$]).

3. Let D be an open bounded symmetric neighborhood of 0 in $\mathbf{R}^N$ and let $\Phi \in C(\overline{D}, \mathbf{R}^N)$ be such that $\Phi(-u) = -\Phi(u)$ $(u \in \partial D)$ and $0 \notin \Phi(\partial D)$. Then
$$d(\Phi, D) = 1 \pmod 2$$
   (Borsuk–Ulam).

   Hint. Prove the result for $\Phi \in C^1(D, \mathbf{R}^N)$ and 0 a regular value by direct computation and use Exercise 5, Weierstrass approximation theorem and invariance of degree.

4. Prove that the $\mathbf{Z}_2$-index defined in Example 2 has the properties (i), (ii), (iii), and (iv) of an index.

5. Let X be a Banach space. Prove that if $A \subset X$ is closed, symmetric with respect to the origin and if $0 \in A$, then $\operatorname{ind}_{\mathbf{Z}_2} A = +\infty$.

6. Let X be a Banach space. Prove that if $A \subset X$ is finite, non-empty, symmetric with respect to the origin and if $0 \notin A$, then $\mathrm{ind}_{\mathbf{Z}_2} A = 1$.

   Hint. Follow the lines of Proposition 5.1.

7. Let Y be a closed subspace with finite codimension of the Banach space X and let $A \subset X$ be closed, symmetric with respect to the origin and such that $A \cap Y = \phi$. Then

$$\mathrm{ind}_{\mathbf{Z}_2} A \leq \mathrm{codim}\, Y.$$

   Hint. Follow the lines of Proposition 5.2.

8. Prove that if $A \subset \mathbf{R}^m$ is a bounded symmetric neighborhood of the origin, then $\mathrm{ind}_{\mathbf{Z}_2} \partial A = m$.

   Hint. Follow the lines of Proposition 5.3 and use Exercise 5.2 and Exercise 5.3.

6

Lusternik–Schnirelman Theory and Multiple Periodic Solutions with Fixed Energy

Introduction

When a function has some symmetry properties, one can expect to obtain more precise information on the set of critical points by using, in the minimax approach, correspondingly symmetric sets which are distinguished by a measure of their "size" given by an index theory.

For example, when φ is an even C^1-mapping, the corresponding *Lusternik–Schnirelman method* consists in finding conditions under which the values c_k defined by

$$c_k = \inf_{A \in \mathcal{A}_k} \max_A \varphi$$

are critical, where $\mathcal{A}_k$ denotes the collection of compact sets symmetric with respect to the origin whose $\mathbf{Z}_2$-index is greater or equal to k. Of course, different values of the c_k for different k will immediately give a multiplicity result for the critical points of φ. The interest of the method lies in the fact that multiplicity conclusions can still be obtained if $c_k = c_j$ for some $k > j$, the "size" $\operatorname{ind}_{\mathbf{Z}_2} K_{c_j}$ of the set of critical points of φ with critical value c_j having, in this case, the lower bound $k - j + 1$. Of course, similar results still hold when φ is invariant with respect to the representation of a topological group for which an index is defined, as it is the case for S^1.

Although a variant of the method of Chapter 4 based upon Ekeland's variational principle could be used as well, we have adopted in this chapter the more classical approach based on *deformations*, which is also at the heart of the Morse theory given in Chapter 8. One considers deformations of the underlying space X along the lines of *steepest descent* associated to φ, i.e. along the trajectories of the differential system

$$\dot{u} = -\nabla\varphi(u)$$

or of an associated *pseudo-gradient vector field* when the above differential system or its solutions are not defined because of the lack of regularity of φ or of structure of X. When φ has some symmetry, the corresponding symmetry of the flow has to be used to preserve the symmetries of the sets used in the minimax process during a deformation.

Some compactness is required in proving the basic properties of the deformation flow, and they follow from a *Palais–Smale-type condition.*

Combined with the natural S^1-invariance of the periodic solutions of an autonomous Hamiltonian system in $\mathbf{R}^{2N}$, this approach provides the existence of at least N distinct periodic orbits on a given convex energy surface which does not differ too much from a sphere, in the sense that it lies between two concentric spheres of radius r and $\sqrt{2}r$.

Local results on the multiplicity of periodic orbits of Hamiltonian systems, namely on *periodic orbits with prescribed energy near an equilibrium,* are also given in Section 6.6. Their obtention relies upon a *variational bifurcation argument* which reduces the problem, through the *Liapunov–Schmidt method,* to a *nonlinear eigenvalue problem* in a finite-dimensional space (Section 6.5). This last problem is studied in Section 6.4 by a variant of the Lusternik–Schnirelman method given in Sections 6.1 and 6.2.

6.1 Equivariant Deformations

The concept of "pseudo-gradient" is required to extend the steepest descent method to a general Banach space.

Definition 6.1. Let X be a Banach space, $\varphi \in C^1(X, \mathbf{R})$ and

$$Y = \{u \in X : \varphi'(u) \neq 0\}. \tag{1}$$

A *pseudo-gradient vector field* for φ on Y is a locally Lipschitz continuous mapping $v : Y \to X$ such that, for every $u \in Y$, one has

$$\|v(u)\| \leq 2\|\varphi'(u)\| \tag{2}$$

$$\langle \varphi'(u), v(u) \rangle \geq \|\varphi'(u)\|^2. \tag{3}$$

Lemma 6.1. *Under the assumptions of Definition* 6.1, *there exists a pseudo-gradient vector field for* φ *on* Y.

Proof. Let $\tilde{u} \in Y$; as $\varphi'(\tilde{u}) \neq 0$, there exists $w \in X$ such that $\|w\| = 1$ and

$$\langle \varphi'(\tilde{u}), w \rangle > \frac{2}{3}\|\varphi'(\tilde{u})\|.$$

Let $v = \frac{3}{2}\|\varphi'(\tilde{u})\|\, w$. Then

$$\|v\| = \frac{3}{2}\|\varphi'(\tilde{u})\| < 2\|\varphi'(\tilde{u})\|$$

$$\langle \varphi'(\tilde{u}), v \rangle = \frac{3}{2}\|\varphi'(\tilde{u})\| < \varphi'(\tilde{u}), w \rangle > \|\varphi'(\tilde{u})\|^2.$$

Since φ' is continuous, there is an open neighborhood $V_{\tilde{u}}$ of $\tilde{u}$ in Y such that, for every $u \in V_{\tilde{u}}$, one has

$$\|v\| < 2\|\varphi'(u)\| \tag{4}$$

$$\langle \varphi'(u), v\rangle > \|\varphi'(u)\|^2. \tag{5}$$

The family $\mathcal{V} = \{V_{\tilde{u}}\}_{\tilde{u}\in Y}$ is an open covering of the metric, and thus, paracompact space Y. Therefore $\mathcal{V}$ has a locally finite refinement $\{W_{\tilde{u}_i}\}_{i\in I}$, i.e. each $W_{\tilde{u}_i}$ is one $V_{\tilde{u}}$ and each $y \in Y$ has a neighborhood U_y such that $U_y \in W_{\tilde{u}_i} \neq \phi$ only for a finite number of values of i. Let us define

$$\rho_i(u) = \operatorname{dist}(u, \complement\, W_{\tilde{u}_i}), \quad i \in I$$

and, for $u \in Y$,

$$v(u) = \sum_{i\in I} \frac{\rho_i(u)}{\sum_{j\in I} \rho_j(u)} v_i$$

where v_i corresponds to $\tilde{u}_i$ in the same way as v corresponds to $\tilde{u}$ above. All the sums are finite as $\{W_{\tilde{u}_i}\}_{i\in I}$ is locally finite, and v is clearly locally Lipschitz continuous. Since ρ_i vanishes outside $W_{\tilde{u}_i}$, $v(u)$ is a convex combination of elements satisfying (4) and (5) and it is then easy to check that (2) and (3) hold for every $u \in Y$. □

Remark 6.1. If X is a Hilbert space and $\varphi : X \to \mathbf{R}$ is differentiable, the gradient $\nabla\varphi$ of φ defined by

$$(\nabla\varphi(u), w) = \langle \varphi'(u), w\rangle, \quad w \in X$$

satisfies (2) and (3).

Definition 6.2. A functional $\varphi : X \to \mathbf{R}$ is *invariant* for the representation $\{T(g)\}_{g\in G}$ of the topological group G if

$$\varphi \circ T(g) = \varphi, \quad g \in G.$$

We shall prove that any invariant $\varphi \in C^1(X, \mathbf{R})$ has an equivariant pseudo-gradient vector field.

Lemma 6.2. *If $\varphi \in C^1(X, \mathbf{R})$ is invariant for the isometric representation $\{T(g)\}_{g\in G}$, then, for all $g \in G$, $u \in X$, $h \in X$,*

$$\langle \varphi'(T(g)u), h\rangle = \langle \varphi'(u), T(-g)h\rangle \tag{6}$$

$$\|\varphi'(T(g)u)\| = \|\varphi'(u)\|. \tag{7}$$

Proof. By definition,

$$\begin{aligned}
\langle \varphi'(T(g)u), h\rangle &= \lim_{t\to 0} \frac{\varphi(T(g)(u + tT(-g)h)) - \varphi(T(g)u)}{t} \\
&= \lim_{t\to 0} \frac{\varphi(u + tT(-g)h) - \varphi(u)}{t} \\
&= \langle \varphi'(u), T(-g)h\rangle.
\end{aligned}$$

Since $T(-g)$ is a surjective isometry, (7) follows from (6). □

Lemma 6.3. *Let X and Y be metric spaces and $T : Y \to X$ a locally Lipschitz continuous function. If $A \subset Y$ is compact, there exists $\delta > 0$ such that T is Lipschitz continuous on A_δ.*

Proof. For each $u \in A$, there is an open ball $B(u, \delta(u))$ and a nonnegative constant $\ell(u)$ such that

$$d(Tv, Tw) \leq \ell(u)\, d(v, w), \quad v, w \in B(u, \delta(u)).$$

The family $\{B(u, \delta(u)/3)\}_{u \in A}$ is an open covering of A and contains a finite covering $\{B(u_i, \delta(u_i)/3)\}_{1 \leq i \leq j}$ of A. Let

$$\delta = \min_{1 \leq i \leq j} \delta(u_i)/3.$$

Then, by construction

$$M = \sup\{d(Tv, Tw) \,:\, v, w \in A_\delta\} < \infty$$

and we shall show that T is Lipschitz continuous on A_δ with Lipschitz constant

$$\ell = \max\{M/\delta, \ell(u_1), \ldots, \ell(u_j)\}.$$

Indeed, take v and w in A_δ; if $d(v, w) < \delta$ and $v \in B(u_i, 2\delta(u_i)/3)$, then $w \in B(u_i, \delta(u_i))$ and

$$d(Tv, Tw) \leq \ell(u_i)\, d(v, w) \leq \ell\, d(v, w).$$

If $d(v, w) \geq \delta$, then

$$d(Tv, Tw) \leq \frac{M}{\delta}\delta \leq \ell\, d(v, w),$$

and the proof is complete. □

Lemma 6.4. *If $\varphi \in C^1(X, \mathbf{R})$ is invariant for the isometric representation $\{T(g)\}_{g \in G}$ of the compact group G, then there exists an equivariant pseudo-gradient vector field for φ and $Y = \{u \in X \,:\, \varphi'(u) \neq 0\}$.*

Proof. Let $w : Y \to X$ be a pseudo-gradient vector field as given by Lemma 6.1 and define $v : Y \to X$ by

$$v(u) = \int_G T(-g)\, w(T(g)u)\, dg,$$

where dg is the (normalized) Haar measure on G. Then v is equivariant because

$$\begin{aligned} v(T(\tilde{g})u) &= \int_G T(-g)\, w(T(g + \tilde{g})u)\, dg \\ &= \int_G T(\tilde{g})\, T(-\tilde{g} - g)\, w(T(g + \tilde{g})u)\, dg \\ &= T(\tilde{g})\, v(u). \end{aligned}$$

Now v satisfies (2) and (3) because, by Lemma 6.2, we obtain

$$\begin{aligned}\|v(u)\| &\leq \int_G \|T(-g)\, w(T(g)u)\|\, dg \\ &= \int_G \|w(T(g)u)\|\, dg \leq 2\int_G \|\varphi'(T(g)u)\|\, dg \\ &= 2\|\varphi'(u)\|,\end{aligned}$$

$$\begin{aligned}\langle \varphi'(u), v(u)\rangle &= \int_G \langle \varphi'(u), T(-g)\, w(T(g)u)\rangle\, dg \\ &= \int_G \langle \varphi'(T(g)u),\, w(T(g)u)\rangle dg \geq \int_G \|\varphi'(T(g)u)\|^2 dg \\ &= \|\varphi'(u)\|^2.\end{aligned}$$

Finally, let $u \in Y$ and $A = \{T(g)u \,:\, g \in G\} \subset Y$. By Lemma 6.3, there exists $\delta > 0$ such that w is Lipschitz continuous, with constant ℓ, on A_δ. Now, if $u_1, u_2 \in B(u, \delta)$, we obtain, since A_δ is invariant,

$$\begin{aligned}\|v(u_1) - v(u_2)\| &\leq \int_G \|T(-g)(w(T(g)u_1) - w(T(g)u_2))\|\, dg \\ &= \int_G \|w(T(g)u_1) - w(T(g)u_2)\|\, dg \\ &\leq \ell \int_G \|T(g)(u_1 - u_2)\|\, dg = \ell\|u_1 - u_2\|,\end{aligned}$$

and v is locally Lipschitz continuous. □

The construction of the required deformations depends on some compactness conditions.

Definition 6.3. The function $\varphi \in C^1(X, \mathbf{R})$ satisfies the *Palais-Smale condition* (PS) if every sequence (u_j) in X such that $(\varphi(u_j))$ is bounded and

$$\varphi'(u_j) \to 0 \quad \text{for } j \to \infty$$

contains a convergent subsequence.

Remark 6.2. The Palais–Smale condition is a compactness condition on φ which replaces the compactness of the manifold in the classical Lusternik–Schnirelman theory.

Remark 6.3. It is, in general, easier to verify (PS) than to find a priori bounds for all possible solutions of $\varphi'(u) = 0$ since in (PS) $(\varphi(u_j))$ has to be bounded.

Remark 6.4. It follows immediately from (PS) that, for each $c \in \mathbf{R}$, the set

$$K_c = \{u \in X \,:\, \varphi'(u) = 0 \ \text{ and } \ \varphi(u) = c\}$$

is compact.

Recall that, for $c \in \mathbf{R}$, $\varphi^c = \{u \in X : \varphi(u) \leq c\}$. The basic deformation result is given by the following.

Lemma 6.5. *If $\varphi \in C^1(X, \mathbf{R})$ satisfies (PS) and if U is an open neighborhood of K_c, then, for every $\bar{\epsilon} > 0$, there exists $\epsilon \in]0, \bar{\epsilon}[$ and $\eta \in C([0,1] \times X, X)$ such that*

(a)

$$\eta(1, \varphi^{c+\epsilon} \setminus U) \subset \varphi^{c-\epsilon},$$

$$\eta(t, u) = u \quad \text{if} \quad u \notin \varphi^{-1}([c - \bar{\epsilon}, c + \bar{\epsilon}]).$$

Moreover, if φ and U are invariant with respect to the isometric representation $\{T(g)\}_{g \in G}$ of the compact group G, then

(b) *$\eta(t, .)$ is equivariant with respect to $\{T(g)\}_{g \in G}$ for every $t \in [0,1]$.*

Proof. If $\bar{\epsilon} > 0$ is given, there is a $\epsilon \in]0, \bar{\epsilon}/2[$ such that if $u \in \varphi^{-1}([c - 2\epsilon, c + 2\epsilon]) \cap (\complement U)_{2\sqrt{\epsilon}}$, then

$$\|\varphi'(u)\| \geq 4\sqrt{\epsilon}. \tag{8}$$

Indeed, if it is not the case, there is a sequence (u_k) such that

$$u_k \in (\complement U)_{2\sqrt{k}}, \quad c - \frac{2}{k} \leq \varphi(u_k) \leq c + \frac{2}{k}, \quad \|\varphi'(u_k)\| < \frac{4}{\sqrt{k}}$$

and, by (PS) we can assume, going if necessary to a subsequence, that $u_k \to u$ for $k \to \infty$. But then $u \in \complement U \cap K_c = \phi$, a contradiction.

Now Lemma 6.1 implies the existence of a pseudo-gradient vector field v for φ on $Y = \{u \in X : \varphi'(u) \neq 0\}$. Let us define

$$A = \varphi^{-1}([c - 2\epsilon, c + 2\epsilon]) \cap (\complement U)_{2\sqrt{\epsilon}} \subset Y,$$

$$B = \varphi^{-1}([c - \epsilon, c + \epsilon]) \cap (\complement U)_{\sqrt{\epsilon}} \subset A,$$

$$\psi(u) = \frac{\operatorname{dist}(u, \complement A)}{d(u, \complement A) + d(u, B)},$$

so that $0 \leq \psi(u) \leq 1$, $\psi(u) = 1$ in B and $\psi(u) = 0$ in $\complement A$. Define the locally Lipschitz continuous vector field f on X by

$$\begin{aligned} f(u) &= -\psi(u)\tfrac{v(u)}{\|v(u)\|} && \text{if } u \in A \\ &= 0 && \text{if } u \notin A. \end{aligned}$$

As f is bounded on X, the Cauchy problem

$$\dot{\sigma} = f(\sigma)$$

$$\sigma(0) = u$$

has, for each $u \in X$, a unique solution $\sigma(.,u)$ defined on $[0,\infty[$ with $\sigma(.,.)$ continuous. Let us define $\eta = C([0,1] \times X, X)$ by

$$\eta(t,u) = \sigma(\sqrt{\epsilon}t, u).$$

For $t \geq 0$, we have

$$\|\sigma(t,u) - u\| = \left\| \int_0^t f(\sigma(\tau,u))\, d\tau \right\| \leq \int_0^t \|f(\sigma(\tau,u))\|\, d\tau \leq t,$$

we have $\sigma(t, \complement\, U) \subset (\complement\, U)_{\sqrt{\epsilon}}$ for $t \in [0,\sqrt{\epsilon}]$. By definition of f, we have for all $u \in X$ and $t \geq 0$,

$$\frac{d}{dt}\varphi(\sigma(t,u)) = \langle \varphi'(\sigma(t,u)), f(\sigma(t,u))\rangle$$

$$= -\psi(\sigma(t,u))\|v(\sigma(t,u))\|^{-1}\langle \varphi'(\sigma(t,u)), v(\sigma(t,u))\rangle \leq 0.$$

Let $u \in \varphi^{c+\epsilon}\backslash U$; if $\varphi(\sigma(t,u)) < c-\epsilon$ for some $t \in [0,\sqrt{\epsilon}[$, then $\varphi(\sigma(\sqrt{\epsilon},u)) < c-\epsilon$ and $\eta(1,u) < c-\epsilon$. If not, then $\sigma(t,u) \in B$ for all $t \in [0,\sqrt{\epsilon}]$, and by (8) and the definition of f we obtain

$$\begin{aligned}
\varphi(\sigma(\sqrt{\epsilon},u)) &= \varphi(u) + \int_0^{\sqrt{\epsilon}} \frac{d}{dt}\varphi(\sigma(t,u))\, dt \\
&= \varphi(u) + \int_0^{\sqrt{\epsilon}} \langle \varphi'(\sigma(t,u)), f(\sigma(t,u))\rangle dt \\
&= \varphi(u) - \int_0^{\sqrt{\epsilon}} \langle \varphi'(\sigma(t,u)), \frac{v(\sigma(t,u))}{\|v(\sigma(t,u))\|}\rangle dt \\
&\leq c+\epsilon - \int_0^{\sqrt{\epsilon}} \frac{\|\varphi'(\sigma(t,u))\|^2}{\|v(\sigma(t,u))\|} dt \\
&\leq c+\epsilon - \frac{1}{2}\int_0^{\sqrt{\epsilon}} \|\varphi'(\sigma(t,u))\| dt \\
&\leq c+\epsilon-2\epsilon = c-\epsilon.
\end{aligned}$$

Thus, $\eta(1,u) \in \varphi^{c-\epsilon}$ and hence $\eta(1, \varphi^{c+\epsilon} \setminus U) \subset \varphi^{c-\epsilon}$.

Finally, if φ is invariant as well as U with respect to $\{T(g)\}_{g\in G}$, we can assume, by Lemma 4, that the pseudo-gradient vector field v is equivariant. Since the invariance of U implies also the invariance of ψ, the vector field f is equivariant. It is then a direct consequence of the uniqueness of the solution of the Cauchy problem that $\sigma(t,.)$ and hence $\eta(t,.)$ are equivariant with respect to $\{T(g)\}_{g\in G}$, and the proof is complete. □

6.2 Existence of Multiple Critical Points

Let φ be a real continuous function on the Banach space X. We shall use the Lusternik–Schnirelman minimax method to construct critical values of

φ. Let $\{T(g)\}_{g\in G}$ be an isometric representation of the compact topological group G. Define, for $j \geq 1$,

$$\mathcal{A}_j = \{A \subset X \,:\, A \text{ is compact, invariant for } \{T(g)\}_{g\in G} \text{ and } \operatorname{ind} A \geq j\}$$

$$c_j = \inf_{A\in\mathcal{A}_j} \max_A \varphi.$$

Clearly, one has $\mathcal{A}_j \subset \mathcal{A}_{j-1}$ $(j \geq 2)$ and hence

$$-\infty \leq c_1 \leq c_2 \leq \ldots < +\infty.$$

Theorem 6.1. *Let $\varphi \in C^1(X,\mathbf{R})$ be an invariant functional with respect to $\{T(g)\}_{g\in G}$ which satisfies* (PS). *If $c_j > -\infty$ for some $j \geq 1$, then c_j is a critical value of φ. Moreover, if $c_k = c_j$, for some $k \geq j$, then*

$$\operatorname{ind} K_{c_j} \geq k - j + 1.$$

Proof. We shall prove that if $-\infty < c_j = c_k = c$ for some $1 \leq j \leq k$, then

$$\operatorname{ind} K_c \geq k - j + 1$$

so that, by property (i) of the index, $K_c \neq \phi$. Notice that K_c is invariant by Lemma 6.2 and is compact by the (PS) condition. Property (iii) of the index implies the existence of a closed invariant neighborhood N of K_c such that $\operatorname{ind} N = \operatorname{ind} K_c$. The interior U of N is an open invariant neighborhood of K_c so that Lemma 6.5 is applicable. Let $\epsilon \in \,]0,1[$ be given by this lemma, $A \in \mathcal{A}_k$ such that

$$\max_A \varphi \leq c + \epsilon$$

and $B = A \setminus U$. We deduce from properties (ii) and (iv) of the index

$$\begin{aligned} k &\leq \operatorname{ind} A \leq \operatorname{ind}(B \cup N) \\ &\leq \operatorname{ind} B + \operatorname{ind} N = \operatorname{ind} B + \operatorname{ind} K_c. \end{aligned} \tag{9}$$

It now follows from conclusion (a) of Lemma 6.5 that $C = \eta(1,B) \subset \varphi^{c-\epsilon}$ and, since B is compact and invariant, the same is true for C by conclusion (b) of Lemma 6.5. But $\max_C \varphi \leq c - \epsilon$ so that, by the definition of $c_j = c$, $\operatorname{ind} C \leq j - 1$. Now property (ii) of the index implies that

$$\operatorname{ind} B \leq \operatorname{ind} C \leq j - 1 \tag{10}$$

and hence, by (9) and (10),

$$\operatorname{ind} K_c \geq k - j + 1. \qquad \square$$

Theorem 6.2. *Let $\varphi \in C^1(X,\mathbf{R})$ be an S^1-invariant functional satisfying* (PS). *Let Y and Z be closed invariant subspaces of X with codim Y and dim Z finite and*

$$\operatorname{codim} Y < \dim Z.$$

Assume that the following conditions are satisfied:

$$\mathrm{Fix}\,(S^1) \subset Y, \quad Z \cap \mathrm{Fix}\,(S^1) = \{0\} \tag{11}$$

$$\inf_Y \varphi > -\infty \tag{12}$$

there exist $r > 0$ *and* $c < 0$ *such that* $\varphi(u) \le c$ *whenever* $u \in Z$ *and* $\|u\| = r$. (13)

$$\textit{If } u \in \mathrm{Fix}\,(S^1) \textit{ and } \varphi'(u) = 0, \textit{ then } \varphi(u) \ge 0. \tag{14}$$

Then there exists at least $\frac{1}{2}(\dim Z - \operatorname{codim} Y)$ *distinct* S^1*-orbits of critical points of* φ *outside of* $\mathrm{Fix}(S^1)$ *with critical values less or equal to* c.

Proof. Let $p = \frac{1}{2}\operatorname{codim} Y$ and let $j \ge p+1$. If $A \in \mathcal{A}_j$, Proposition 5.2 and (11) imply that $A \cap Y \ne \phi$. By assumption (12),

$$\max_A \varphi \ge \inf_Y \varphi > -\infty$$

so that $c_j \ge \inf_Y \varphi > -\infty$. Let $q = \frac{1}{2}\dim Z$ and $D = \{u \in Z \,:\, \|u\| < r\}$. Proposition 5.3 and (11) imply that $\operatorname{ind} \partial D = q$. For $j \le q$ we deduce from (13) that

$$c_j \le \max_{\partial D} \varphi \le c.$$

Consequently,

$$-\infty < c_{p+1} \le \ldots \le c_q \le c.$$

By Theorem 6.1, each c_j $(p+1 \le j \le q)$ is a critical value of φ. If all the c_j are distinct, the proof is complete because of (14). If $c_j = c_k$ for some $p+1 \le j < k \le q$, then, by Theorem 6.1,

$$\operatorname{ind} K_{c_j} \ge k - j + 1 \ge 2.$$

By (14), $K_{c_j} \cap \mathrm{Fix}\,(S^1) = \phi$ and hence, by Proposition 5.1, K_{c_j} contains necessarily infinitely many S^1-orbits and the proof is complete. □

6.3 Multiple Periodic Solutions with Prescribed Energy of Autonomous Hamiltonian Systems

We study the existence of multiple periodic solutions of the autonomous Hamiltonian system

$$J\dot{u}(t) + \nabla H(u(t)) = 0 \tag{15}$$

on a convex energy surface $H^{-1}(c)$.

Theorem 6.3. *Let* $H \in C^1(\mathbf{R}^{2N}, \mathbf{R})$ *and* $c \in \mathbf{R}$ *be such that* $\nabla H(u) \ne 0$ *for every* $u \in S = H^{-1}(c)$. *Assume that* S *is the boundary of a strictly* *convex compact set C and that there exists $r > 0$ and $R \in]r, \sqrt{2}r[$ such that*

$$B[0,r] \subset C \subset B[0,R]. \tag{16}$$

Then there exists at least N periodic orbits of (15) *on S.*

The proof of Theorem 6.3 requires some preliminary results. Let $F = j^{3/2}$, where j is the gauge of C, so that F is strictly convex and satisfies properties (i) to (iv) of Lemma 3.2. Assumption (16) implies that

$$(|u|/R)^{3/2} \leq F(u) \leq (|u|/r)^{3/2}.$$

By relation (2.4), we obtain

$$\eta r^3|v|^3 \leq F^*(v) \leq \eta R^3|v|^3 \tag{17}$$

where $\eta = 4/27$. Theorem 2.3 implies that the dual action φ defined by

$$\varphi(v) = \int_0^T [(1/2)(J\dot{v}(t), v(t)) + F^*(\dot{v}(t))]\, dt$$

is continuously differentiable over $X = \tilde{W}_T^{1,3}$.

The function φ is invariant under the representation of $S^1 \approx \mathbf{R}/\mathbf{Z}$ defined over X by the translation in time t. In order to apply Theorem 6.2, we need some easy estimates.

Lemma 6.6. *If $p \geq 2$ and $v \in W_T^{1,p}$, then*

$$\int_0^T (J\dot{v}(t), v(t))\, dt \geq -(1/2\pi)T^{(2-2/p)} \left(\int_0^T |\dot{v}(t)|^p dt\right)^{2/p}.$$

Proof. By Proposition 3.2, we have, for $v \in W_T^{1,p}$,

$$\int_0^T (J\dot{v}(t), v(t))\, dt \geq -(T/2\pi)||\dot{v}||_{L^2}^2.$$

Hölder's inequality implies that

$$||\dot{v}||_{L^2} \leq T^{(1/2-1/p)}||\dot{v}||_{L^p}. \qquad \square$$

As $p = 3$, Lemma 6.6 suggests to make the convenient choice $T = (2\pi)^{3/4}$ for the period.

Lemma 6.7. *If $v \in X$, then*

$$\varphi(v) \geq \eta r^3||\dot{v}||_{L^3}^3 - (1/2)||\dot{v}||_{L^3}^2.$$

Proof. It follows from (17) that

$$\int_0^T F^*(\dot{v}(t))\, dt \geq \eta r^3||\dot{v}||_{L^3}^3$$

and from Lemma 6.6 that

$$\int_0^T (J\dot{v}(t), v(t))\,dt \geq -\|\dot{v}\|_{L^3}^2. \qquad \square$$

Let

$$V = \{v \in X \,:\, v \text{ is } (T/k)\text{-periodic for some integer } k \geq 2\}.$$

Lemma 6.8. *If $v \in V$, then*

$$\varphi(v) \geq m = -(1/12)(6\eta r^3)^{-2}.$$

Proof. If $v \in X$ and is T/k periodic for some integer $k \geq 2$, Lemma 6.6 implies that

$$\int_0^T (J\dot{v}(t), v(t))\,dt = k\int_0^{T/k} (J\dot{v}(t), v(t))\,dt \geq -k^{-1/3}\left(\int_0^T |\dot{v}(t)|^3 dt\right)^{2/3}$$

$$= -k^{-1}\|\dot{v}\|_{L^3}^2 \geq -\frac{1}{2}\|\dot{v}\|_{L^3}^2.$$

Then, using (17), we obtain, for $v \in V$,

$$\varphi(v) \geq \eta r^3 \|\dot{v}\|_{L^3}^3 - (1/4)\|\dot{v}\|_{L^3}^2,$$

and the right hand membre of this inequality is minimum when $\|\dot{v}\|_{L^3} = (6\eta r^3)^{-1}$. $\square$

Let

$$Z = \left\{\left(\cos\frac{2\pi t}{T}\right)e + \left(\sin\frac{2\pi t}{T}\right)Je \,:\, e \in \mathbf{R}^{2N}\right\}.$$

Lemma 6.9. *If $v \in Z$ and $\|\dot{v}\|_{L^3} = \rho \equiv (3\eta r^3)^{-1}$, then*

$$\varphi(v) \leq c \equiv -\rho^2/6.$$

Proof. If $v \in Z$, then $J\dot{v} = -(2\pi/T)v$, so that

$$\int_0^T (J\dot{v}(t), v(t))\,dt = (-T/2\pi)\int_0^T |J\dot{v}(t)|^2 dt = (-T/2\pi)\|\dot{v}\|_{L^2}^2$$

$$\geq (-T/2\pi)T^{1/3}\|\dot{v}\|_{L^3}^2 = -\|\dot{v}\|_{L^3}^2.$$

Consequently, (17) implies that, for $v \in Z$, with $\|\dot{v}\|_{L^3} = \rho$,

$$\varphi(v) \leq \eta r^3\|\dot{v}\|_{L^3}^3 - (1/2)\|\dot{v}\|_{L^3}^2 = -\rho^2/6. \qquad \square$$

Proof of Theorem 6.3. We apply Theorem 6.2 to the invariant functional φ on X. It is convenient to use the norm $\|v\| = \|\dot{v}\|_{L^3}$ on X.

1) Let (v_j) be a sequence in X such that $(\varphi(v_j))$ is bounded and $\varphi'(v_j) \to 0$ as $j \to \infty$. Lemma 6.7 implies that (v_j) is bounded in X. As in the proof of Lemma 4.5, this implies that (v_j) contains a convergent subsequence, so that φ satisfies the (PS)-condition.

2) The space $Y = X$ and the space Z introduced in Lemma 6.9 satisfy the conditions (11) and (14) since $\text{Fix}\,(S^1) = \{0\}$ and $\varphi(0) = 0$. Moreover, condition (12) follows from Lemma 6.7 and condition (13) from Lemma 6.9. We also notice that

$$\operatorname{codim} Y = \operatorname{codim} X = 0 < \dim Z = 2N.$$

Thus, by Theorem 6.2, there exists at least N distinct S^1-orbits of critical points of φ, namely $\{T(\theta)v_j \, : \, \theta \in S^1\}$ outside $\text{Fix}\,(S^1) = \{0\}$ with $\varphi(v_j) \leq c$ $(j = 1, 2, \ldots, N)$.

By Theorem 2.3, the function u_j defined by

$$u_j(t) = \nabla F^*(\dot{v}_j(t))$$

is a T-periodic solution of

$$J\dot{u}(t) + \nabla F(u(t)) = 0. \tag{18}$$

Now condition $R \in \,]r, \sqrt{2}r[$ is equivalent to $m > c$, where m is defined in Lemma 6.8 and c in Lemma 6.9. Since $\varphi(v_j) \leq c$ $(1 \leq j \leq N)$, we have $v_j \notin V$ $(1 \leq j \leq N)$, i.e. T is the minimal period of u_j $(1 \leq j \leq N)$.

3) If $d_j \equiv F(u_j) > 0$, then w_j, defined by

$$w_j(t) = d_j^{-2/3} u_j(d_j^{1/3} t),$$

is a solution of (18) with minimal period $T/d_j^{1/3}$ and energy $F(w_j(t)) = 1$. If w_j and w_k describe the same orbit on $S = F^1(1)$, then $w_k = T(\theta)w_j$ for some θ. Thus, the minimal periods of w_j and w_k are the same, i.e. $d_j = d_k$. But then u_j and u_k also describe the same orbit, i.e. $u_k = T(\theta)u_j$ for some θ, so that

$$\dot{v}_k = \nabla F(u_k) = \nabla F(T(\theta)u_j) = T(\theta)\nabla F(u_j) = T(\theta)\dot{v}_j$$

and hence $v_k = T(\theta)v_j$ as v_k and v_j have mean value zero. Thus, v_k and v_j describe the same orbit, which implies $j = k$. Thus, there exists at least N distinct periodic orbits of (18) on S. By Lemma 3.1, the proof is complete. □

Exercise 3.5 shows that the estimate N is optimal.

The *Poincaré integral invariant* around a closed oriented curve Γ in $\mathbf{R}^{2N}$ is the line integral

$$\frac{1}{2}\int_\Gamma (Ju, du).$$

Theorem 6.4. *Let $H \in C^1(\mathbf{R}^{2N}, \mathbf{R})$, $c \in \mathbf{R}$ and $r > 0$ be such that $S = H^{-1}(c)$ is the boundary of a compact convex set C enclosing the closed ball $B[0,r]$. Assume that $\nabla H(u) \neq 0$ whenever $u \in S$. Then every oriented periodic orbit Γ of (15) on S satisfies the inequality*

$$\frac{1}{2}\int_\Gamma (Ju, du) \geq \pi r^2.$$

Proof. Let $v : [0,T] \to \mathbf{R}^{2N}$ be the parametrization of the oriented periodic orbit Γ by the arclength. Thus $|\dot{v}(t)| = 1$ for all $t \in [0,T]$ and

$$J\dot{v}(t) = -N(v(t)),$$

where N is defined on S by

$$N(u) = \nabla H(u)/|\nabla H(u)|.$$

Since $(u, N(u))$ is equal to the distance from the origin to the tangent hyperplane to S at u, we have, by assumption

$$(u, N(u)) \geq r$$

for every $u \in S$. By using Proposition 3.2, we obtain

$$T^2/2\pi = (T/2\pi)\int_0^T |\dot{v}(t)|^2 dt \geq -\int_0^T (v(t), J\dot{v}(t))\, dt$$

$$= \int_0^T (v(t), N(v(t)))\, dt \geq rT.$$

In particular, we have $T \geq 2\pi r$ and

$$(1/2)\int_\Gamma (Ju, du) = (1/2)\int_0^T (Jv(t), \dot{v}(t))\, dt$$

$$= (-1/2)\int_0^T (v(t), J\dot{v}(t))\, dt \geq rT/2 \geq \pi r^2. \qquad \square$$

Example. The Henon–Heiles Hamiltonian $H : \mathbf{R}^4 \to \mathbf{R}$ given by

$$H(q,p) = (1/2)(p_1^2 + p_2^2 + q_1^2 + q_2^2) + \mu(p_1^2 q_2 - q_2^3/3)$$

occurs in the simulation of a three atoms solid and in the Hartree averaged field seen by a star moving in the galaxy. When $c > 0$ is sufficiently small, $H^{-1}(c)$ is the boundary of a strictly convex compact set of $\mathbf{R}^4$ and Theorem 6.3 can be applied.

6.4 Nonlinear Eigenvalue Problems

In this section, we denote by $T(\theta)$ an isometric representation of S^1 over $\mathbf{R}^{2N}$ such that $\mathrm{Fix}(S^1) = \{0\}$. Let D be an invariant open subset of $\mathbf{R}^{2k}$ and let $\varphi, \chi \in C^1(D, \mathbf{R})$ be two invariant functions. We consider the *nonlinear eigenvalue problem*

$$\nabla\varphi(u) = \mu\nabla\chi(u), \quad u \in Z_a, \tag{19}$$

where

$$Z_a = \{u \in D : \chi(u) = a\}.$$

By definition, a *critical point of φ restricted to Z_a* is a solution of (19).

The invariance of φ and χ implies the invariance of the set of critical points of φ restricted to Z_a.

Theorem 6.5. *If there exists an equivariant diffeomorphism $h : D \to h(D)$ such that $h(Z_a) = S^{2k-1}$, then there exists at least k S^1-orbits of critical points of φ restricted to Z_a.*

Proof. Let us define $\tilde{\varphi}$ and $\tilde{\chi}$ by $\tilde{\varphi} = \varphi \circ h^{-1}$ and $\tilde{\chi} = \chi \circ h^{-1}$. It suffices, therefore, to prove the existence of at least k S^1-orbits of critical points of $\tilde{\varphi}$ restricted to

$$\tilde{Z}_a = \{v \in h(D) : \tilde{\chi}(v) = a\} = S^{2k-1}$$

Now, for every $v \in \tilde{Z}_a$, $\nabla\tilde{\chi}(v)$ and v are normal to $\tilde{Z}_a$, and hence there exists $\lambda(v) \in \mathbf{R}$ such that

$$\nabla\tilde{\chi}(v) = \lambda(v)v.$$

Therefore, it suffices to prove the existence of k S^1-orbits of solutions of

$$\nabla\tilde{\varphi}(v) = \mu v, \quad v \in S^{2k-1},$$

so that the proof of Theorem 6.5 is reduced to that of the following lemma. □

Lemma 6.10. *Let D be an open invariant neighborhood of S^{2k-1} and let $\varphi \in C^1(D, \mathbf{R})$ be an invariant function. Then there exists at least k S^1-orbits of critical points of φ restricted to S^{2k-1}.*

Proof. Let us define the equivariant vector field w on S^{2k-1} by

$$w(u) = \nabla\varphi(u) - (\nabla\varphi(u), u)u.$$

As in Section 6.1, it is easy to define on

$$Y = \{u \in S^{2k-1} : w(u) \neq 0\}$$

a pseudo-gradient vector field v such that

(a) $\|v(u)\| \leq 2\|w(u)\|$

(b) $(w(u), v(u)) \geq \|w(u)\|^2$

(c) $(v(u), u) = 0$

(d) v is locally Lipschitzian and equivariant.

Let us define, for $c \in \mathbf{R}$, K_c and φ^c by

$$K_c = \{u \in S^{2k-1} \ : \ w(u) = 0 \text{ and } \varphi(u) = c\}$$

$$\varphi^c = \{u \in S^{2k-1} \ : \ \varphi(u) \leq c\}.$$

Let U be an open invariant neighborhood of K_c. Using the pseudo-gradient vector field v, it is easy to prove, as in Lemma 6.5, the existence of $\epsilon > 0$ and of $\eta \in C([0,1] \times S^{2k-1}, S^{2k-1})$ such that:

(a) $\eta(1, \varphi^{c+\epsilon} \setminus U) \subset \varphi^{c-\epsilon}$

(b) $\eta(t, .)$ is equivariant for every $t \in [0,1]$.

By Proposition 5.3, $\operatorname{ind} S^{2k-1} = k$. Define, for $1 \leq j \leq k$, $\mathcal{A}_j$ and c_j by

$$\mathcal{A}_j = \{A \subset S^{2k-1} \ : \ A \text{ is closed, invariant and } \operatorname{ind} A \geq j\},$$

$$c_j = \inf_{A \in \mathcal{A}_j} \sup_A \varphi.$$

It then suffices to prove, like in Theorem 6.1, that if $c_p = c_q = c$ for some $1 \leq p \leq q \leq k$, one has

$$\operatorname{ind} K_c \geq q - p + 1. \qquad \square$$

Remark 6.5. When $\chi'(u) \neq 0$ on Z_a, the notion of critical point of φ restricted to Z_a has a simple geometric interpretation. By the preimage theorem, Z_a is a C^1-manifold of $\mathbf{R}^{2k}$ of dimension $2k - 1$. Hence, for every $z \in Z_a$, there exists an open neighborhood A of 0 in $\mathbf{R}^{2k}$, an open neighborhood B of z in $\mathbf{R}^{2k}$ and a diffeomorphism $\Phi : A \to B$ such that $\Phi(A \cap \mathbf{R}^{2k-1}) = B \cap Z_a$. The *tangent space* of Z_a at z is defined by

$$T_z Z_a = \{\Phi'(\Phi^{-1}(z))v \ : \ v \in \mathbf{R}^{2k-1}\}.$$

By the definition of Z_a, we obtain

$$\chi(\Phi(y)) = a$$

for all $y \in A \cap \mathbf{R}^{2k-1}$. Hence, we have

$$\chi'(\Phi(y))\, \Phi'(y)v = 0$$

for all $y \in A \cap \mathbf{R}^{2k-1}$ and $v \in \mathbf{R}^{2k-1}$. In particular, this gives

$$\chi'(z)\Phi'(\Phi^{-1}(z))\, v = 0$$

for all $v \in \mathbf{R}^{2k-1}$, so that $T_z Z_a \subset \ker \chi'(z)$. Since

$$\operatorname{codim} T_z Z_a = 1 = \operatorname{codim} \ker \chi'(z),$$

we have, necessarily,

$$T_z Z_a = \ker \chi'(z).$$

By definition, z is a critical point of φ restricted to Z_a if and only if

$$\ker \chi'(z) \subset \ker \varphi'(z),$$

i.e. if and only if

$$T_z Z_a \subset \ker \varphi'(z).$$

6.5 Application to Bifurcation Theory

This section is devoted to the nonlinear eigenvalue problem

$$\nabla \alpha(u) + \lambda \nabla \beta(u) = 0 \tag{20}$$

where α and β are functionals of class C^2 on a Hilbert space X. We assume that

$$\nabla \alpha(0) = \nabla \beta(0) = 0.$$

Thus $\mathbf{R} \times \{0\}$ is a branch of trivial solutions of (20). Using the Liapunov–Schmidt method and an elementary variational argument, we shall construct two distinct one-parameter families of non trivial solutions of (20). We shall also prove a stronger multiplicity result when α and β are S^1-invariant.

Our basic assumptions are as follows:

(H_1) $\nabla \alpha(0) = \nabla \beta(0) = 0$.

(H_2) $L = \alpha''(0)$ is a Fredholm operator, i.e. the dimension of ker L and the codimension of $R(L)$ are finite.

(H_3) dim ker $L \geq 2$ and $M = \beta''(0)$ is positive definite on ker L.

Remarks 6.6.

1) Since L is a Fredholm operator, $R(L)$ is closed. The symmetry of L then implies that X is the orthogonal direct sum of $R(L)$ and $\ker L$.

2) Without loss of generality, we can assume that

$$\alpha(0) = \beta(0) = 0.$$

3) We shall denote by P (resp. Q) the orthogonal projector on $\ker L$ (resp. $R(L)$), so that $Q = I - P$. We shall use the following notations:

$$A = \nabla\alpha, \quad B = \nabla\beta, \quad R = A - L, \quad S = B - M.$$

Theorem 6.6. *Under assumptions* ($\mathrm{H}_{1,2,3}$), *equation* (20) *has, for each sufficiently small* $\epsilon > 0$, *at least two solutions* $(\lambda(\epsilon), u(\epsilon))$ *such that* $\beta(u(\epsilon)) = \epsilon$. *Moreover,*

$$\lambda(\epsilon) \to 0 \quad \text{as} \quad \epsilon \to 0.$$

Proof. 1) *Liapunov–Schmidt reduction.* Equation (20) is equivalent to the system

$$\begin{aligned} P[R(v+w) + \lambda B(v+w)] &= 0 \\ Lw + Q[R(v+w) + \lambda B(v+w)] &= 0 \end{aligned} \tag{21}$$

where $v = Pu$, $w = Qu$. Since $L : R(Q) \to R(Q)$ is invertible, it follows from the implicit function theorem that (21) defines near $\lambda = 0$, $v = 0$, $w = 0$ a C^1-function $w = w^*(\lambda, v)$. Since $\mathbf{R} \times \{0\}$ is a branch of solutions, we have $w^*(\lambda, 0) = 0$. Differentiating the identity

$$Lw^*(\lambda, v) + Q[R(v + w^*(\lambda, v)) + \lambda B(v + w^*(\lambda, v))] = 0 \tag{22}$$

with respect to v at $[0,0]$ and using the fact that $R'(0) = 0$, we obtain

$$LD_v w^*(0,0) = 0,$$

i.e. $D_v w^*(0,0) = 0$. Thus,

$$\|w^*(\lambda, v)\|/\|v\| \to 0 \tag{23}$$

as $v \to 0$ uniformly for λ near zero. Thus, equation (20) in the neighborhood of $\lambda = 0$, $v = 0$, $w = 0$ is equivalent to the finite-dimensional system

$$P[R(v + w^*(\lambda, v)) + \lambda B(v + w^*(\lambda, v))] = 0. \tag{24}$$

2) *Deparametrization.* Taking the inner product of (24) with v, we obtain

$$(R(v + w^*(\lambda, v)), v) + \lambda(B(v + w^*(\lambda, v)), v) = 0. \tag{25}$$

By Lemma 6.11 below, equation (25) defines a C^1-function $\lambda = \lambda^*(v)$ for $v \neq 0$ and sufficiently small. Moreover, λ^* is extended continuously at zero by setting $\lambda^*(0) = 0$. If we define f near the origin by

$$f(v) = w^*(\lambda^*(v), v),$$

it follows from (23) that

$$\|f(v)\|/\|v\| \to 0 \quad \text{as} \quad v \to 0. \tag{26}$$

Equation (20) is now equivalent to the equation

$$P[R(v + f(v)) + \lambda^*(v)B(v + f(v))] = 0. \tag{27}$$

3) *Constrained extremisation.* For $\rho > 0$ small enough, the function χ defined by

$$\chi(v) = \beta(v + f(v))$$

is continuous on $B(0,\rho)$ and of class C^1 on $B(0,\rho) \setminus \{0\}$. Since

$$\beta(u) = (1/2)(Mu, u) + \int_0^T (S(t,u), u)\, dt,$$

we have

$$\chi(v) = (1/2)(Mv, v) + (Mv, f(v)) + (1/2)(Mf(v), f(v)) + \int_0^T (S(tv + tf(v)), v + f(v))\, dt. \tag{28}$$

Assumption (H_3) implies the existence of $c > 0$ such that

$$(Mv, v) \geq c\|v\|^2, \quad v \in \ker L. \tag{29}$$

Using (26), (28), (29) and the fact that $S(0) = 0$, we can choose ρ small enough so that

$$\chi(v) \geq (c/4)\|v\|^2 \tag{30}$$

for all $v \in B(0,\rho)$. Furthermore,

$$\begin{aligned}(\nabla\chi(v), v) &= (\nabla\beta(v + f(v)), v + f'(v)v)\\ &= ((M + S)(v + f(v)), v + f'(v)v).\end{aligned} \tag{31}$$

By Lemma 6.12 below, we have

$$\|f'(v)\|/\|v\| \to 0 \quad \text{as} \quad v \to 0. \tag{32}$$

It then follows from (26), (29), and (32) that

$$(\nabla\chi(v), v) \geq \frac{c}{2}\|v\|^2 \tag{33}$$

on $B(0,\rho)$ for $\rho > 0$ small enough.

By (30) and (33),

$$Z_\epsilon = \{v \in \ker L : \chi(v) = \epsilon\}$$

will be a compact subset of $B(0,\rho) \setminus \{0\}$ when $\epsilon > 0$ is small enough, and moreover,

$$(\nabla\chi(v), v) \neq 0 \tag{34}$$

when $v \in Z_\epsilon$. The function φ defined by

$$\varphi(v) = \alpha(v + f(v)) + \lambda^*(v)(\chi(v) - \epsilon)$$

achieves its minimum on Z_ϵ at a point v_ϵ. Since $\nabla\chi(v_\epsilon) \neq 0$ because of (34), the Lagrange multiplies rule implies the existence of μ such that

$$\nabla\varphi(v_\epsilon) = \mu\nabla\chi(v_\epsilon),$$

i.e.

$$(A(v_\epsilon + f(v_\epsilon)) + \lambda^*(v_\epsilon)B(v_\epsilon + f(v_\epsilon)), h + f'(v_\epsilon)h) = \mu(\nabla\chi(v_\epsilon), h)$$

for every $k \in \ker L$. Since $f'(v_\epsilon)h \in R(Q)$, the definition of $f(v_\epsilon) = w^*(\lambda(v_\epsilon, v_\epsilon)$ implies that

$$(A(v_\epsilon + f(v_\epsilon)) + \lambda^*(v_\epsilon)B(v_\epsilon + f(v_\epsilon)), f'(v_\epsilon)h) = 0.$$

Using the fact that $PA = PR$, we obtain, for $h \in \ker L$,

$$(R(v_\epsilon + f(v_\epsilon)) + \lambda^*(v_\epsilon)B(v_\epsilon + f(v_\epsilon)), h) = \mu(\nabla\chi(v_\epsilon), h). \tag{35}$$

It follows now from the definition of $\lambda^*(v_\epsilon)$ that

$$0 = (R(v_\epsilon + f(v_\epsilon)) + \lambda^*(v_\epsilon)B(v_\epsilon + f(v_\epsilon)), v_\epsilon) = \mu(\nabla\chi(v_\epsilon), v_\epsilon),$$

so that relation (34) implies that $\mu = 0$. So, we finally have by (35)

$$(R(v_\epsilon + f(v_\epsilon)) + \lambda^*(v_\epsilon)B(v_\epsilon + f(v_\epsilon)), h) = 0$$

for all $h \in \ker L$, i.e. v_ϵ is a solution of (27). Thus $(\lambda(\epsilon), \mu_\epsilon)$, with

$$\lambda(\epsilon) = \lambda^*(v_\epsilon), \quad u_\epsilon = v_\epsilon + f(v_\epsilon)$$

is a solution of (20) such that

$$\beta(u_\epsilon) = \chi(v_\epsilon) = \epsilon.$$

By (30), $v_\epsilon \to 0$ as $\epsilon \to 0$, so that $\lambda(\epsilon) = \lambda^*(v_\epsilon) \to \lambda(0) = 0$ as $\epsilon \to 0$. The other solution is obtained by maximizing φ on Z_ϵ. □

We now state and prove the technical lemmas used in the proofs of Theorem 6.6.

Lemma 6.11. *Equation* (25) *defines for small nonzero v a C^1-function $\lambda = \lambda^*(v)$ such that*

$$\lambda^*(v) \to 0 \quad \textit{as} \quad v \to 0.$$

Proof. Since $M = B'(0)$, it follows from (23) that

$$B(v + w^*(\lambda, v), v) = (Mv, v) + 0(|v|^2) \text{ as } v \to 0,$$

uniformly for λ near zero. Assumption (H_3) implies the existence of $c > 0$ such that

$$(Mv, v) \geq c\|v\|^2$$

for each $v \in \ker L$. Thus, we have

$$B(v + w^*(\lambda, v), v) \geq \frac{c}{2}\|v\|^2$$

on a neighborhood of $\lambda = 0$, $v = 0$, and the function g given by

$$\begin{aligned} g(\lambda, v) &= \lambda + \frac{(R(v + w^*(\lambda, v)), v)}{(B(v + w^*(\lambda, v)), v)} && \text{if } f \neq 0, \\ &= \lambda && \text{if } v = 0, \end{aligned}$$

is well defined. Since $R'(0) = 0$, we have

$$R(v + w^*(\lambda, v), v) = 0(|v|^2) \quad \text{as} \quad v \to 0 \tag{36}$$

uniformly for λ near zero, so that g is continuous.

Let us prove that $D_\lambda g$ is also continuous. Differentiating (22) with respect to λ, we obtain

$$[L + QR'(v + w^*) + \lambda QB'(v + w^*)]\, D_\lambda w^* + QB(v + w^*) = 0.$$

Now, $L : R(Q) \to R(Q)$ is invertible,

$$QR'(v + w^*) + \lambda QB'(v + w^*) \to 0$$

as $[\lambda, v] \to 0$, and

$$B(v + w^*) = Mv + 0(v) \quad \text{as} \quad v \to 0$$

uniformly for λ near zero. Thus, there exists $c_1 > 0$ such that

$$\|D_\lambda w^*\| / \|v\| \leq c_1 \tag{37}$$

for $[\lambda, v]$ near zero. When $v \neq 0$, we have

$$D_\lambda g = 1 + \frac{(R'(v + w^*)D_\lambda w^*, v)}{(B(v + w^*), v)} - \frac{(R(v + w^*), v)(B'(v + w^*)D_\lambda w^*, v)}{((B(v + w^*, v)))2}.$$

Using (35), (36), and (37) in this formula, it is easy to verify that

$$D_\lambda g(\lambda, v) \to 1 \quad \text{as} \quad v \to 0,$$

and $D_\lambda g$ is continuous. Since

$$g(0,0) = 0, \quad D_\lambda g(0,0) = 1,$$

we deduce from the implicit function theorem that the equation

$$g(\lambda, v) = 0 \tag{38}$$

defines near $\lambda = 0$, $v = 0$ a continuous function $\lambda = \lambda^*(v)$. Moreover, since $D_v g(\lambda, v)$ exists and is continuous for $v \neq 0$, λ^* is of class C^2 for small nonzero v. For $v \neq 0$, equations (25) and (38) are equivalent, so that the proof is complete. □

Lemma 6.12. *The function f defined by*

$$f(v) = w^*(\lambda^*(v), v)$$

is such that

$$\lim_{\substack{v \to 0 \\ v \neq 0}} \|f'(v)v\| / \|v\| = 0.$$

Proof. For $v \neq 0$, it follows from the definition of f that

$$f'(v)v = D_\lambda w^*(\lambda^*(v), v) D\lambda^*(v)v + D_v w^*(\lambda^*(v), v)\, v.$$

Since $D_v w^*$ and λ^* are continuous, we have

$$D_v w^*(\lambda^*(v), v) \to D_v w^*(0,0) = 0$$

as $v \to 0$, so that

$$\|D_v w^*(\lambda^*(v), v)v\| / \|v\| \to 0 \quad \text{as } v \to 0.$$

Since, by (37),

$$\|D_\lambda w^*(\lambda^*(v), v)\| / \|v\| \leq c_1,$$

it remains only to prove that $D\lambda^*(v)v \to 0$ as $v \to 0$. Differentiating the identity $g(\lambda^*(v), v) = 0$, we obtain

$$D\lambda^*(v)v = -\frac{D_v g(\lambda^*(v), v)v}{D_\lambda g(\lambda^*(v), v)}.$$

The continuity of $D_\lambda g$ implies that $D_\lambda g(\lambda^*(v), v) \to 1$ as $v \to 0$. It thus suffices to show that

$$D_v g(\lambda^*(v), v)v \to 0 \quad \text{as} \quad v \to 0.$$

When $v \neq 0$, we have, omitting the argument $(\lambda^*(v), v)$ in w^*,

$$D_v g(\lambda^*(v), v)v = \frac{(R'(v + w^*)(v + D_v w^* v), v)}{(B(v + w^*), v)}$$

$$- \frac{(R(v + w^*), v)(B'(v + w^*)(v + d_v w^* v), v)}{[(B(v + w^*), v)]^2}.$$

Using (35) and (36), it is easy to complete the proof. □

Now let $T(\theta)$ be an isometric representation of S^1 over the Hilbert space X such that the following condition holds.

(H_4) α and β are S^1-invariant and

$$\ker L \cap \mathrm{Fix}(S^1) = \{0\}.$$

This assumption implies that A, B, L, M and S are equivariant. In particular, $\ker L$ is invariant and, by the last part of (H_4), the dimension of $\ker L$ will be even.

Theorem 6.7. *Under the assumptions of Theorem 6.6 and (H_4), equation (20) has, for each sufficiently small $\epsilon > 0$, at least $(1/2)\dim\ker L$ S^1-orbits $\{[\lambda(\epsilon), T(\theta)u_\epsilon] : \theta \in S^1\}$ of solutions such that $\beta(u(\epsilon)) = \epsilon$. Moreover, $\lambda(\epsilon) \to 0$ as $\epsilon \to 0$.*

Proof. 1. Let $k = (1/2)\dim\ker L$, so that $\ker L \approx \mathbf{R}^{2k}$. We shall apply Theorem 6.5 to the functions φ and χ defined in the proof of Theorem 6.6. Let us prove that α and χ are invariant. The invariance of P and Q. It then follows from equation (22) that

$$LT(\theta)w^* + Q[R(T(\theta)v + T(\theta)w^*) + \lambda B(T(\theta)v + T(\theta)w^*)] = 0.$$

From the uniqueness of w^*, we, therefore, get the equivariance property

$$w^*(\lambda, T(\theta)v) = T(\theta)w^*(\lambda, v). \tag{39}$$

$T(\theta)$ being an isometry, we deduce from equation (25) with $\lambda = \lambda^*$, and (39) that

$$(R(T(\theta)v + w^*(\lambda^*, T(\theta)v), T(\theta)v) + \lambda^*(B(T(\theta)v$$
$$+ w^*(\lambda^*, T(\theta)v), T(\theta)v) = 0.$$

Thus the uniqueness of λ^* implies its invariance. In particular, we have

$$f(T(\theta)v) = w^*(\lambda^*(T(\theta)v), T(\theta)v) = T(\theta)w^*(\lambda^*(v), v) = T(\theta)f(v),$$

i.e. f is equivariant. The invariance of α and β then implies that of χ and φ.

2. As in the proof of Theorem 6.6, there exists $\rho > 0$ such that $(\nabla\chi(u), u) > 0$ whenever $0 < |u| \le \rho$. Thus, for $\epsilon > 0$ small, there exists a diffeomorphism h defined on a neighborhood of

$$Z_\epsilon = \{u \in \ker L : \chi(u) = \epsilon\}$$

such that $h(Z_\epsilon) = S^{2k-1}$. Thus, condition (H_4) and Theorem 6.5 imply the existence of k S^1-orbits of critical points of φ restricted to Z_ϵ, namely

$$\{T(\theta)v_\epsilon^j : \theta \in S^1\}, \quad j = 1, 2, \ldots, k.$$

It then suffices to verify, as in Theorem 6.6, that if $\lambda_j(\epsilon) = \lambda^*(v_\epsilon^j)$, $u_\epsilon^j = v_\epsilon^j + f(v_\epsilon^j)$, then

$$\{[\lambda_j(\epsilon), T(\theta)u_\epsilon^j] : \theta \in S^1\}$$

is an S^1-orbit of solutions of (20) such that $\beta(u_\epsilon^j) = \epsilon$ and $\lambda_j(\epsilon) \to 0$ as $\epsilon \to 0$. □

6.6 Multiple Periodic Solutions with Prescribed Energy Near an Equilibruim

Let $H \in C^2(\mathbf{R}^{2N}, \mathbf{R})$ be such that $H(0) = 0$, $\nabla H(0) = 0$. We consider the existence of periodic solutions of the Hamiltonian system

$$J\dot{u}(t) + \nabla H(u(t)) = 0 \tag{40}$$

on the energy surface $H^{-1}(\epsilon)$ for small $\epsilon > 0$. If $C = H''(0)$, we assume that the linearized system

$$J\dot{v}(t) + Cv(t) = 0 \tag{41}$$

has $2k$ linearly independent solutions with (not necessarily minimal) period T. Moreover, we assume that

$$(Cv(t), v(t)) \equiv (Cv(0), v(0)) > 0$$

for every nonzero solution v of (41) with (not necessarily minimal) period T.

Theorem 6.8. *Under the above assumptions, equation* (40) *has, for each sufficiently small* $\epsilon > 0$, *at least* k *periodic orbits on* $H^{-1}(\epsilon)$ *whose periods are near* T.

Proof. After the change of variable $s = \tau^{-1}t$, equation (40) becomes

$$J\dot{z}(s) + \tau\nabla H(z(s)) = 0$$

and any 1-periodic solution of this equation corresponds to a τ-periodic solution of (40). Setting $\tau = T + \lambda$, we obtain the bifurcation problem

$$\nabla\alpha(z) + \lambda\nabla\beta(z) = 0 \tag{42}$$

where the functionals α and β given by

$$\alpha(z) = \int_0^1 [(J\dot{z}(s), z(s)) + TH(z(s))]\, ds,$$

$$\beta(z) = \int_0^1 H(z(s))\, ds,$$

are defined on the Hilbert space $X = H_1^1$. The functionals α and β are of class C^2 and are invariant under the representation of $S^1 \approx \mathbf{R}/\mathbf{Z}$ defined over X by the translations in time.

Let $L = \alpha''(0)$, so that $z \in \ker L$ if and only if

$$J\dot{z}(s) + TCz(s) = 0$$

$$z(0) = z(1).$$

By assumption, dim ker $L = 2k$ and $M = \beta''(0)$ are positive definite on ker L. Since $\alpha''(0)$ is symmetric, dim ker L = codim $R(L)$ and $\alpha''(0)$ is a Fredholm operator. Since $H''(0)$ is nonsingular, ker $L \cap \mathrm{Fix}(S^1) = \{0\}$.

By Theorem 6.7, equation (42) has, for every sufficiently small $\epsilon > 0$, at least k S^1-orbits

$$\{[\lambda_j(\epsilon), T(\theta)z_\epsilon^j] : \theta \in S^1\}, \quad j = 1, \ldots, k,$$

such that

$$\beta(z_\epsilon^j) = \int_0^1 H(z_\epsilon^j(s))\, ds = \epsilon,$$

and $\lambda_j(\epsilon) \to 0$ as $\epsilon \to 0$. The corresponding solutions of (40) on $H^{-1}(\epsilon)$ are given by

$$u_\epsilon^j(t) = z_\epsilon^j((T + \lambda_j(\epsilon))^{-1}t). \quad \square$$

Corollary 6.1. *Let $H \in C^2(\mathbf{R}^{2N}, \mathbf{R})$ be such that $H(0) = 0$, $\nabla H(0) = 0$ and $H''(0)$ is positive definite. Then, for each sufficiently small $\epsilon > 0$, equation (40) has at least N periodic orbits with energy ϵ.*

Proof. Since $C = H''(0)$ is positive definite, any solution of (41) is bounded on $\mathbf{R}$. Thus (41) has $2N$ linearly independent periodic solutions. The periodic solutions of (41) split into n families with incommensurable periods $T_1, \ldots, T_n$ and dimensions $k_1, \ldots, k_n$. Theorem 6.8 applied to each of those families implies the existence of $\frac{k_1}{2} + \ldots + \frac{k_n}{2} = N$ periodic orbits on $H^{-1}(\epsilon)$ where $\epsilon > 0$ is small enough. These periodic orbits are distinct from one family to another because they have no common period for $\epsilon > 0$ sufficiently small. $\square$

Historical and Bibliographical Notes

For surveys on the mathematical work of Lusternik, see [AVS$_1$, AVD$_1$], [Ale$_1$] and on minimax methods see [Pal$_2$], [Rab$_6$]. The Lusternik–Schnirelmann theory [LJS$_1$] generalizes to smooth functions on a compact manifold the minimax theory of eigenvalues due to Raleigh, Poincaré, Fischer, Courant, Weyl, etc. The basic topological invariant is not the homology of the manifold as in Morse theory but the category. The category $\mathrm{cat}_M A$ of a closed subset A of a compact manifold M is the least number of closed sets, each contractible in M, needed to cover A. By an elementary minimax argument, it is proved in [LJS$_1$] that a smooth function φ on M has at least $\operatorname{cat}_M M$ critical points. Since $\operatorname{cat}_{\mathbf{R}(P_N)} \mathbf{R}(P_N) = N + 1$ where $\mathbf{R}(P_N) = S^N/\mathbf{Z}_2$ is the real projective N-space, every smooth, even function on S^N will have at least $N + 1$ pairs of critical points. This result generalizes the linear theory of eigenvalues. If $A \subset S^N$ is closed and symmetric with respect to the origin, the following relation

$$\operatorname{cat}_{\mathbf{R}(P_N)}(A/\mathbf{Z}_2) = \operatorname{ind} A$$

between the category and the $\mathbf{Z}_2$-index has been proved by Rabinowitz [Rab$_8$]. For other group actions, the quotient space is not, in general, a manifold, so that the index is more flexible than the category.

The concept of pseudo-gradient vector field was introduced by Palais [Pal$_3$] in order to extend the classical Lusternik–Schnirelmann theory to infinite-dimensional Banach manifolds. The compactness of the domain is replaced by the Palais–Smale condition on the function ([Pal$_1$], [PaS$_1$], [Sma$_1$]). The next basic step toward the approach developed in this Chapter was the obtention by Clark [Clk$_1$] of strong multiplicity results for an even function defined on a Banach space X. Since X is contractible, the multiplicity follows from geometric conditions on the function itself as in Theorem 6.2. The $\mathbf{Z}_2$-versions of Lemma 6.4, Lemma 6.5, Theorem 6.1, and Theorem 6.2 are due to Clark [Clk$_1$]. Related results are due to Ambrosetti [Amb$_1$].

Theorem 6.2 is a generalization, due to Costa–Willem [CoW$_1$] of a result of Ekeland–Lasry [EkL$_1$] (dealing with the case where $Y = X$ and $\operatorname{Fix}(S^1) = \{0\}$). Situations where $\operatorname{Fix}(S^1)$ is finite dimensional have been considered by Benci [Ben$_2$].

Theorem 6.3 is a result of Ekeland–Lasry [EkL$_1$]. Other proofs have been given by Hofer [Hof$_2$] and Ambrosetti–Mancini [AmM$_2$]. An extension which replaces the strict convexity of C by starshapeness properties has been given by Berestycki–Lasry–Mancini–Ruf [BLM$_1$]. Local results in the line of Theorem 6.3 had been proved earlier by Weinstein [Wei$_2$] and Moser [Mos$_1$], following the pioneering work of Lyapunov [Lya$_1$] and Horn [Hor$_1$]. The concept of an integral invariant is due to Poincaré [Poi$_1$] and Theorem 6.4 to Croke–Weinstein [CrW$_1$].

Theorem 6.5 is due to Krasnosel'skii [Kra$_1$] and Theorem 6.6, due to Stuart [Stu$_1$], generalizes classical results of Krasnosel'skii, Böhme, and Marino. Theorem 6.8 is due to Moser [Mos$_1$] and Corollary 6.1 is a result of Weinstein [Wei$_2$].

Further applications of the Lusternik–Schnirelmann type arguments with S^1 — or other indices or pseudo-indices can be found in [BCP$_{1,2}$], [BeF$_{1,2}$], [Cap$_{1,2}$], [CaF$_1$], [CFS$_2$], [CaS$_{1,2}$], [CWL$_1$], [Ckk$_3$], [Cor$_3$], [DCF$_1$], [Gir$_1$], [GiM$_{1,2,3}$], [Lov$_1$], [Mnc$_2$], [Rab$_{18}$], [Sal$_1$], [VGr$_{5-10}$].

For other papers on Lusternik–Schnirelmann theory, see [Amb$_6$], [CoP$_1$], [Rab$_{22}$], [Wu$_3$].

Multiplicity results for the fixed energy problem can also be found in [Ben$_5$], [LasV$_1$], [Szu$_4$], [Vi$_{1,2}$], [HofZ$_1$], [Rab$_{23}$]. See also [Bere$_2$] and [BL$_1$] for conservative systems and [Ben$_6$] for Lagrangian systems.

Exercises

1. Let $\varphi \in C^1(X, \mathbf{R})$ be an even functional satisfying (PS). Let Y and Z be closed subspaces of X with codim Y and dim Z finite and

$$\operatorname{codim} Y < \dim Z.$$

   Assume that the following conditions are satisfied:

   (i) $\inf_Y \varphi > -\infty$,

   (ii) there exists $r > 0$ such that $\varphi(u) < 0$ whenever $u \in Z$ and $\|u\| = r$.

   (iii) if $\varphi'(0) = 0$, then $\varphi(0) \geq 0$.

   Then there exists at least $\dim Z - \operatorname{codim} Y$ distinct pairs of nonzero critical points of φ.

   Hint. Follow the line of Theorem 6.2 using $\mathbf{Z}_2$-index.

2. Let $H \in C^1(\mathbf{R}^{2N}, \mathbf{R})$ be strictly convex and such that its Fenchel transform $H^* \in C^1(\mathbf{R}^{2N}, \mathbf{R})$. If $v_1, v_2 \in C^1([0,T],)$ are T-periodic and such that $u_1 = \nabla H^*(\dot{v}_1)$ and $u_2 = \nabla H^*(\dot{v}_2)$ belong to the same S^1-orbit, then v_1 and v_2 belong to the same S^1-orbit.

3. Let $H \in C^1(\mathbf{R}^{2N}, \mathbf{R})$ be strictly convex, such that $H(0) = 0$, $\nabla H(0) = 0$,

$$\underline{\lim}_{u \to 0} 2|u|^{-2} H(u) > \alpha$$

   and such that its Fenchel transform $H^* \in C^1(\mathbf{R}^{2N}, \mathbf{R})$. Then there exists $\nu > \alpha$ and $\rho > 0$ such that

$$H^*(v) \leq (2\nu)^{-1}|v|^2$$

   whenever $|v| \leq \rho$.

4. Let $H \in C^1(\mathbf{R}^{2N}, \mathbf{R})$ be strictly convex and such that

$$H(0) = 0, \quad \nabla H(0) = 0$$

   and let $T > 0$. Assume that there exists $n \in \mathbf{N}$, $\gamma \in]2\pi n/T, 2\pi(n+1)/T[$, $c \geq 0$, and $\delta \in]0, \min(\gamma - (2\pi n/T), (2\pi(n+1)/T - \gamma)[$ such that

$$|\nabla H(u) - \gamma u| \leq \delta|u| + c$$

for all $u \in \mathbf{R}^{2N}$ and

$$\underline{\lim}_{u\to 0} 2|u|^{-2}H(u) > (2\pi/T)(n+k)$$

for some $k \geq 1$. Then the problem

$$J\dot{u} + \nabla H(u) = 0$$

$$u(0) = u(T)$$

has at least Nk S^1-orbits of nontrivial solutions.

Hint. Use Exercises 5.5 and 5.6. Show that the same is true for $-J\dot{u} + \nabla H(u) = 0$.

5. ([BHR1]). Let $H, K \in C^1(\mathbf{R}^{2N}, \mathbf{R})$ and let U be a compact neighborhood of $S = H^{-1}(1)$. Assume that there exists $a, b \geq 0$ with $a + b > 0$ such that

$$a(q, D_q H(q,p)) + b(p, D_q H(q,p)) + (\nabla K(q,p), J\nabla H(q,p)) > 0$$

for all $(q,p) \in U$. Then there exists $\alpha, \beta > 0$ such that every T-periodic solution of $J\dot{u} + \nabla H(u) = 0$ on $S_\epsilon = H^{-1}(1+\epsilon) \subset U$ satisfies

$$\alpha T \leq \int_0^T (Ju, \dot{u}) \leq \beta T.$$

7

Morse–Ekeland Index and Multiple Periodic Solutions with Fixed Period

Introduction

An autonomous Hamiltonian system

$$J\dot{u}(t) + \nabla H(u(t)) = 0$$

is called ***asymptotically linear*** if there exist symmetric matrices A_0 and A_∞ such that

$$\nabla H(u) = A_0 u + o(|u|) \quad \text{as} \quad |u| \to 0$$

and

$$\nabla H(u) = A_\infty u + o(|u|) \quad \text{as} \quad |u| \to \infty.$$

Such a system necessarily has the trivial periodic solution $u = 0$ and it is interesting to find conditions for A_0 and A_∞ which guarantee the existence of ***nontrivial periodic solutions of a given period*** and provide information about their multiplicity.

A rough condition follows from degree theory by requiring that some topological degrees associated to the linearizations at zero and at infinity would be different. A more precise tool would be provided by the ***Morse index*** of the Hamiltonian actions associated to those linearized systems but this index is equal to infinity in both cases because of the strongly indefinite character of the action.

When A_0 and A_∞ are positive definite, the corresponding dual actions are well defined and have finite Morse indices, which we can denote by $i(A_0, T)$ and $i(A_\infty, T)$, respectively, where T is the period. Combining their properties with the Lusternik–Schnirelman theory applied to the dual action, it is then possible to prove, under a nonresonance condition at infinity, that the asymptotically linear Hamiltonian system has at least

$$(1/2)[i(A_0, T) - i(A_\infty, T)]$$

nontrivial T-periodic solutions when $i(A_0, T) > i(A_\infty, T)$. This condition is reminiscent of the twist condition in the ***Poincaré–Birkhoff fixed point theorem*** for area-preserving mapping of an annulus into itself.

7.1 The Index of a Linear Positive Definite Hamiltonian System

Let A be a continuous mapping from $\mathbf{R}$ into the space of symmetric positive definite matrices of order $2N$. We consider the periodic boundary value problem

$$\begin{aligned} J\dot{u}(t) + A(t)u(t) &= 0 \\ u(0) &= u(T) \end{aligned} \tag{1}$$

where $T > 0$ is fixed. The corresponding Hamiltonian is given by

$$H(t,u) = (1/2)(A(t)u,u).$$

It is easy to verify that its Legendre transform (with respect to u) $H^*(t,\cdot)$ is of the form

$$H^*(t,v) = (1/2)(B(t)v,v)$$

where $B(t) = (A(t))^{-1}$, so that the corresponding dual action is defined on H_T^1 by

$$\chi_T(v) = \int_0^T (1/2)[(J\dot{v}(t),v(t)) + (B(t)\dot{v}(t),\dot{v}(t))]\,dt.$$

Definition 7.1. The index $i(A,T)$ is the *Morse index* of χ_T, i.e., the supremum of the dimensions of the subspaces of H_T^1 on which χ_T is negative definite.

Notice that the Hamiltonian action associated to (1) is given by

$$\varphi(u) = \int_0^T (1/2)[(J\dot{u}(t),u(t)) + (A(t)u(t),u(t))]\,dt$$

and hence is strongly indefinite because of the spectral properties of $u \to J\dot{u}$ seen in Chapter 3. Therefore defining the index $i(A,T)$ as the Morse index of φ would always give the value $+\infty$ and would be useless.

Since $\chi_T(v+w) = \chi_T(v)$ for every constant function w, it is sufficient to consider the restriction of χ_T to the subspace

$$\tilde{H}_T^1 = \left\{ v \in H_T^1 \; : \; \int_0^T v(t)\,dt = 0 \right\}.$$

From our assumptions follows the existence of $\delta_T > 0$ such that

$$(B(t)v,v) \geq \delta_T |v|^2$$

for all $t \in [0,T]$ and $v \in \mathbf{R}^{2N}$. Thus Wirtinger's inequality implies that the symmetric bilinear form given by

$$((v,w)) = \int_0^T (B(t)\dot{v}(t),\dot{w}(t))\,dt$$

defines an inner product on $\tilde{H}^1_T$. The corresponding norm $\|\cdot\|$ is such that

$$\|v\|^2 \geq \delta_T |\dot{v}|^2_{L^2}. \tag{2}$$

Let us define the linear operator K on $\tilde{H}^1_T$ by the formula (using the Riesz theorem)

$$((Kv, w)) = \int_0^T (Jv(t), \dot{w}(t))\, dt.$$

It is easy to check that K is self-adjoint and compact. Moreover

$$2\chi_T(v) = \int_0^T [-(Jv(t), \dot{v}(t)) + (B(t)\dot{v}(t), \dot{v}(t))]\, dt = ((v - Kv, v)). \tag{3}$$

It follows from the spectral theory that $\tilde{H}^1_T$ will be the orthogonal sum of $\ker(I - K)$, H^+ and H^- with $I - K$ positive definite (resp. negative definite) on H^+ (resp. H^-). Since K has at most finitely many eigenvalues (with finite multiplicity) greater than one,

$$i(A, T) = \dim H^- < +\infty$$

i.e., *the index* $i(A,T)$ *is finite.* On the other hand, there exists $\tilde{\delta} > 0$ such that

$$((v - Kv, v)) \geq \tilde{\delta}\|v\|^2, \quad v \in H^+$$

and

$$((v - Kv, v)) \leq -\tilde{\delta}\|v\|^2, \quad v \in H^-.$$

Setting $\delta = \tilde{\delta}\delta_T > 0$ we deduce from (2) and (3) the estimates

$$\chi_T(v) \geq (\delta/2)|\dot{v}|^2_{L^2}, \quad v \in H^+ \tag{4}$$

and

$$\chi_T(v) \leq -(\delta/2)|\dot{v}|^2_{L^2}, \quad v \in H^-. \tag{5}$$

We now state and prove some preliminary results which will allow us to prove a geometrical interpretation of the index.

Proposition 7.1. *The dimension of* $\ker(I - K)$ *is equal to the number of linearly independent solutions of* (1).

Proof. By a Fourier series argument, $v \in \ker(I - K)$ if and only if $v \in \tilde{H}^1_T$ and

$$B(t)\dot{v}(t) = Jv(t) + c \tag{6}$$

for some $c \in \mathbf{R}^{2N}$ and a.e. $t \in [0, T]$. As $(B(t))^{-1} = A(t)$ is invertible, v is of class C^1, (6) holds for all $t \in [0, T]$, and $B(\cdot)\dot{v}(\cdot) \in H^1_T$. Setting

$$u(t) = (\Phi v)(t) = B(t)\dot{v}(t),$$

we have, for a.e. $t \in [0,T]$,

$$\dot{u}(t) = \frac{d}{dt}(B(t)\dot{v}(t)) = J\dot{v}(t),$$

and u is a solution of (1).

Conversely, assume now that u is a solution of (1). Then

$$\int_0^T A(t)u(t)\,dt = 0$$

and hence there will exist a unique $v \in \tilde{H}_T^1$ such that $u = \Phi v$. Thus

$$-J\dot{u}(t) = A(t)u(t) = \dot{v}(t)$$

for a.e. $t \in [0,T]$, so that

$$u(t) = Jv(t) + c$$

for some $c \in \mathbf{R}^{2N}$ and all $t \in [0,T]$. Consequently

$$A(t)[Jv(t) + c] = \dot{v}(t)$$

for a.e. $t \in [0,T]$, which is equivalent to (6) and shows that $v \in \ker(I - K)$. Thus Φ is an isomorphism between $\ker(I - K)$ and the space of solutions of (1). $\square$

According to the Poincaré–Weyl–Fisher–Courant principle, the (possible) positive eigenvalues of K,

$$\lambda_1 \geq \lambda_2 \geq \lambda_3 \geq \cdots,$$

are given by the formulas

$$\lambda_k = \lambda_k(T) = \max_{H_{T,k}} \min_{\substack{v \in H_{T,k} \\ \|v\|=1}} ((Kv, v))$$

$$= \max_{H_{T,k}} \min_{\substack{v \in H_{T,k} \\ \int_0^T (B(t)\dot{v}(t),\dot{v}(t))dt=1}} \int_0^T (Jv(t), \dot{v}(t))\,dt, \tag{7}$$

where the maximum is taken over all subspaces $H_{T,k}$ of $\tilde{H}_T^1$ having dimension k. By Proposition 3.1, the bilinear form

$$v \longrightarrow \int_0^T (Jv(t), \dot{v}(t))\,dt = -\int_0^T (J\dot{v}(t), v(t))\,dt$$

is positive on the space spanned by

$$\left(\cos\frac{2\pi kt}{T}\right)c + \left(\sin\frac{2\pi kt}{T}\right)Jc, \quad k \in \mathbf{N}^*, \; c \in \mathbf{R}^{2N}$$

and hence K has infinitely many positive eigenvalues.

Lemma 7.1. *Let Y, Z and W be subspaces of a vector space X. If X is the direct sum of Y and Z and if $W \cap Y = \{0\}$, then*

$$\dim W \leq \dim Z.$$

Proof. Let P be the projection on Z along Y. Since $W \cap Y = \{0\}$, $Pu = 0$ implies that $u = 0$ for every $u \in W$. Hence $P : W \to Z$ is one to one and the result follows. □

Proposition 7.2. *The eigenvalues $\lambda_k(T)$ are increasing functions of T.*

Proof. Let $0 < S < T$. There exists a k-dimensional subspace V of $\tilde{H}^1_S$ such that

$$\lambda_k(S) = \min_{\substack{v \in V \\ \int_0^S (B(t)\dot{v}(t), \dot{v}(t))\,dt = 1}} \int_0^S (Jv(t), \dot{v}(t))\,dt.$$

If $v \in V$, let us extend it to $[0, T]$ by setting

$$v(t) = v(S), \quad S < t \leq T$$

and let

$$\hat{v}(t) = v(t) - \frac{1}{T} \int_0^T v(s)\,ds, \quad 0 \leq t \leq T.$$

Then the space

$$W = \{\hat{v} : v \in V\} \subset \tilde{H}^1_T$$

is such that $\dim W = k$. Moreover, for each $v \in V$, we have

$$\int_0^S (B(t)\dot{v}(t), \dot{v}(t))\,dt = \int_0^T (B(t)\dot{\hat{v}}, \dot{\hat{v}}(t))\,dt$$

and

$$\int_0^S (Jv(t), \dot{v}(t))\,dt = \int_0^T (J\hat{v}(t), \dot{\hat{v}}(t))\,dt.$$

Hence, (7) implies that

$$\lambda_k(S) = \min_{\substack{w \in W \\ \|w\| = 1}} ((Kw, w)) \leq \lambda_k(T). \tag{8}$$

If $\lambda_k(S)$ is not an eigenvalue of K on $\tilde{H}^1_T$, then $\lambda_k(S) < \lambda_k(T)$ and the proof is complete. Let us assume that $\lambda = \lambda_k(S)$ is an eigenvalue of K on $\tilde{H}^1_T$. According to the spectral theory, $\tilde{H}^1_T$ is the orthogonal sum of H_1, H_2, and H_3, with $H_2 = \ker(K - \lambda I)$, $K - \lambda I$ negative definite on H_1 and positive definite on H_3. We show that $W \cap (H_1 \oplus H_2) = \{0\}$. If $w \in W \cap (H_1 \oplus H_2)$, then $w = w_1 + w_2$ with $w_i \in H_i$ $(i = 1, 2)$. If $w_1 \neq 0$, then

$$\left(\left(K\left(\frac{w}{\|w\|}\right), \frac{w}{\|w\|}\right)\right) = \left(\left(K\left(\frac{w_1}{\|w\|}\right), \frac{w_1}{\|w\|}\right)\right) + \left(\left(\lambda\left(\frac{w_2}{\|w\|}\right), \frac{w_2}{\|w\|}\right)\right)$$
$$< \lambda\frac{\|w_2\|^2}{\|w\|^2} + \|w_1\|^2 \leq \lambda = \lambda_k(S),$$

a contradiction with (8). Thus $w = w_2 \in \ker(K - \lambda I)$, i.e.,

$$\lambda B(t)\dot{w}(t) = Jw(t) + c$$

for some $c \in \mathbf{R}^{2N}$ and a.e. $t \in [0, T]$. Since by construction $\dot{w}(t) = 0$ on $]S, T]$, we have $w(t) = Jc$ on the same interval. By the uniqueness of the solution of the Cauchy problem, $w(t) = Jc$ on $[0, T]$ and hence $w = 0$ as $w \in \tilde{H}_T^1$. Now it follows from Lemma 7.1 that

$$k = \dim W \leq \dim H_3.$$

Let $H_{T,k}$ be a k-dimensional subspace of H_3. We have, by definition of H_3 and (7),

$$\lambda_k(S) = \lambda < \min_{\substack{v \in H_{T,k} \\ \|v\|=1}} ((Kv, v)) \leq \lambda_k(T),$$

and the proof is complete. □

Proposition 7.3. *The eigenvalues $\lambda_k(T)$ are continuous functions of T.*

Proof. Let $T > 0$ be fixed and let $S > 0$. Define φ on $\tilde{H}_T^1$ by

$$\varphi_S(v) = \frac{T}{S}\int_0^T \left(B\left(\frac{S}{T}t\right)\dot{v}(t), \dot{v}(t)\right) dt.$$

Then, the change of variable $t = \frac{S}{T}\tau$ easily implies that

$$\lambda_k(S) = \max_{H_{T,k}} \min_{\substack{v \in H_{T,k} \\ \varphi_S(v)=1}} \int_0^T (Jv(\tau), \dot{v}(\tau))\, d\tau, \tag{9}$$

where the maximum is taken over all subspaces $H_{T,k}$ of $\tilde{H}_T^1$ having dimension k.

Let $\epsilon > 0$; there will exist $\delta > 0$ such that

$$\frac{1}{1+\epsilon}B(t) \leq \frac{T}{S}B\left(\frac{S}{T}t\right) \leq \frac{1}{1-\epsilon}B(t)$$

whenever $t \in [0,T]$ and $|S - T| \leq \delta$. Thus

$$\frac{1}{1+\epsilon}\|v\|^2 \leq \varphi_S(v) \leq \frac{1}{1-\epsilon}\|v\|^2 \tag{10}$$

whenever $|S-T| \leq \delta$ and $v \in \tilde{H}^1_T$. It follows then from (9) and (10) that

$$\begin{aligned}
\lambda_k(S) &= \max_{H_{T,k}} \min_{\substack{v \in H_{T,k} \\ v \neq 0}} \int_0^T (Jv(t), \dot{v}(t))\, d)/\varphi_S(v) \\
&\leq \max_{H_{T,k}} \min_{\substack{v \in H_{T,k} \\ v \neq 0}} (1+\epsilon) \left(\int_0^T (Jv(t), \dot{v}(t))\, dt \right) /\|v\|^2 \\
&= (1+\epsilon) \max_{H_{T,k}} \min_{\substack{v \in H_{T,k} \\ \|v\|=1}} \int_0^T (Jv(t), \dot{v}(t)) dt = (1+\epsilon)\lambda_k(T)
\end{aligned}$$

whenever $|S-T| \leq \delta$. Similarly, $\lambda_k(S) \geq (1-\epsilon)\lambda_k(T)$ when $|S-T| \leq \delta$, and the proof is complete. □

Definition 7.2. A point $T > 0$ is *conjugate* to 0 for

$$J\dot{u}(t) + A(t)u(t) = 0 \tag{11}$$

with multiplicity m if the periodic boundary value problem (1) has m linearly independent solutions.

We can now state and prove our geometric interpretation of $i(A,T)$.

Theorem 7.1. *The index $i(A,T)$ is equal to the sum of the multiplicities of the conjugate points to* 0 *for* (11) *situated in* $]0,T[$.

Proof. By definition of the index and the relation

$$\chi_T(v) = (1/2)((v - Kv, v))$$

we see immediately that the index $i = i(A,T)$ is characterized by the relation

$$\lambda_i(T) > 1 \geq \lambda_{i+1}(T).$$

It follows easily from Wirtinger's inequality that, for $S > 0$ sufficiently small, we have

$$\ldots \lambda_2(S) \leq \lambda_1(S) = \max_{\substack{v \in \tilde{H}^1_S \\ \int_0^S (B(t), \dot{v}(t), \dot{v}(t)) dt = 1}} \int_0^T (Jv(t), \dot{v}(t))\, dt < 1.$$

Thus, by Propositions 2 and 3, i is equal to the number of eigenvalues which have crossed the value one when S increases from 0 to T. But, if for some $S > 0$,

$$\lambda_k(S) > 1 = \lambda_{k+1}(S) = \ldots = \lambda_{k+m}(S) > \lambda_{k+m+1}(S),$$

Proposition 7.1 implies that S is conjugate to 0 for (11) with multiplicity m, and hence the proof is complete. □

7.2 Linear Autonomous Positive Definite Hamiltonian Systems

Let A be a symmetric positive definite matrix of order $2N$. Then the energy integral

$$(Au(t), u(t)) = c$$

implies that all the solutions of the equation

$$J\dot{u}(t) + Au(t) = 0$$

are bounded on $\mathbf{R}$. Consequently, all the eigenvalues of JA must be pure imaginary, so that

$$\sigma(JA) = \{i\alpha_k : \alpha_k > 0, \alpha_{N+k} = -\alpha_k, k = 1, \ldots, N\}$$

and there exists a basis of $\mathbf{C}^{2N}$ of the form

$$x_1 + iy_1, \ldots, x_N + iy_N, \quad x_1 - iy_1, \ldots, x_N - iy_N$$

such that

$$JA(x_k + iy_k) = i\alpha_k(x_k + iy_k) \quad (i \leq k \leq N).$$

Hence

$$JAx_k = -\alpha_k y_k, \quad JAy_k = \alpha_k x_k \quad (1 \leq k \leq N)$$

so that a fundamental system of solutions is given by

$$(\sin \alpha_k t)x_k + (\cos \alpha_k t)y_k, \quad (\cos \alpha_k t)x_k - (\sin \alpha_k t)y_k \quad k = 1, \ldots, N.$$

It follows immediately from Proposition 7.1 that

$$\dim \ker(I - K) = 2\sum_{j=0}^{\infty} \dim \ker \left(JA - \frac{2i\pi j}{T}\right) \tag{12}$$

and from Theorem 7.1 that

$$i(A, T) = 2\sum_{k=1}^{N} \# \left\{j \in \mathbf{N}^* : \frac{2\pi j}{T} < \alpha_k\right\}. \tag{13}$$

Proposition 7.4. *If* $\sigma(JA) \cap \frac{2i\pi}{T}\mathbf{N} = \phi$, *then* $I - K$ *is invertible and* $\operatorname{codim} H^+ = i(A, T)$.

Proof. By (12), $\ker(I - K) = \{0\}$ so that $I - K$ is invertible and $\tilde{H}_T^{\frac{1}{2}}$ is the orthogonal direct sum of H^+ and H^-. Hence

$$\operatorname{codim} H^+ = \dim H^- = i(A, T). \qquad \square$$

7.3 Periodic Solutions of Convex Asymptotically Linear Autonomous Hamiltonian Systems

We consider the existence of multiple periodic solutions for the autonomous Hamiltonian system

$$J\dot{u}(t) + \nabla H(u(t)) = 0 \tag{14}$$

where $H \in \mathbf{C}^1(\mathbf{R}^{2N}, \mathbf{R})$ is strictly convex and satisfies the conditions

$$\nabla H(u) = A_0 u + o(|u|) \quad \text{as} \quad |u| \to 0 \tag{15}$$

and

$$\nabla H(u) = A_\infty u + o(|u|) \quad \text{as} \quad |u| \to \infty \tag{16}$$

with symmetric positive definite matrices A_0 and A_∞.

Theorem 7.2. *Assume that $T > 0$ is such that*

A$_1$. $\sigma(JA_\infty) \cap \frac{2i\pi}{T}\mathbf{N} = \phi$

A$_2$. $i(A_0, T) > i(A_\infty, T)$.

Then system (14) *has at least*

$$\frac{1}{2}[i(A_0, T) - i(A_\infty, T)]$$

nonzero T-periodic orbits.

Remarks.

1) It follows from (A$_1$) that the linear system

$$J\dot{u}(t) + A_\infty u(t) = 0$$

has no nontrivial T-periodic solution. Thus (A$_1$) is a nonresonance condition "at infinity."

2) Assumption (A$_2$), which requires a distinct behavior of ∇H "at the origin" and "at infinity," is similar to the twist condition in the Poincaré–Birkhoff theorem on the invariant points of a self-mapping on an annulus.

3) Since H is strictly convex and $\nabla H(0) = 0$ by (15), 0 is the unique equilibrium point of (14).

4) Without loss of generality, we can assume that $H(0) = 0$. Since $\nabla H(0) = 0$, this implies that

$$H^*(0) = 0.$$

5) Since H is strictly convex and, because of (16) such that

$$H(u)/|u| \to +\infty \quad \text{as} \quad |u| \to \infty,$$

Proposition 2.4 implies that $H^* \in \mathbf{C}^1(\mathbf{R}^{2N}, \mathbf{R})$. Moreover, (16) implies also that

$$\nabla H^*(v) = B_\infty v + o(|v|) \quad \text{as} \quad |v| \to \infty \tag{17}$$

where $B_\infty = A_\infty^{-1}$. Indeed, by duality, if $v = \nabla H(u)$,

$$\nabla H^*(v) - B_\infty v = u - B_\infty \nabla H(u) = -B_\infty(\nabla H(u) - A_\infty u).$$

Moreover, as A_∞ is positive definite, there exists $\delta > 0$ such that $|A_\infty u| \geq \delta|u|$ for $u \in \mathbf{R}^{2N}$ and hence, by (16), we have, for sufficiently large $|u|$,

$$\begin{aligned} [|A_\infty| + (\delta/2)]|u| &\geq |A_\infty u - (A_\infty u - \nabla H(u))| = |v| \\ &\geq |A_\infty u| - |A_\infty u - \nabla H(u)| \geq (\delta/2)|u| \end{aligned}$$

so that $|u| \to \infty$ whenever $|v| \to \infty$, and hence (17) holds. By Theorem 2.3, the dual action φ defined by

$$\chi(v) = \int_0^T \left[\frac{1}{2}(J\dot{v}(t), v(t)) + H^*(\dot{v}(t))\right] dt$$

is continuously differentiable on $\tilde{H}_T^1$.

Since φ is invariant for the representation of $S^1 \cong \mathbf{R}/T\mathbf{Z}$ defined over $\tilde{H}_T^1$ by the translations in time

$$(T(\theta)v)(t) = v(t + \theta),$$

we are in a position to apply Theorem 6.2.

6) It is convenient in this section to use the inner product

$$((v, w)) = \int_0^T (B_\infty \dot{v}(t), \dot{w}(t))\, dt$$

and the corresponding norm $\|\cdot\|$ in $\tilde{H}_T^1$. By Wirtinger's inequality and the positive-definiteness of B_∞, this norm is equivalent to the standard norm of $\tilde{H}_T^1$.

The proof of Theorem 6.2 will depend on the following lemmas. The first one will imply that φ satisfies the (PS)-condition.

Lemma 7.2. *Every sequence (v_j) in $\tilde{H}_T^1$ such that $\varphi'(v_j) \to 0$ contains a convergent subsequence.*

Proof. Let us define the operators K and N over $\tilde{H}_T^1$, using the Riesz theorem, by the formulas

$$((Kv, w)) = \int_0^T (Jv(t), \dot{w}(t))\, dt,$$

$$((Nv, w)) = \int_0^T (\nabla H^*(\dot{v}(t)) - B_\infty \dot{v}(t), \dot{w}(t))\, dt.$$

Since

$$\langle \varphi'(v), w\rangle = \int_0^T (\nabla H^*(\dot v(t)) - Jv(t), \dot w(t))\, dt = ((v - Kv + Nv, w)),$$

we have, by assumption,

$$v_j - Kv_j + Nv_j = f_j, \quad j \in \mathbf{N}^*, \tag{18}$$

with $f_j \to 0$ in $\tilde H^1_T$. In particular, $\|f_j\| \le R$ for some $R > 0$ and all $j \in \mathbf{N}^*$. Assumption (A_1) and Proposition 7.4 imply that $L = I - K$ is invertible. Thus it follows from (16) that there exists some $c > 0$ such that

$$\|Nv\| \le (1/2)\|L^{-1}\|^{-1}\,\|v\| + c$$

for all $v \in \tilde H^1_T$. Therefore (18) implies that

$$\|v_j\| \le \|L^{-1}\|(\|Nv_j\| + \|f_j\|) \le (1/2)\|v_j\| + \|L^{-1}\|(c + R), \quad j \in \mathbf{N}^*,$$

so that (v_j) is bounded. The rest of the proof uses the same argument as that of the end of the proof of Lemma 4.5. □

We now verify the first geometric condition of Theorem 6.2 for φ.

Lemma 7.3. *The functional φ is bounded from below on a closed invariant subspace Y of $\tilde H^1_T$ of codimension $i(A_\infty, T)$.*

Proof. By Assumption (A_1), Proposition 7.4 and formula (4), there exists a closed invariant subspace $Y = H^+$ of $\tilde H^1_T$ with codimension $i(A_\infty, T)$ and there exists $\delta > 0$ such that, for each $v \in Y$, one has

$$\chi_T^\infty(v) \equiv \int_0^T (1/2)[J\dot v(t), v(t)) + (B_\infty \dot v(t), \dot v(t))]\, dt \ge (\delta/2)|\dot v|^2_{L^2}.$$

It follows from (17) that there exists $c > 0$ such that

$$|\nabla H^*(v) - B_\infty v| \le (\delta/2)|v| + c$$

for each $v \in \mathbf{R}^{2N}$. Hence, by the mean value theorem,

$$\begin{aligned} |H^*(v) - (1/2)(B_\infty v, v)| &\le \int_0^T |\nabla H^*(tv) - B_\infty(tv), v)|dt \\ &\le \int_0^1 [(\delta/2)t\,|v|^2 + c\,|v|]\, dt = (\delta/4)|v|^2 + c\,|v|. \end{aligned}$$

Consequently, we have, for $v \in Y$,

$$\begin{aligned} \varphi(v) &= \chi_T^\infty(v) + \int_0^T [H^*(\dot v(t) - (1/2)(B_\infty \dot v(t), \dot v(t))]\, dt \\ &\ge (\delta/2)|\dot v|^2_{L^2} - \int_0^T [(\delta/2)|\dot v(t)|^2 - c\,|\dot v(t)|]\, dt \\ &= (\delta/4)|\dot v|^2_{L^2} - c\,|\dot v|_{L^1} \ge (\delta/4)|\dot v|^2_{L^2} - cT\,(1/2)|\dot v|_{L^1}, \end{aligned}$$

and φ is bounded from below on Y. □

We show now that the second geometric condition of Theorem 6.2 holds for φ.

Lemma 7.4. *There exists an invariant subspace Z of $\tilde{H}^1_T$ with dimension $i(A_0, T)$ and some $r > 0$ such that $\varphi(v) < 0$ whenever $v \in Z$ and $\|v\| = r$.*

Proof. Assumption (15) and the reasoning of Remark 5 imply that

$$\nabla H^*(v) = B_0 v + o(|v|) \quad \text{as} \quad |v| \to 0 \tag{19}$$

where $B_0 = A_0^{-1}$. By (5) there exists an invariant subspace $Z = H^-$ of $\tilde{H}^1_T$ with dimension $i(A_0, T)$ and some $\delta > 0$ such that

$$\chi^0_T(v) \equiv \int_0^T (1/2)[J\dot{v}(t), v(t)) + (B_0\dot{v}(t), \dot{v}(t))]\,dt \leq -(\delta/2)|\dot{v}|^2_{L^2}$$

whenever $v \in Z$. By (19), there exists $\rho > 0$ such that

$$|\nabla H^*(v) - B_0 v| \leq (\delta/2)\,|v|$$

for $v \in \mathbf{R}^{2N}$ with $|v| \leq \rho$. Hence, by the mean value theorem, we have

$$\begin{aligned} |H^*(v) - (1/2)(B_0 v, v) &\leq \int_0^1 |(\nabla H^*(tv) - B_0(tv), v)|\,dt \\ &\leq \int_0^1 (\delta/2)t\,|v|^2 dt = (\delta/4)\,|v|^2 \end{aligned}$$

whenever $|v| \leq \rho$. Consequently, if $v \in Z$ and $0 < |v|_\infty < \rho$, we get

$$\begin{aligned} \varphi(v) &= \chi^0_T(v) + \int_0^T [H^*(\dot{v}(t)) - (1/2)(B_0\dot{v}(t), \dot{v}(t))]\,dt \\ &\leq -(\delta/2)|\dot{v}|^2_{L^2} + (\delta/4)\int_0^T |\dot{v}(t)|^2 dt = -(\delta/4)|\dot{v}|^2_{L^2}, \end{aligned}$$

and the proof is complete since Z is finite-dimensional. □

Proof of Theorem 7.2. We apply Theorem 6.2 to φ which is invariant and satisfies the (PS)-condition by Lemma 7.2. The spaces Y and Z introduced respectively in Lemmas 7.3 and 7.4 satisfy the assumption

$$i(A_\infty, T) = \operatorname{codim} Y < \dim Z = i(A_0, T)$$

and the conditions (11), (12), (13), and (14) in Chapter 6 since Fix $(S^1) = \{0\}$ and $\varphi(0) = 0$. Thus Theorem 6.2 implies the existence of at least

$$(1/2)[i(A_0, T) - i(A_\infty, T)]$$

distinct orbits $\{T(\theta)v_j \,:\, \theta \in S^1\}$ of critical points of φ outside of Fix $(S^1) = \{0\}$. By Theorem 2.3, u_j defined by

$$u_j(t) = \nabla H^*(\dot{v}_j(t))$$

is a T-periodic solution of (14). Clearly, $v_j \neq 0$ implies $u_j \neq 0$ as v_j has a mean value of zero. Finally, if u_j and u_k describe the same orbit, then $u_k = T(\theta)u_j$ for some θ, so that

$$\dot{v}_k = \nabla H(u_k) = \nabla H(t(\theta)u_j) = T(\theta)\nabla H(u_j) = T(\theta)\dot{v}_j$$

and hence

$$v_k = T(\theta)v_j$$

and v_k and v_j have a mean value of zero. Thus v_k and v_j describe the same orbit, which implies $j = k$ and completes the proof. □

Historical and Bibliographical Notes

The Morse index of a quadratic form was introduced by Morse [$\text{Mrs}_{1,2}$] in finite dimension and by Lichtenstein for an infinite dimensional situation.

The index of a linear positive definite Hamiltonian is a concept due to Ekeland [Eke_7], the present definition being taken from [Eke_8] as well as the material of Sections 7.1 and 7.2.

Multiplicity results for the periodic solutions of asymptotically linear autonomous systems of second order equations with odd nonlinearities were already obtained in 1978 by Clark [Clk_2].

In the case of Hamiltonian systems, the first multiplicity results are due to Amann–Zehnder [$\text{AmZ}_{1,2}$] who combine a Liapunov–Schmidt-type of finite-dimensional reduction [Ama_1] with the Z_2- and S_1-indices. The approach of Section 7.3 is due to Costa–Willem [CoW_2].

For other results of the Amann–Zehder type see also [Ama_2], [AmZ_3], [BMi_1], [BCF_3], and [Hof_5].

For results related to the index of Hamiltonian systems, see [Bro_1], [$\text{EkH}_{2,3}$], [$\text{ELa}_{1,2}$], and [$\text{GiM}_{4,5}$].

Exercises

1. Let $H \in C^2(\mathbf{R}^{2N}, \mathbf{R})$ be such that $\nabla H(0) = 0$ and $H''(x)$ is positive definite for all $x \in \mathbf{R}^{2N} \setminus \{0\}$ (in particular, H is strictly convex). Define the index i_u of an non-constant solution u of

$$J\dot{u}(t) + \nabla H(u(t)) = 0,$$

$$u(0) = u(T),$$

by $i_u = i(H'' \circ u, T)$. Prove that if u is T/k-periodic, then $k \leq i_u + 1$. In particular, if $i_u = 0$, then T is the minimal period of u.

2. Consider the linear Hamiltonian systems

$$J\dot{u}(t) + A_1(t)u(t) = 0, \quad J\dot{u}(t) + A_2(t)u(t) = 0,$$

with A_1 and A_2 continuous mappings from $\mathbf{R}$ into the space of positive definite matrices of order $2N$. If

$$A_1(t) \leq A_2(t)$$

for all $t \in [0, T]$, then

$$i(A_1, T) \leq (A_2, T).$$

3. Let $H \in C^1(\mathbf{R}^{2N}, \mathbf{R})$ be strictly convex and such that

$$H(0) = 0, \quad \nabla H(0) = 0,$$

and let $T > 0$. Assume that there exists $n \in \mathbf{N}$, $\gamma \in]2\pi n/T, 2\pi(n+1)/T[$ such that

$$\nabla H(u) = \gamma u + o(|u|) \quad \text{as} \quad |u| \to \infty$$

and some $k \geq 1$ and $\beta > (2\pi/T)(n+k)$ such that

$$\nabla H(u) = \beta u + o(|u|) \quad \text{as} \quad |u| \to 0.$$

Then the problem

$$J\dot{u} + \nabla H(u) = 0,$$
$$u(0) = u(T)$$

has at least Nk S^1-orbits of nontrivial solutions. Compare this result with the Exercise 7 in Chapter 6.

Hint. Use Theorem 7.2 and the results of Section 7.2.

8

Morse Theory

Introduction

Morse theory's object is the relation between the topological type of critical points of a function φ and the topological structure of the manifold on which the function is defined.

The topological type of a critical point u is described by the *critical groups* of Morse $C_n(\varphi, u)$ (see Section 8.2) which exhibit the following properties:

a) in the nondegenerate case, the critical groups are computable by linearization (see Section 8.6);

b) the critical groups are stable under small perturbations of the function φ (see Sections 8.9 and 8.10).

If $\varphi'(u) = 0$ and $\varphi''(u)$ is invertible, then

$$\dim C_n(\varphi, u) = \delta_{n,k}$$

where k is the *Morse index* of $\varphi''(u)$. Recall that this Morse index is an integer measuring the maximal dimension of the spaces on which $\varphi''(u)$ is negative definite (see Section 8.6). We also present some results in the degenerate case when $\varphi''(u)$ is a Fredholm operator.

The topological structure of the manifold M is described by its *Betti numbers* B_n. Intuitively, B_n is the maximal number of n-dimensional surfaces without boundaries on M which are not the boundaries of a $(n+1)$ dimensional surface on M (see Section 8.1). For example B_0 is the number of path connected components of M. In the case of a sphere, every closed curve is a boundary and $B_1 = 0$. On the other hand, $B_1 = 2$ for the two-dimensional torus.

To illustrate Morse theory, let us consider the classical situation of the function $\varphi(x, y, z) = z$ defined on a two-dimensional torus $M \subset \mathbf{R}^3$ tangent to the plane Oxy (see Figure 8.1). The function φ has a critical point u_1 with Morse index zero, two critical points u_2 and u_3 with Morse index one, and one critical point u_4 with Morse index 2. If M_k denotes the number of critical points of φ with Morse index k, we have $B_k = M_k$ $(k = 0, 1, 2)$.

In general, for an N-dimensional compact manifold, the following relation is valid

$$\sum_{k=0}^{N} M_k t^k = \sum_{k=0}^{N} B_k t^k + (1+t)Q(t),$$

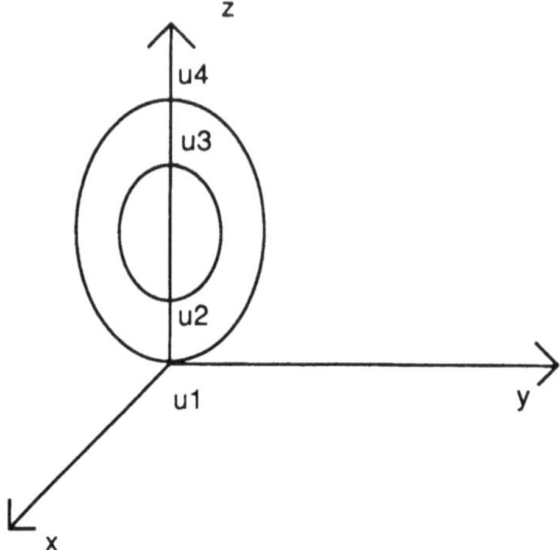

FIGURE 8.1.

where $Q(t)$ is a polynomial with nonnegative integer coefficients (see Section 8.5). In particular, for $t = -1$, we obtain the *Poincaré–Hopf formula* for a gradient vector field

$$\sum_{k=0}^{N}(-1)^k M_k = \sum_{k=0}^{N}(-1)^k B_k.$$

It is important to notice that a single degenerate critical point can contribute to different numbers M_k.

The proof of these results makes use of a deformation technique along the paths of steepest descent along $\nabla\varphi$. The corresponding tools and results are developed in Sections 8.3 and 8.4.

A first application of Morse theory deals with the bifurcation of solutions of equations depending upon a parameter. Loosely speaking, a change of critical groups of the trivial solution implies bifurcation. This result, which corresponds, in the context of Morse theory, to Krasnosel'skii's bifurcation theorem in degree theory, is given in Section 8.9.

8.1 Relative Homology

Let B be a subspace of a topological space A. For every integer n, we denote by $H_n(A, B)$ the nth *singular homology group* of the pair (A, B) over a field F. For $n \leq -1$, $H_n(A, B) = \{0\}$. For any map $f : (A, B) \to (A', B')$ (i.e., any continuous map $f : A \to A'$ such that $f(B) \subset B'$) there is a homomorphism

$$f_{n*} : H_n(A, B) \to H_n(A', B')$$

called the *induced homomorphism*. Let $C \subset B$; there is a homomorphism

$$\partial_n : H_n(A,B) \to H_{n-1}(B,C)$$

called the *boundary homomorphism*. We shall frequently write f_* and ∂, omitting the subindex. The *Eilenberg–Steenrod axioms* are satisfied:

(a) $id_* = id$.

(b) $(g \circ f)_* = g_* \circ f_*$.

(c) The following diagram commutes:

$$\begin{array}{ccc} H_n(A,B) & \xrightarrow{f_*} & H_n(A',B') \\ \partial \downarrow & & \downarrow \partial \\ H_{n-1}(B,C) & \xrightarrow{(f|_B)_*} & H_{n-1}(B',C'). \end{array}$$

(d) (*Exactness*). Let

$$i : (B,C) \to (A,C)$$

$$j : (A,C) \to (A,B)$$

be the inclusion maps. The homology sequence

$$\ldots \to H_{n+1}(A,B) \xrightarrow{\partial} H_n(B,C) \xrightarrow{i_*} H_n(A,C) \xrightarrow{j_*} H_n(A,B) \to \ldots$$

is exact (i.e., the image of any homomorphism is equal to the kernel of the next one).

(e) (*Homotopy invariance*). If f and g are homotopic (i.e., $f = F(0,\cdot)$, $g = F(1,\cdot)$ for some continuous mapping $F : [0,1] \times A \to A'$ such that $F([0,1] \times B) \subset B'$), then $f_* = g_*$.

(f) (*Excision*). Assume that C is an open subset of A such that the closure of C is contained in the interior of B. Let $i : (A \setminus C, B \setminus C) \to (A,B)$ be the inclusion map. Then i_* is an isomorphism.

(g) If u is a point, then $H_n(\{u\},\phi) = \delta_{n,0}F$, where $\delta_{n,0}$ is the Kronecker symbol.

We shall also need the following results.

(h) (*Decomposition theorem*). If $(A,B) = \cup_{i=1}^{j}(A_i,B_i)$, where the A_i are closed and disjoint, then

$$H_n(A,B) = \oplus_{i=1}^{j} H_n(A_i,B_i).$$

(i) (*Mayer–Vietoris sequence*). Assume that X_1, X_2 are open in $X = X_1 \cup X_2$ and that $Y_1 \subset X_1$, $Y_2 \subset X_2$ are open in $Y = Y_1 \cup Y_2$. If $X_1 \cap X_2 \neq \phi$, there is an exact sequence

$$\ldots \to H_n(X_1,Y_1) \oplus H_n(X_2,Y_2) \xrightarrow{\Psi} H_n(X,Y) \xrightarrow{\Delta} H_{n-1}(X_1 \cap X_2, Y_1 \cap Y_2)$$

$$\xrightarrow{\Phi} H_{n-1}(X_1,Y_1) \oplus H_{n-1}(X_2,Y_2) \to \ldots$$

called the Mayer–Vietoris sequence of $\{(X_1, Y_1), (X_2, Y_2)\}$.

Since F is a field, the homology groups are vector spaces and the *Betti numbers* $B_n(A, B)$ of the pair (A, B) are defined by

$$B_n(A, B) = \dim H_n(A, B).$$

Let $R_n(A, B, C)$ be the rank of ∂_n. By exactness, we obtain

$$\begin{aligned} B_n(A, B) &= \dim R(j_{n*}) + R_n(A, B, C) \\ B_n(B, C) &= R_{n+1}(A, B, C) + \dim R(i_{n*}) \\ B_n(A, C) &= \dim R(i_{n*}) + \dim R(j_{n*}). \end{aligned} \tag{1}$$

A pair (A, B) is *admissible* if $B_n(A, B)$ is finite for each n and zero for all sufficiently large n. The *Poincaré polynomial* of an admissible pair (A, B) is defined by

$$P(t, A, B) = \sum_{n=0}^{\infty} B_n(A, B)\, t^n.$$

Let us also define

$$Q(t, A, B, C) = \sum_{n=0}^{\infty} R_{n+1}(A, B, C)\, t^n.$$

If (A, B) and (B, C) are admissible, then (1) implies that (A, C) is also admissible and that

$$B_n(A, B) + B_n(B, C) = B_n(A, C) + R_n(A, B, C) + R_{n+1}(A, B, C).$$

Multiplying this equation by t^n and adding over n, we get

$$P(t, A, B) + P(t, B, C) = P(t, A, C) + (1 + t)Q(t, A, B, C), \tag{2}$$

where we have used the fact that $R_0(A, B, C) = 0$.

Assume that $A_1 \supset A_2 \supset \ldots \supset A_j$ are such that (A_i, A_{i+1}) is admissible for $i = 1, \ldots, j-1$. Applying equation (2) to (A_1, A_i, A_{i+1}), we obtain

$$P(t, A_1, A_i) + P(t, A_i, A_{i+1}) = P(t, A_1, A_{i+1}) + (1 + t)Q(t, A_1, A_i, A_{i+1}).$$

Adding those equations, we find

$$\sum_{i=1}^{j-1} P(t, A_i, A_{i+1}) = P(t, A_1, A_j) + (1 + t)Q(t), \tag{3}$$

where $Q(t)$ is a polynomial with nonnegative integer coefficients (by exactness, $P(t, A_1, A_1) = 0$).

A subset A' of A is a *strong deformation retract of* A if there exists $h \in C([0,1] \times A, A)$ such that

$$h(t,u) = u \quad \text{whenever} \quad u \in A' \quad \text{and} \quad t \in [0,1],$$

$$h(0,u) = u \quad \text{and} \quad h(1,u) \in A' \quad \text{whenever} \quad u \in A.$$

Assume that $A \supset A' \supset C$ with A' a strong deformation retract of A. Define $r : (A, C) \to (A', C')$ by

$$r(u) = h(1,u)$$

and let

$$i : (A', C) \to (A, C)$$

be the inclusion map. By homotopy invariance, we obtain

$$i_* \circ r_* = (i \circ r)_* = id_* = id,$$

and, on the other hand, by definition of r, we have

$$r_* \circ i_* = (r \circ i)_* = id_* = id,$$

so that r_* is an isomorphism between $H_n(A, C)$ and $H_n(A', C)$. In particular, if $C = A'$, we see that

$$H_n(A, A') \approx H_n(A', A') \approx \{0\}.$$

Now, if $A \supset B \supset B'$, with B' a strong deformation retract of B, we have, by the above result and exactness,

$$\{0\} = H_n(B, B') \xrightarrow{i_*} H_n(A, B') \xrightarrow{j_*} H_n(A, B) \xrightarrow{\partial} H_{n-1}(B, B') \approx \{0\},$$

and hence

$$\{0\} = \operatorname{Im} i_* = \ker j_*, \ \operatorname{Im} j_* = \ker \partial = H_n(A, B).$$

Thus $j_* : H_n(A, B') \to H_n(A, B)$ is one to one and

$$H_n(A, B') \approx H_n(A, B).$$

Let A be a subset of $\mathbf{R}^p$ containing 0 and let B^k be the k-ball. Then, for $k \geq 1$,

$$H_n(A \times B^k, (A \times B^k) \setminus \{0\}) \approx H_{n-k}(A, A \setminus \{0\}). \tag{4}$$

Proof. If $k \geq 2$, we obtain, after the identification $B^k \approx [-1,1]^k$,

$$(A \times B^k, (A \times B^k) \setminus \{0\}) = (A \times B^{k-1} \times [-1,1], (A \times B^{k-1} \times [-1,1]) \setminus \{0\}).$$

Thus the result follows by induction from the case $k = 1$. Let us assume that $k = 1$. The suspension ΣA of A is obtained from $A \times [-1,1]$ by identifying the pair of sets $(A\times\{-1\}, A\times\{1\})$ to a pair of points (w_-, w_+). By excision,

$$H_n(A \times [-1,1], (A \times [-1,1]) \setminus \{0\}) \approx H_n(\Sigma A, \Sigma A \setminus \{0\}).$$

Let us define the sets

$$X_+ = \Sigma A \setminus \{w_-\}, \quad X_- = \Sigma A \setminus \{w_+\},$$

$$Y_+ = X_+ \setminus (\{0\}\times\,]-1,0]), \quad Y_- = X_- \setminus (\{0\} \times [0,1[),$$

so that

$$X_+ \cup X_- = \Sigma A, \quad X_+ \cap X_- = A\times\,]-1,1[,$$
$$Y_+ \cup Y_- = \Sigma A \setminus \{0\}, \quad Y_+ \cap Y_- = (A \setminus \{0\})\times\,]-1,1[.$$

Since X_+ and Y_+ (resp. X_- and Y_-) are contractible to w_+ (resp. w_-) we have, by homotopy invariance and exactness

$$H_n(X_\pm, Y_\pm) \approx H_n(\omega_\pm, \omega_\pm) \approx \{0\}.$$

The exactness of the Mayer–Vietoris sequence of $\{(X_+, Y_+), (X_-, Y_-)\}$ implies that

$$H_n(X_+ \cup X_-, Y_+ \cup Y_-) \approx H_{n-1}(X_+ \cap X_-, Y_+ \cap Y_-),$$

i.e.,

$$\begin{aligned} H_n(\Sigma A, \Sigma A \setminus \{0\}) &\approx H_{n-1}(A\times\,]-1,1[, (A \setminus \{0\})\times\,]-1,1[) \\ &\approx H_{n-1}(A, A \setminus \{0\}), \end{aligned}$$

and the proof is complete. □

In particular, if $A = \{0\}$, we obtain

$$\begin{aligned} H_n(B_k, B_k \setminus \{0\}) &\approx H_n(\{0\} \times B^k, (\{0\} \times B^k) \setminus \{0\}) \\ &\approx H_{n-k}(\{0\}, \phi) = \delta_{n-k,0} F = \delta_{n,k} F. \end{aligned}$$

Let B^∞ (resp. S^∞) be the unit ball (resp. the unit sphere) in an infinite-dimensional normed space. Then, since S^∞ is a strong deformation retract of B^∞, we have

$$H_n(B^\infty, B^\infty \setminus \{0\}) \approx H_n(B^\infty, S^\infty) \approx H_n(S^\infty, S^\infty) \approx \{0\}.$$

8.2 Manifolds

Let M be a set and V a Banach space. A *chart* is a bijection $x : D(x) \subset M \to R(x) \subset V$ such that $R(x)$ is open. An *atlas of class* C^k $(k \geq 0)$ on M is a set $\mathcal{A}$ of charts such that

(AT1) $\cup_{x \in \mathcal{A}} D(x) = M$.

(AT2) $x(D(x) \cap D(y))$ is an open subset of V whenever $x \in \mathcal{A}$ and $y \in \mathcal{A}$.

(AT3) the mapping

$$y \circ x^{-1} : x(D(x) \cap D(y)) \to y(D(x) \cap D(y))$$

is a C^k-diffeomorphism for each $x \in \mathcal{A}$ and $y \in \mathcal{A}$.

A *manifold of class* C^k *modeled on* V (or briefly a C^k*-manifold*) is a pair $(M, \mathcal{A})$ where M is a set and $\mathcal{A}$ is an atlas of class C^k on M. We will use the same symbol M to denote the C^k-manifold $(M, \mathcal{A})$ and the underlying set M. The topology of the manifold M is, by definition, the unique topology on M such that the domain of each chart is open and each chart is an homeomorphism.

Example 8.1. The singleton $\{Id : V \to V\}$ is an atlas of class C^∞ on the Banach space V.

Example 8.2. Let $\mathcal{A}$ be an atlas of class C^k on M and let N be an open subset of the manifold M. The restriction to N of the charts in $\mathcal{A}$ is an atlas of class C^k on N.

Example 8.3. Let G be a discrete subgroup of V and $\pi : V \to V/G$ the canonical projection. Then

$$\{\pi^{-1} : \pi(U) \to U,\ U \text{ is open and } \pi : U \to V/G \text{ is injective}\}$$

is an atlas of class C^∞ in V/G.

An important example of a manifold is given by the tangent bundle of a C^1-manifold. If x and y are two charts on M whose domains contain a point u, and if $v \in V$ and $w \in V$, let us introduce the equivalence relation (verify it!)

$$(u, x, v) \sim (u, y, w) \Leftrightarrow w = (y \circ x^{-1})'(x(u))v$$

and define the equivalent class

$$[u, x, v] = \{(u, y, w) : u \in D(y) \text{ and } (u, y, w) \sim (u, x, v)\}.$$

The *tangent space of* M at u is the set $T_u M$ of the equivalence classes $[u, x, v]$ such that $u \in D(x)$ and $v \in V$. A vector space structure is defined on $T_u M$ by the formulas

$$[u, x, v] + [u, x, w] = [u, x, v + w],$$

$$s[u, x, v] = [u, x, sv].$$

The chain rule shows that this definition is independent of the chart x.

The *tangent bundle* TM *of* M is defined by

$$TM = \bigcup_{u \in M} T_u M,$$

and the projection $\pi : TM \to M$ is defined by

$$\pi : [u, x, v] \to u.$$

Let M and N be C^k-manifolds modeled on Banach spaces V and W, respectively. A mapping $f : M \to N$ is *locally Lipschitzian* (resp. of *class* C^k) if $y \circ f \circ x^{-1}$ is locally Lipschitzian (resp. of class C^k) for every chart x on M and every chart y on N. If $f : M \to N$ is of class C^1, the *differential* of f is the mapping $df : TM \to TN$ defined by

$$df([u, x, v]) = [f(u), y, (y \circ f \circ x^{-1})'(x(u))v],$$

where x is a chart at u and y a chart at $f(u)$. One can check that this definition is independent of x and y and that the following diagram commutes

$$\begin{array}{ccc} TM & \xrightarrow{df} & TN \\ \pi \downarrow & & \downarrow \pi \\ M & \xrightarrow{f} & N. \end{array}$$

If N is a Banach space W, then $TW \approx W^2$ and $df : TM \to W^2$ is defined by (taking $y = Id$ on W)

$$df([u, x, v]) = (f(u), (f \circ x^{-1})'(x(u))v).$$

In particular, if x is a chart on M,

$$dx : \pi^{-1}(D(x)) \to V^2$$

is a chart and $\{dx : x \in \mathcal{A}\}$ is an atlas of class C^{k-1} on TM, such that

$$dx([u, x, v]) = (x(u), v).$$

A *critical point* of $\varphi \in C^1(M, \mathbf{R})$ is a point $u \in M$ such that $d\varphi|_{T_u M} = 0$. The change in topology near an isolated critical point is described by the critical groups. We assume that the C^1-manifold M is regular. (Recall that a topological space is regular if every neighborhood of a point contains a closed neighborhood.) Let u be an isolated critical point of $\varphi \in C^1(M, \mathbf{R})$. The *critical groups* (over a field F) of u are defined by

$$C_n(\varphi, u) = H_n(\varphi^c \cap U, \varphi^c \cap U \setminus \{u\}), \quad n = 0, 1, \ldots,$$

where $c = \varphi(u)$ and U is a closed neighborhood of u. By excision, the critical groups are independent of U.

Let us complete the critical groups in a trivial but important case, namely when u is an isolated local minimum point. Then there exists a closed neighborhood U of u such that

$$\varphi(v) > c = \varphi(u)$$

whenever $v \in U \setminus \{u\}$. We obtain therefore

$$C_n(\varphi, u) = H_n(\{u\}, \phi) = \delta_{n,0}F, \quad n = 0, 1, \ldots.$$

8.3 Vector Fields

In this section, M will denote a Hausdorff manifold of class C^2 modeled on a Banach space V. A *vector field* on M is a mapping $f : M \to TM$ such that $\pi \circ f = Id$. If $\sigma :]a,b[\to M$ is a C^1-mapping, then we define $\dot{\sigma}(t)$ for $t \in]a,b[$ by

$$\dot{\sigma}(t) = [\sigma(t), x, (x \circ \sigma)'(t)(1)] = d\sigma[t, id, 1],$$

where x is a chart at $\sigma(t)$.

Proposition 8.1. *If f is a locally Lipschitzian vector field on M, then, for every $u \in M$, the Cauchy problem*

$$\begin{cases} \dot{\sigma}(t) = f(\sigma(t)) \\ \sigma(0) = u \end{cases} \tag{5}$$

has a solution defined on some open interval containing 0. *Moreover, if $\sigma_1 : I_1 \to M$ and $\sigma_2 : I_2 \to M$ is a pair of solutions of* (5) *defined on open intervals I_j ($j = 1, 2$), then $\sigma_1 = \sigma_2$ on $I_1 \cap I_2$.*

Proof. Let x be a chart at u; near u, the Cauchy problem is equivalent to

$$dx(\dot{\sigma}(t)) = dx(f(\sigma(t)))$$

$$x(\sigma(0)) = x(u)$$

or

$$\begin{aligned} \dot{\eta}(t) &= (P \circ dx \circ f \circ x^{-1})(\eta(t)) \\ \eta(0) &= x(u) \end{aligned} \tag{6}$$

where $\eta = x \circ \sigma$ and $P : V \times V \to V$ is defined by $P(v, w) = w$. Since f is locally Lipschitzian, the same is true for $dx \circ f \circ x^{-1}$. The local theory of differential equations in a Banach space implies the existence of a solution $\eta :]-\epsilon, \epsilon[\to V$ of (6), and each solution of (6) defined on $]-\epsilon, \epsilon[$ is equal to η.

Let $I = \{t \in I_1 \cap I_2 : \sigma_1(t) = \sigma_2(t)\}$; this set contains 0 and is closed in $I_1 \cap I_2$ since M is Hausdorff. Using the local uniqueness result, it is easy to verify that I is open in $I_1 \cap I_2$, so that $I = I_1 \cap I_2$, and the proof is complete. □

Proposition 8.1 implies that the union of the graphs of all solutions of (5) defined on open intervals is a solution of (5) defined on an interval $]\omega_-(u), \omega_+(u)[$ with

$$-\infty \leq \omega_-(u) < 0 < \omega_+(u) \leq +\infty.$$

This solution is called the *maximal solution* of (5) and is denoted by $\sigma(\cdot, u)$.

As in the Banach space theory, the set

$$\mathcal{D} = \{(t, u) \, : \, \omega_-(u) < t < \omega_+(u)\}$$

is open in $\mathbf{R} \times M$ and the *flow*

$$\sigma : \mathcal{D} \to M, \quad (t, u) \to \sigma(t, u)$$

is continuous.

8.4 Riemannian Manifolds

Let M be a manifold of class C^k ($k \geq 1$) modeled on a Hilbert space V. A *Riemannian metric of class* C^{k-1} on M is a mapping which associates to each pair (u, x), with $u \in M$ and x a chart at u, a positive definite invertible symmetric operator $M_x(u) \, : \, V \to V$ such that the following properties hold.

(RM1) The mapping M_x

$$D(x) \to \mathcal{L}(V) \, : \, u \to M_x(u)$$

is of class C^{k-1} for each chart x.

(RM2) If x and y are two charts at $u \in M$, then

$$[(y \circ x^{-1})'(x(u))]^* M_y(u)[(y \circ x^{-1})'(x(u))] = M_x(u).$$

It follows from (RM2) that the relation

$$([u, x, v], [u, x, w]) = (M_x(u)v, w)$$

defines an inner product on $T_u M$, and the corresponding norm is given by

$$\|[u, x, v]\| = (M_x(u)v, v)^{1/2}.$$

A *Riemannian manifold of class* C^k is a regular connected manifold of class C^k modeled on a Hilbert space and equipped with a Riemannian metric of class C^{k-1}.

Let M be a Riemannian manifold of class C^1. A *piecewise C^1 path from $u \in M$ to $v \in M$* is a piecewise C^1 mapping $\sigma : [a,b] \to M$ such that $\sigma(a) = u$ and $\sigma(b) = v$. We shall denote by C_v^u the set of all piecewise C^1 paths from u to v and define the *length* of $\sigma \in C_v^u$ by

$$L(\sigma) = \int_a^b |\dot{\sigma}(t)|\, dt.$$

Proposition 8.2. *For each $u \in M$ and $v \in M$, the set C_v^u is non-empty.*

Proof. For each $u \in M$, define $A = \{v \in M : C_v^u \neq \phi\}$. Since M is connected and $A \neq \phi$, it suffices to prove that A is open and closed.

If $v \in A$, there is a path $\sigma : [a,b] \to M$ in C_v^u. Let x be a chart at $v = \sigma(b)$. There is a $r > 0$ such that $B = x^{-1}(B(x(v), r))$ is an open subset of $D(x)$, and thus of M. For $w \in B$, the path $\tilde{\sigma} : [a, b+1] \to M$ defined by

$$\tilde{\sigma}(t) = \sigma(t), \quad a \leq t \leq b$$

$$\tilde{\sigma}(t) = x^{-1}((1 - (t-b))x(v) + (t-b)x(w)), \quad b \leq t \leq b+1$$

is in C_u^w. Thus $B \subset A$ and A is open.

Now let v be in the closure of A and x be a chart at v. Define B as before; there will exist $w \in A \cap B$ and then a path $\sigma : [a,b] \to M$ in C_u^w. The path $\tilde{\sigma} : [a, b+1]$ defined by

$$\tilde{\sigma}(t) = \sigma(t), \quad a \leq t \leq b$$

$$\tilde{\sigma}(t) = x^{-1}((1 - (t-b))x(w) + (t-b)x(v)), \quad b \leq t \leq b+1$$

is in C_u^v. Thus $v \in A$ and A is closed. □

Proposition 8.2 justifies the following definition of the *geodesic distance* d on M

$$d(u,v) = \inf\{L(\sigma) : \sigma \in C_u^v\}.$$

Proposition 8.3. *The geodesic distance d is a distance on M whose topology is compatible with the manifold topology.*

Proof. Clearly d is symmetric and verifies the triangle inequality. Let x be a chart at $u \in M$. By definition, there exists $0 < \alpha \leq \beta$ such that

$$\alpha |h|^2 \leq (M_x(u)h, h) \leq \beta |h|^2, \quad h \in V.$$

By continuity, there exists $r > 0$ such that $B = x^{-1}(B(x(u), r))$ is an open subset of $D(x)$, and hence of M and such that

$$(\alpha/2)|h|^2 \leq (M_x(v)h, h) \leq 2\beta |h|^2, \quad v \in B,\ h \in V.$$

For every piecewise C^1 path $\sigma : [a,b] \to B$, we have

$$L(\sigma) = \int_a^b (M_x(\sigma(t))(x \circ \sigma)'(t), (x \circ \sigma)'(t))^{1/2}\, dt$$

$$\geq (\alpha/2)^{1/2} \int_a^b |(x \circ \sigma)'(t)|\, dt \geq (\alpha/2)^{1/2} \left| \int_a^b (x \circ \sigma)'(t)\, dt \right| \tag{7}$$
$$= (\alpha/2)^{1/2} |(x \circ \sigma)(b) - (x \circ \sigma)(a)|.$$

For every $v \in B$, the path $\tilde{\sigma}$ defined by

$$\tilde{\sigma}(t) = x^{-1}((1-t)x(u) + tx(v)), \quad 0 \leq t \leq 1,$$

is such that

$$L(\tilde{\sigma}) \leq (2\beta)^{1/2} \int_0^1 |x(v) - x(u)|\, dt = (2\beta)^{1/2} |x(v) - x(u)|. \tag{8}$$

Let A be a neighborhood of u in M. Since M is regular, there exists a closed neighborhood C of u such that $C \subset A \cap B$. Define $\delta > 0$ by

$$\delta = \inf\{|x(w) - x(u)| \,:\, w \in \partial C\}. \tag{9}$$

Let $v \in M$. If $\sigma : [a,b] \to M$ belongs to C_u^v, then either $\sigma([a,b]) \subset C$ or there is a $c \in]a,b[$ such that $\sigma([a,c]) \subset C$ and $\sigma(c) \in \partial C$. In the first case, it follows from (7) that

$$L(\sigma) \geq (\alpha/2)^{1/2} |x(v) - x(u)|.$$

In the second case, (7) and (9) imply that

$$L(\sigma) \geq (\alpha/2)^{1/2} |x(\sigma(c)) - x(u)| \geq (\alpha/2)^{1/2} \delta.$$

In particular, $d(u,v) > 0$ for $v \neq u$ and d is a distance.

On the other hand, if $v \in M \setminus C$, then $L(\sigma) \geq (\alpha/2)^{1/2}\delta$ so that

$$\{v \in M \,:\, d(u,v) < (\alpha/2)^{1/2}\delta\} \subset C \subset A.$$

A being arbitrary, this implies that the topology induced by d is stronger than the manifold topology. Now (8) implies that

$$x^{-1}(B(x(u), (2\beta)^{-1/2}R)) \subset \{v \in M \,:\, d(u,v) \leq R\}$$

whenever $R \in]0, (2\beta)^{1/2} r[$, showing that the topology induced by d is weaker than the manifold topology. □

A subset of a Riemannian manifold of class C^1 will be said to be *complete* if it is complete for the geodesic distance.

Let M be a Riemannian manifold of class C^2 and let $\varphi \in C^{2-0}(M, \mathbf{R})$. The *gradient* of φ is the vector field defined on M by

$$\nabla\varphi(u) = [u, x, M_x^{-1}(u)\, J(\varphi \circ x^{-1})'(x(u))]$$

where $J : V^* \to V$ is the inverse duality mapping.

If $\varphi \in C^{2-0}(M,\mathbf{R})$, the Cauchy problem

$$\begin{cases} \dot{\sigma}(t) = -\nabla\varphi(\sigma(t)) \\ \sigma(0) = u \end{cases} \tag{10}$$

has a unique maximal solution $\sigma(.) = \sigma(.,u)$. Since

$$\begin{aligned}
\frac{d}{dt}(\varphi\circ\sigma)(t) &= \frac{d}{dt}(\varphi\circ x^{-1}\circ x\circ\sigma)(t) \\
&= \langle(\varphi\circ x^{-1})'(x(\sigma(t))), \frac{d}{dt}(x\circ\sigma)(t)\rangle \\
&= (M_x(\sigma(t))M_x^{-1}(\sigma(t))J(\varphi\circ x^{-1})'(x(\sigma(t))), \frac{d}{dt}(x\circ\sigma)(t)) \\
&= -(\nabla\varphi(\sigma(t)), \nabla\varphi(\sigma(t))) \\
&= -|\nabla\varphi(\sigma(t))|^2,
\end{aligned}$$

where x is a chart at $\sigma(t)$, either $\varphi(\sigma(t)) = \varphi(u)$ for all $t \geq 0$ or $\varphi\circ\sigma$ is decreasing. Moreover we have

$$\varphi(\sigma(t)) = \varphi(\sigma(s)) - \int_s^t |\nabla\varphi(\sigma(r))|^2 dr, \quad \omega_-(u) \leq s \leq t \leq \omega_+(u). \tag{11}$$

Proposition 8.4. *Under the above assumptions, if $\omega_+(u)$ is finite and the set $\{\sigma(t) : t \in [0,\omega_+(u)[\}$ is contained in a complete subset of M, then $\varphi(\sigma(t)) \to -\infty$ when $t \to \omega_+(u)$.*

Proof. For $0 \leq s \leq t < \omega_+(u)$, the definition of d and (11) imply that

$$\begin{aligned}
d(\sigma(t),\sigma(s)) \leq \int_s^t |\nabla\varphi(\sigma(r))|\,dr &\leq (t-s)^{1/2}\left(\int_s^t |\nabla\varphi(\sigma(r))|^2 dr\right)^{1/2} \\
&= (t-s)^{1/2}(\varphi(\sigma(s)) - \varphi(\sigma(t)))^{1/2}.
\end{aligned} \tag{12}$$

Since $\omega_+(u) < \infty$, $\sigma(t)$ does not converge as $t \to \omega_+(u)$, and hence does not verify the corresponding Cauchy condition. Since $\varphi\circ\sigma$ is non-increasing, (12) implies that $\varphi(\sigma(t)) \to -\infty$ as $t \to \omega_+(u)$. □

8.5 Morse Inequalities

Let us consider the following framework:

i) M is a Riemannian manifold of class C^2 and $\varphi \in C^{2-0}(M,\mathbf{R})$;

ii) $X \subset M$ is positively invariant for the flow σ defined by (10) (i.e., $\sigma(t,u) \in X$ whenever $u \in X$ and $t \in]0,\omega_+(u)[$);

iii) $a < b$ are real numbers such that the critical points of φ in $\varphi^{-1}([a,b]) \cap X$ are isolated and contained in the interior of $\varphi^{-1}([a,b]) \cap X$;

iv) $\varphi^{-1}([a,b]) \cap X$ is complete;

v) the Palais–Smale condition over $\varphi^{-1}([a,b]) \cap X$ is satisfied, i.e., every sequence (u_j) in $\varphi^{-1}([a,b]) \cap X$ such that $|\nabla\varphi(u_j)| \to 0$ contains a convergent subsequence.

More generally, we shall say that φ satisfies the *Palais–Smale condition over a closed subset* S of M if every sequence $(u_j) \subset S$ such that $(\varphi(u_j))$ is bounded and $|\nabla\varphi(u_j)| \to 0$ contains a convergent subsequence.

Lemma 8.1. *Let M be a Riemannian manifold of class C^2 and let v be an isolated critical point of $\varphi \in C^{2-0}(M, \mathbf{R})$. If the Palais–Smale condition is satisfied over a closed neighborhood A of v, then there exists $\epsilon > 0$ and a neighborhood B of v such that, if $u \in B$, either $\sigma(t,u)$ stays in A for $0 < t < \omega_+(u)$, or $\sigma(t,u)$ stays in A until $\varphi(\sigma(t,u))$ becomes less than $\varphi(v) - \epsilon$.*

Proof. Let $\rho > 0$ be such that $B[v,\rho] \subset A$, φ is bounded on $B[v,\rho]$, and $C = \{u \in M \,:\, \rho/2 \le d(u,v) \le \rho\}$ is free of critical points. The Palais–Smale condition implies that

$$\delta = \inf\{|\nabla\varphi(u)| \,:\, u \in C\} > 0.$$

Let us define $B = B[v,\rho/2] \cap \varphi^{c+\delta\rho/4}$ where $c = \varphi(v)$. If $u \in B$ is such that $\sigma(t,u)$ does not stay in A for all $0 < t < \omega_+(u)$, then there exists $0 \le t_1 < t_2 < \omega_+(u)$ such that $\sigma(t,u) \in C$ for $t_1 \le t \le t_2$, $d(\sigma(t_1,u),v) = \rho/2$ and $d(\sigma(t_2,u),v) = \rho$. It follows from (11) that

$$\begin{aligned}
\varphi(\sigma(t_2,u)) &\le \varphi(\sigma(t_1,u)) - \delta \int_{t_1}^{t_2} |\nabla\varphi(\sigma(r,u))|\, dr \\
&\le \varphi(u) - \delta \int_{t_1}^{t_2} |\dot\sigma(r,u)|\, dr \\
&\le \varphi(u) - \delta d(\sigma(t_1,u),\sigma(t_2,u)) \\
&\le c + \delta\rho/4 - \delta(d(\sigma(t_2,u),v) - d(\sigma(t_1,u),v)) \\
&= c + \delta\rho/4 - \delta\rho/2 \\
&= \varphi(v) - \delta\rho/4,
\end{aligned}$$

and the proof is complete with $\epsilon = \delta\rho/4$. □

Lemma 8.2. *If Assumptions* (A) *hold, then, for every $u \in \varphi^{-1}([a,b]) \cap X$, either there is a (unique) $t \ge 0$ such that $\varphi(\sigma(t,u)) = a$ or $\omega_+(u) = +\infty$ and there is a critical point v of φ in $\varphi^{-1}([a,b]) \cap X$ such that $\sigma(t,u) \to v$ when $t \to +\infty$.*

Proof. If $\varphi(\sigma(t,u)) > a$ for all $t \in]0,\omega_+(u)[$, Proposition 8.4 implies that $\omega_+(u) = +\infty$, and hence $\varphi(\sigma(t,u)) \to c \ge a$ when $t \to +\infty$. By (11),

$$\int_0^\infty |\nabla\varphi(\sigma(r,u))|^2 dr < \infty.$$

Consequently, $\liminf_{t\to+\infty} |\nabla\varphi(\sigma(t,u))|^2 = 0$ and the (PS) condition implies the existence of a sequence (t_j) tending to $+\infty$ and of a critical point v such that $\sigma(t_j,u) \to v$ as $j \to \infty$. In particular, $v \in X$ and $c = \varphi(v)$. It follows then from Lemma 8.1 that $\sigma(t,u) \to v$ as $t \to +\infty$. □

Let us define, for $c \in [a,b]$,

$$X^c = \{u \in X : \varphi(u) \le c\}$$

$$K_c = \{u \in X : \varphi(u) = c, d\varphi(u) = 0\}.$$

Lemma 8.3. *Under assumptions* (A), *let* $a \le \alpha < \beta \le b$ *be such that* $\varphi^{-1}(]\alpha,\beta[) \cap X$ *is free of critical points. Then* X^α *is a strong deformation retract of* $X^\beta \setminus K_\beta$. *Moreover,* φ *is non-increasing during the deformation.*

Proof. By Lemma 8.2, if $u \in X^\beta \setminus K_\beta$ and $\varphi(u) > \alpha$, either there is a unique $t(u)$ such that $\varphi(\sigma(t(u),u)) = \alpha$ or $\varphi(\sigma(t,u)) \to \alpha$ as $t \to +\infty$. If $\psi(t,u) = \varphi(\sigma(t,u))$, then $D_t\psi(t(u),u) = -|\nabla(\sigma(t(u),u)|^2 \neq 0$, and $t(u)$ is continuous by the implicit function theorem. Define the function ρ by

$$\begin{aligned} \rho(t,u) &= \sigma(t,u) && \text{if } 0 \le t \le t(u) \\ &= \sigma(t(u),u) && \text{if } t(u) < t < \infty \end{aligned}$$

in the first case and by

$$\rho(t,u) = \sigma(t,u), \quad 0 \le t < +\infty$$

in the second case. Moreover define ρ by

$$\rho(t,u) = u, \quad 0 \le t < +\infty$$

whenever $u \in X^\alpha$. The continuity of the flow σ implies the continuity of ρ. Now define the deformation on $[0,1] \times (X^\beta \setminus K_\beta)$ by

$$\eta(t,u) = \rho\left(\frac{t}{1-t}, u\right), \quad 0 \le t < 1$$

$$\eta(1,u) = \lim_{t\to\infty} \rho(t,u).$$

The continuity of η follows from Lemma 8.1 and from the continuity of ρ. By construction, $\varphi(\eta(.,u))$ is non-increasing, and the proof is complete. □

It is easy to verify that, under assumptions (A), $\varphi^{-1}([a,b]) \cap X$ contains at most a finite number of critical points $u_1, \ldots, u_j$. The *Morse numbers* of the pair (X^b, X^a) are defined by

$$M_n(X^b, X^a) = \sum_{i=1}^{j} \dim C_n(\varphi, u_i), \quad n = 0, 1, \ldots.$$

If $M_n(X^b, X^a)$ is finite for every n and is equal to zero for n sufficiently large, the *Morse polynomial* of the pair (X^b, X^a) is defined by

$$M(t, X^b, X^a) = \sum_{n=0}^{\infty} M_n(X^b, X^a) t^n.$$

Theorem 8.1. *Under assumptions* (A) *if, every critical point in* $\varphi^{-1}([a,b] \cap X$ *corresponds to the same critical value* $c \in]a,b[$, *then*

$$M_n(X^b, X^a) = B_n(X^b, X^a), \quad n = 0, 1, \ldots.$$

Proof. Lemma 8.3 implies that

$$H_n(X^b, X^a) \approx H_n(X^c, X^a) \approx H_n(X^c, X^c \setminus K_c).$$

Since $K_c = \{u_1, \ldots, u_j\}$ is contained in the interior of $\varphi^{-1}([a,b]) \cap X$, the critical points have disjoint closed neighborhoods $U_1, \ldots, U_j$ such that

$$U = \bigcup_{i=1}^{j} U_i \subset \varphi^{-1}([a,b]) \cap X.$$

Therefore we obtain, by the excision and decomposition properties,

$$\begin{aligned} H_n(X^c, X^c \setminus K_c) &\approx H_n(X^c \cap U, (X^c \setminus K_c) \cap U) \\ &= H_n(\varphi^c \cap U, (\varphi^c \setminus K_c) \cap U) \\ &\approx \bigoplus_{i=1}^{j} H_n(\varphi^c \cap U_i, \varphi^c \cap U_i \setminus \{u_i\}) \\ &= \bigoplus_{i=1}^{j} C_n(\varphi, u_i), \end{aligned}$$

and the result follows from the definitions. □

Theorem 8.2. *Under assumptions* (A), *if* $M_n(X^b, X^a)$ *is finite for every* n *and equal to zero for* n *sufficiently large, then there exists a polynomial* $Q(t)$ *with nonnegative integer coefficients such that*

$$M(t, X^b, X^a) = P(t, X^b, X^a) + (1+t) Q(t).$$

Proof. Let $a < c_1 < \ldots < c_j < b$ be the critical values corresponding to the critical points in $\varphi^{-1}([a,b]) \cap X$. If we take real numbers a_i such that

$$a = a_0 < c_1 < a_1 < c_2 < \ldots < a_{j-1} < c_j < a_j = b,$$

Theorem 8.1 implies that the pairs $(X^{a_{i+1}}, X^{a_i})$ are admissible and that

$$\sum_{i=0}^{j-1} B_n\left(X^{a_{i+1}}, X^{a_i}\right) = \sum_{i=0}^{j-1} M_n\left(X^{a_{i+1}}, X^{a_i}\right) = M_n(X^b, X^a).$$

It follows then from formula (3) that

$$M(t, X^b, X^a) = \sum_{i=0}^{j-1} P\left(t, X^{a_{i+1}}, X^{a_i}\right) = P(t, X^b, X^a) + (1+t)\,Q(t),$$

where $Q(t)$ is a polynomial with nonnegative integer coefficients. □

Remarks.

1. Theorem 8.2 implies that

$$M_n(X^b, X^a) \geq B_n(X^b, X^a), \quad n = 0, 1, \ldots,$$

and that

$$\sum_{n=0}^{\infty}(-1)^n M_n(X^b, X^a) = \sum_{n=0}^{\infty}(-1)^n B_n(X^b, X^a).$$

The second relation is an extension of the Poincaré–Hopf formula.

2. If $M_n(X^b, X^a)$, $M_{n+1}(X^b, X^a) \equiv 0$ for every n, then necessarily

$$M(t, X^b, X^a) = P(t, X^b, X^a).$$

The above observation is called the *Morse lacunary principle.*

Let us now extend Theorem 8.2 to the case of an unbounded interval $[a, +\infty[$.

Lemma 8.4. *Let M be a Riemannian manifold of class C^2, let $\varphi \in C^{2-0}(M, \mathbf{R})$, and let X be a subset of M positively invariant with respect to the flow σ defined by* (10). *If for every $d > b$, $\varphi^{-1}([b, d]) \cap X$ is complete and free of critical points, and if φ satisfies* (PS) *over $\varphi^{-1}([b, d]) \cap X$, then X^b is a strong deformation retract of X.*

Proof. Let $u \in X$ be such that $\varphi(u) > b$. If $\varphi(\sigma(t, u)) > b$ for every $t \in]0, \omega^+(u)[$ then, as in the first part of the proof of Lemma 8.2, there exists a critical point v of φ in X such that $\varphi(v) \geq b$. But this is not possible by assumption. Thus there exists a unique $t(u)$ such that $\varphi(\sigma(t(u), u)) = b$. The deformation can then be given on $[0, 1] \times X$ by

$$\eta(s, u) = \sigma(t(u)s, u), \quad 0 \leq s \leq 1$$

if $u \in X \setminus X^b$ and by

$$\eta(s, u) = u, \quad 0 \leq s \leq 1$$

if $u \in X^b$. □

Let us suppose that, in addition to assumption (A), the following condition holds.

(B) For every $d \geq b$, $\varphi^{-1}([b, d[) \cap X$ is complete and free of critical points and φ satisfies (PS) over $\varphi^{-1}([b, d[) \cap X$.

The *Morse numbers* of the pair (X, X^a) are defined by

$$M_n(X, X^a) = M_n(X^b, X^a).$$

If $M_n(X, X^a)$ is finite for every n and equal to zero for n sufficiently large, the *Morse polynomial* of the pair (X, X^a) is defined by

$$M(t, X, X^a) = M(t, X^b, X^a).$$

Corollary 8.1. *Under assumptions* (A) *and* (B), *if* $M_n(X, X^a)$ *is finite for every* n *and equal to zero for* n *sufficiently large, there exists a polynomial* $Q(t)$ *with nonnegative integer coefficients such that*

$$M(t, X, X^a) = P(t, X, X^a) + (1 + t)\,Q(t).$$

Proof. By Lemma 8.4, X^b is a strong deformation retract of X so that

$$P(t, X, X^a) = P(t, X^b, X^a).$$

The result then follows from Theorem 8.2 and from the definition of $M(t, X, X^a)$. □

Corollary 8.2. *Let* M *be a complete Riemannian manifold of class* C^2 *and let* $\varphi \in C^{2-0}(M, \mathbf{R})$. *If*

i) φ *satisfies the Palais–Smale condition over* M,

ii) φ *is bounded from below on* M,

iii) φ *has only a finite number of critical points* $u_1, \ldots, u_j$ *and* $\dim C_n(\varphi, u_i)$ *is finite for every* n *and zero for* n *sufficiently large,* $i = 1, \ldots, j$,

then there exists a polynomial $Q(t)$ *with nonnegative integer coefficients such that*

$$\sum_{n=0}^{\infty} \sum_{i=1}^{j} \dim C_n(\varphi, u_i) t^n = P(t, M, \phi) + (1 + t)\,Q(t).$$

Proof. Let $a < \inf_M \varphi$ and $b > \sup\{\varphi(u) : \nabla\varphi(u) = 0\}$. It suffices to apply Corollary 8.1 with $X = M$. □

8.6 The Generalized Morse Lemma

The generalized Morse lemma, also called the splitting theorem, is the basic tool for the effective computation of critical groups. The theory of Fredholm operators provides a natural setting for this lemma.

A linear continuous operator L between two Banach spaces is called a *Fredholm operator* if the dimension of $\ker L$ and the codimension of $R(L)$ are finite. This implies that $R(L)$ is closed.

Let V be a Hilbert space, U an open neighborhood of $u \in V$, and let $\varphi \in C^2(U, \mathbf{R})$. Define implicitly the linear operator $L : V \to V$ by

$$(Lv, w) = \varphi''(u)(v, w).$$

Then L is self-adjoint and we shall identify L with $\varphi''(u)$. If $\varphi''(u)$ is a Fredholm operator, V is the orthogonal sum of $R(\varphi''(u))$ and $\ker(\varphi''(u))$.

Assume now that u is a critical point of φ. The *Morse index of* u is defined as the supremum of the dimensions of the vector subspaces of V on which $\varphi''(u)$ is negative definite. The *nullity* of u is defined as the dimension of $\ker \varphi''(u)$. Finally, the critical point u will be said to be *non-degenerate* if $\varphi''(u)$ is invertible.

Theorem 8.3. *Let U be an open neighborhood of 0 in a Hilbert space V and let $\varphi \in C^2(U, \mathbf{R})$. Suppose that 0 is a critical point of φ with positive nullity and that $L = \varphi''(0)$ is a Fredholm operator, so that V is the orthogonal direct sum of* $\ker(L)$ *and* $R(L)$. *Let $w + v$ be the corresponding decomposition of $u \in V$. Then there exists an open neighborhood A of 0 in V, an open neighborhood B of 0 in* $\ker(L)$, *a local homeomorphism h from A into U, and a function $\hat{\varphi} \in C^2(B, \mathbf{R})$ such that*

$$h(0) = 0, \quad \hat{\varphi}'(0) = 0, \quad \hat{\varphi}''(0) = 0$$

and

$$\varphi(h(u)) = (1/2)(Lv, v) + \hat{\varphi}(w)$$

on the domain of h.

Proof. 1) Let $Q : V \to V$ be the orthogonal projection onto $R(L)$. By the implicit function theorem, we can find $r_1 > 0$ and a C^1-mapping

$$g : B(0, r_1) \cap \ker L \to R(L)$$

such that $g(0) = 0$, $g'(0) = 0$ and

$$Q\nabla\varphi(w + g(w)) = 0. \tag{13}$$

Let us define $\hat{\varphi}$ on $B = B(0, r_1) \cap \ker L$ by

$$\hat{\varphi}(w) = \varphi(w + g(w))$$

so that, by direct computation and (13),

$$\nabla\hat{\varphi}(w) = (I - Q)\nabla\varphi(w + g(w))$$

and

$$\hat{\varphi}''(w) = (I - Q)\varphi''(w + g(w))(Id + g'(w)).$$

In particular

$$\nabla\hat{\varphi}(0) = (I - Q)\nabla\varphi(0) = 0$$

and

$$\hat{\varphi}''(0) = (I - Q)\varphi''(0) = (I - Q)L = 0.$$

Let us define, near $[0,1] \times \{0\}$, the function

$$\Phi(t, v, w) = (1 - t)(\hat{\varphi}(w) + (1/2)(Lv, v)) + t\varphi(v + w + g(w))$$

and the vector field

$$\begin{aligned} f(t, v, w) &= 0, & \text{if } v = 0, \\ &= -\Phi_t(t, v, w)\,|\Phi_v(t, v, w)|^{-2}\Phi_v(t, v, w), & \text{if } v \neq 0. \end{aligned}$$

If $\eta(t) = \eta(t, v, w)$ is a solution of the Cauchy problem

$$\dot{\eta} = f(t, \eta, w)$$

$$\eta(0) = v$$

we have

$$\begin{aligned} \frac{d}{dt}\Phi(t, \eta(t), w) &= \Phi_t(t, \eta(t), w) + (\Phi_v(t, \eta(t), w), \dot{\eta}(t)) \\ &= 0 \end{aligned}$$

and, in particular,

$$\begin{aligned} \hat{\varphi}(w) + (1/2)(Lv, v) &= \Phi(0, v, w) \\ &= \Phi(1, \eta(1, v, w), w) \\ &= \varphi(\eta(1, v, w) + w + g(w)). \end{aligned}$$

Let us assume that the flow $\eta(t, v, w)$ is well defined and continuous on $[0,1] \times A$, where A is an open neighborhood of 0 in V. Then the local homeomorphism h is given by

$$h(u) = h(v, w) = w + g(w) + \eta(1, v, w).$$

The local invertibility of h follows from the local invertibility of $\eta(1, \cdot, w)$.

2) It remains to prove that η is well defined and continuous. Let us define Ψ by

$$\Psi(v, w) = \varphi(v + w + g(w)) - \hat{\varphi}(w) - (1/2)(Lv, v).$$

We obtain, using (13),

$$\Psi(0, w) = 0, \quad \Psi_v(0, w) = 0, \quad \Psi_v''(0, 0) = 0;$$

and, consequently

$$\Psi(v, w) = \int_0^1 (1 - s)(\Psi_v''(sv, w)v, v)\, ds$$

$$\Psi_v(v,w) = \int_0^1 \Psi_v''(sv,w)v\,ds.$$

Thus, for each $\epsilon > 0$, there exists $\delta(\epsilon) \in]0, r_1[$ such that

$$|\Psi(v,w)| \leq \epsilon|v|^2, \quad |\Psi_v(v,w)| \leq \epsilon|v| \tag{14}$$

whenever $|v+w| \leq \delta(\epsilon)$. Since $L : R(L) \to R(L)$ is continuous and invertible, there exists $c > 0$ such that

$$c^{-1}|v| \leq |Lv| \leq c\,|v| \tag{15}$$

for $v \in R(L)$. We have, for $v \neq 0$,

$$f(t,v,w) = -\Psi(v,w)|Lv + t\Psi_v(v,w)|^{-2}(Lv + t\Psi_v(v,w)).$$

Let $\epsilon = (2c)^{-1}$. Using (14) and (15), we obtain, for $|v+w| \leq \delta(\epsilon)$,

$$|f(t,v,w)| \leq 2c(c+\epsilon)\epsilon\,|v|. \tag{16}$$

Since $f(t,0,w) = 0$, f is continuous. Let $\rho \in]0, \delta(\epsilon)[$ be such that

$$|\Psi_v''(v,w)| \leq 1 \tag{17}$$

for $|v+w| \leq \rho$ and $v \neq 0$. Using (14), (15), and (17), it is easy to verify the existence of $c_1 > 0$ such that

$$|f_v(t,v,w)| \leq c_1$$

for $|v+w| \leq \rho$ and $v \neq 0$. By the mean value theorem and (16), there exists $c_2 > 0$ such that

$$|f(t,v_1,w) - f(t,v_2,w)| \leq c_2|v_1 - v_2|$$

for $|v_i + w| \leq \rho$, $i = 1,2$. Thus the flow η is locally well defined and continuous. Moreover, since $\eta(t,0,w) = 0$, η is well defined on $[0,1] \times A$ where A is an open neighborhood of 0 in V. □

Remarks. 1) It is easy to verify that h restricted to $R(L)$ is a local diffeomorphism since $f_v(t,v,0)$ is continuous.

2) A similar but simpler proof gives the following result, which is called the *Morse lemma.*

Theorem 8.3bis. *Let U be a neighborhood of 0 in a Hilbert space V and let $\varphi \in C^2(U,\mathbf{R})$ be such that 0 is a non-degenerate critical point of φ. Then there exists an open neighborhood A of 0 in V and a local diffeomorphism h from A into U such that $h(0) = 0$ and*

$$\varphi(h(u)) = \varphi(0) + (1/2)(\varphi''(0)u, u).$$

Let now M be a regular C^2-manifold modelled on a Hilbert space V and let u be an isolated critical point of $\varphi \in C^2(M, \mathbf{R})$. The *Morse index* (resp. the nullity) of u is defined as the Morse index (resp. the *nullity*) of $x(u)$ as a critical point of $\varphi \circ x^{-1}$, where x is a chart at u. The critical point u is called *non-degenerate* if $x(u)$ is a non-degenerate critical point of $\varphi \circ x^{-1}$.

Remarks. 1. If y is another chart at u, then, since

$$\varphi \circ x^{-1} = \varphi \circ y^{-1} \circ y \circ x^{-1}$$

on $D(x) \cap D(y)$, it is easy to verify that

$$(\varphi \circ x^{-1})''(x(u)) = (\varphi \circ y^{-1})''(y(u))[(y \circ x^{-1})'(x(u)), (y \circ x^{-1})'(x(u))].$$

The invertibility of $(y \circ x^{-1})'(x(u))$ implies that the above definitions are independent of the chart x.

2. In the non-degenerate case, the Morse index is the supremum of the dimensions of the subspaces along which φ is decreasing near the critical point u.

3. By the implicit function theorem (or by the Morse lemma) any non-degenerate critical point is isolated.

We now show that the critical groups of a non-degenerate critical point depend only upon its Morse index.

Corollary 8.3. *Let M be a regular C^2 manifold modeled on a Hilbert space V and let u be a non-degenerate critical point of $\varphi \in C^2(M, \mathbf{R})$ with Morse index k. Then*

$$C_n(\varphi, u) = \delta_{n,k} F, \quad n = 0, 1, \ldots.$$

Proof. 1) Let x be a chart at u and let $U \subset D(x)$ be a closed neighborhood of u. Since, by definition

$$C_n(\varphi, u) = H_n(\varphi^c \cap U, \varphi^c \cap U \setminus \{0\})$$

with $c = \varphi(u)$, it is sufficient to consider the case where M is an open subset of V.

2) We can assume without loss of generality that $u = 0$ and $c = 0$. By Theorem 8.3bis, there exists an open neighborhood A of 0 in M and a local homeomorphism h from A into V such that $h(0) = 0$ and

$$\varphi(h(u)) = \psi(u) \equiv (1/2)(\varphi''(0)u, u)$$

whenever $u \in A$. Let $B \subset A$ be a closed ball centered at 0. We have

$$\begin{aligned} C_n(\varphi, 0) &= H_n(\varphi^0 \cap h(B), \varphi^0 \cap h(B) \setminus \{0\}) \\ &\approx H_n(\psi^0 \cap B, \psi^0 \cap B \setminus \{0\}). \end{aligned}$$

From the invertibility of $\varphi''(0)$ it follows that V is the orthogonal sum of V^- and V^+ with ψ negative (resp. positive) definite on V^- (resp. V^+). Let

$v = v^- + v^+$ be the corresponding decomposition of any $v \in V$. Define the deformation η of B by

$$\eta : [0,1] \times B \to B, \quad (t,v) \to v^- + (1-t)v^+$$

so that

$$\psi(\eta(t,v)) = \psi(v^-) + (1-t)^2 \psi(v^+).$$

Thus $V^- \cap B \setminus \{0\}$ is a deformation retract of $\psi^0 \cap B \setminus \{0\}$ and $V^- \cap B$ is a deformation retract of $\psi^0 \cap B$. Since, by definition, $k = \dim V^-$, we obtain, for $k \geq 1$,

$$H_n(\psi^0 \cap B, \psi^0 \cap B \setminus \{0\}) \approx H_n(\psi^0 \cap B, V^- \cap B \setminus \{0\})$$

$$\approx H_n(V^- \cap B, V \cap B \setminus \{0\}) \approx H_n(B^k, S^{k-1}) \approx \delta_{n,k} F,$$

and for $k = 0$,

$$H_n(\psi^0 \cap B, \psi^0 \cap B \setminus \{0\}) \approx H_n(\{0\}, \phi) = \delta_{n,0} F. \quad \square$$

Remarks. 1. If the Morse index of a nondegenerate critical point u is infinite, then all the critical groups of φ at u are isomorphic to 0.

2. Under the assumptions of Theorem 8.2, if the critical points of φ in $\varphi^{-1}([a,b]) \cap X$ are non-degenerate, then $M_n(X^b, X^a)$ is equal to the number of critical points of φ with Morse index n in $\varphi^{-1}([a,b]) \cap X$.

8.7 Computation of the Critical Groups

The use of Morse inequalities depends on the effective computation of the critical groups in the degenerate case.

Lemma 8.5. *Let U be an open neighborhood of v in a Hilbert space V and let $\varphi \in C^{2-0}(U; \mathbf{R})$. If v is the only critical point of φ, and if the Palais–Smale condition is satisfied over a closed ball $B[v,r] \subset U$, then there exists $\epsilon > 0$ and $X \subset U$ such that:*

i) *X is a neighborhood of v, closed in U;*

ii) *X is positively invariant for the flow σ defined by* (10);

iii) *$\varphi^{-1}([c-\epsilon, c+\epsilon]) \cap X$ is complete, where $c = \varphi(v)$;*

iv) *the Palais–Smale condition is satisfied over $\varphi^{-1}([c-\epsilon, c+\epsilon]) \cap X$.*

Proof. Let $\epsilon > 0$ and $B \subset U$ be given by Lemma 8.1 applied to $A = B[v,r]$ and let X be the closure in U of the set

$$Y = \{\sigma(t,u) : u \in B, 0 \leq t < \omega_+(u)\}.$$

By construction, X satisfies i) and ii). Lemma 8.1 implies that

$$\varphi^{-1}([c-\epsilon, c+\epsilon]) \cap Y \subset B[v, r].$$

Since $B[v,r]$ is closed in U, the set $\varphi^{-1}([c-\epsilon, c+\epsilon]) \cap X$ is contained in $B[v,r]$ and closed in $B[v,r]$; hence it is complete. By our Palais–Smale assumption, iv) then follows from iii). □

We shall prove that, in the setting of Theorem 8.3, the critical groups depend on the Morse index and on the "degenerate part" of the functional. Thus the computation of the critical groups is reduced to a finite dimensional problem. This result is called the Shifting theorem.

Theorem 8.4. *Under the assumptions of Theorem 8.3, if 0 is the only critical point of φ, and if the Morse index k of 0 is finite, then*

$$C_n(\varphi, 0) \approx C_{n-k}(\hat{\varphi}, 0), \quad n = 0, 1, \ldots.$$

Proof. 1) With the notations of Theorem 8.3, let $C \subset A$ be a closed neighborhood of 0. Setting $c = \varphi(0) = \hat{\varphi}(0)$ and $\psi(u) = \psi(v+w)\ (1/2)(Lv, v) + \hat{\varphi}(w)$, we obtain

$$\begin{aligned} C_n(\varphi, 0) &= H_n(\varphi^c \cap h(C), \varphi^c \cap h(C) \setminus \{0\}) \\ &\approx H_n(\psi^c \cap C, \psi^c \cap C \setminus \{0\}) = C_n(\psi, 0). \end{aligned}$$

2) By assumption, $0 \in \ker L$ is the only critical point of $\hat{\varphi} \in C^2(B, \mathbf{R})$. Since $\dim \ker L$ is finite, the Palais–Smale condition is satisfied over any closed ball $B[0,r] \subset B$. Let $\epsilon > 0$ and $X \subset B$ be given by Lemma 8.5 applied to $\hat{\varphi}$. Lemma 8.3 implies that X^c is a strong deformation retract of $X^{c+\epsilon}$. Moreover, $\hat{\varphi}$ is non-increasing during the corresponding deformation η. Define the deformation Δ over $D = R(L) \times X^{c+\epsilon}$ by

$$\Delta(t, v, w) = v^- + (1-t)v^+ + \eta(t, w).$$

It is easy to verify that $V^- \times X^c$ is a strong deformation retract of $\psi^c \cap D$ and that $(V^- \times X^c) \setminus \{0\}$ is a strong deformation retract of $\psi^c \cap D \setminus \{0\}$. Therefore we obtain

$$\begin{aligned} C_n(\psi, 0) &= H_n(\psi^c \cap D, \psi^c \cap D \setminus \{0\}) \\ &\approx H_n(V^- \times X^c, (V^- \times X^c) \setminus \{0\}). \end{aligned}$$

3) If $k = \dim V^- = 0$, we have

$$\begin{aligned} C_n(\psi, 0) &= H_n(X^c, X^c \setminus \{0\}) \\ &= H_n(\hat{\varphi}^c \cap X, \hat{\varphi}^c \cap X \setminus \{0\}) = C_n(\hat{\varphi}, 0), \end{aligned}$$

and the proof is complete. If $k \geq 1$, relation (4) implies that

$$\begin{aligned} C_n(\psi, 0) &\approx H_n(\mathbf{R}^k \times X^c, (\mathbf{R}^k \times X^c) \setminus \{0\}) \\ &\approx H_n(B^k \times X^c, (B^k \times X^c) \setminus \{0\}) \\ &\approx H_{n-k}(X^c, X^c \setminus \{0\}) = C_{n-k}(\hat{\varphi}, 0). \quad \square \end{aligned}$$

Lemma 8.6. *Let U be an open subset of $\mathbf{R}^p$ and let v be the only critical point of $\varphi \in C^2(U,\mathbf{R})$. Then, for every $\rho > 0$, there exists $\tilde{\varphi} \in C^2(U,\mathbf{R})$ such that the following hold:*

a) *The critical points of $\tilde{\varphi}$, if any, are finite in number and non-degenerate.*

b) *If $|u-v| \geq \rho$, then $\tilde{\varphi}(u) = \varphi(u)$.*

c) *If $u \in U$, then*

$$|\tilde{\varphi}(u) - \varphi(u)| + |\tilde{\varphi}'(u) - \varphi'(u)| + |\tilde{\varphi}''(u) - \varphi''(u)| \leq \rho.$$

Proof. We can assume that the closed ball $B[v,\rho]$ is contained in U. Let $\omega \in C^2(U,\mathbf{R})$ be such that

$$\omega(u) = \begin{cases} 1 & \text{if } |u-v| \leq \rho/2 \\ 0 & \text{if } |u-v| \geq \rho \end{cases}$$

and let $e \in \mathbf{R}^p$. The function $\tilde{\varphi} \in C^2(U,\mathbf{R})$ defined by

$$\tilde{\varphi}(u) = \varphi(u) - \omega(u)(u,e)$$

satisfies b). It is easy to verify the existence of $\alpha > 0$ such that c) is satisfied for $|e| \leq \alpha$. Since

$$\nabla\tilde{\varphi}(u) = \nabla\varphi(u) - \omega(u)e - \nabla\omega(u)(u,e),$$

we obtain

$$|\nabla\tilde{\varphi}(u)| \geq |\nabla\varphi(u)| - |e|\,|\omega(u)| - |\nabla\omega(u)|\,|u|\,|e|.$$

But

$$\delta = \inf\{|\nabla\varphi(u)| \,:\, \rho/2 \leq |u-v| \leq \rho\} > 0.$$

Thus there exists $\beta \in]0,\alpha]$ such that, for $|e| \leq \beta$,

$$\inf\{|\nabla\tilde{\varphi}(u)| \,:\, \rho/2 \leq |u-v| \leq \rho\} \geq \delta/2.$$

By Sard's theorem, we can assume that e is a regular value of $\nabla\varphi$ such that $|e| \leq \beta$. If $|u-v| \geq \rho$, $\tilde{\varphi}(u) = \varphi(u)$, so that $\nabla\tilde{\varphi}(u) \neq 0$. If $\rho/2 \leq |u-v| \leq \rho$, we have $|\nabla\tilde{\varphi}(u)| \geq \delta/2$. If $|u-v| < \rho/2$, then, by definition

$$\nabla\tilde{\varphi}(u) = 0 \quad \text{if and only if} \quad \nabla\varphi(u) = e.$$

Since e is a regular value of $\nabla\varphi$, the critical points of $\tilde{\varphi}$ are non degenerate and, consequently, isolated. Being contained in $B[v,\rho/2]$, they must be finite in number. □

Let U be an open subset of $\mathbf{R}^p$ and let v be an isolated zero of $f \in C(\overline{U}, \mathbf{R}^p)$. Assume that $r > 0$ is such that the ball $B[v, r]$ is contained in $\overline{U}$ and v is the unique zero of f in $B[v, r]$. Then the *topological index* $i(f, v)$ of f at v is defined by

$$i(f, v) = d(f, B(v, r)).$$

By the excision property of the topological degree, the right-hand member is independent of r.

The following theorem gives a relation between the topological index and the critical groups.

Theorem 8.5. *Let U be an open subset of $\mathbf{R}^p$ and let v be an isolated critical point $\varphi \in C^2(U, \mathbf{R})$. Then $\dim C_n(\varphi, v)$ is finite for every n and is zero for $n \geq p+1$. Moreover*

$$i(\nabla\varphi, v) = \sum_{n=0}^{p} (-1)^n \dim C_n(\varphi, v).$$

Proof. 1) By diminishing U if necessary, we can assume that v is the only critical point of φ lying in U. Moreover, the Palais–Smale condition is satisfied over any closed ball $B[v, r] \subset U$. Let $\epsilon > 0$ and $X \subset U$ be given by Lemma 8.5. The definition of the Morse numbers and Theorem 8.1 imply that

$$\dim C_n(\varphi, v) = M_n(X^{c+\epsilon}, X^{c-\epsilon}) = B_n(X^{c+\epsilon}, X^{c-\epsilon}) \tag{18}$$

where $c = \varphi(v)$.

2) There exists $\rho \in]0, \epsilon/3]$ such that

$$B[v, 2\rho] \subset \varphi^{-1}\left(\left[c - \frac{\epsilon}{3}, c + \frac{\epsilon}{3}\right]\right) \cap X.$$

Let $\tilde{\varphi} \in C^2(U, \mathbf{R})$ be given by Lemma 8.6. Properties b) and c) of $\tilde{\varphi}$ imply that $\tilde{\varphi}^{c\pm\epsilon} = \varphi^{c\pm\epsilon}$. Thus $\tilde{\varphi}^{-1}([c-\epsilon, c+\epsilon]) \cap X = \varphi^{-1}([c-\epsilon, c+\epsilon]) \cap X$ is complete. In particular, $\tilde{\varphi}$ satisfies the Palais–Smale condition over $\tilde{\varphi}^{-1}([c-\epsilon, c+\epsilon]) \cap X$. Since $B[v, \rho]$ is contained in the interior of X, property b) of $\tilde{\varphi}$ implies that X is positively invariant for the flow $\tilde{\sigma}$ defined by

$$\dot{\tilde{\sigma}}(t) = -\nabla\tilde{\varphi}(\tilde{\sigma}(t))$$

$$\tilde{\sigma}(0) = u.$$

By a), $\tilde{\varphi}$ has only a finite number of critical points $u_1, \ldots, u_j$, all non degenerate. By b), the critical points are contained in $B[v, \rho]$, and, hence, in the interior of $\tilde{\varphi}^{-1}([c-\epsilon, c+\epsilon]) \cap X$.

3) Let $k_i \in \{0, 1, \ldots, p\}$ be the Morse index of u_i, $i = 1, \ldots, j$. If we denote by $\tilde{M}_n(X^{c+\epsilon}, X^{c-\epsilon})$ the Morse numbers corresponding to $\tilde{\varphi}$, Corollary 8.2 implies that

$$\tilde{M}_n(X^{c+\epsilon}, X^{c-\epsilon}) = \sum_{i=1}^{j} \delta_{n,k_i}. \tag{19}$$

In particular, $\tilde{M}_n(X^{c+\epsilon}, X^{c-\epsilon})$ is finite for every n and equal to zero for $n \geq p+1$. It follows from Theorem 8.2 that

$$\tilde{M}_n(X^{c+\epsilon}, X^{c-\epsilon}) \geq B_n(X^{c+\epsilon}, X^{c-\epsilon})$$

and that

$$\sum_{n=0}^{p}(-1)^n \tilde{M}_n(X^{c+\epsilon}, X^{c-\epsilon}) = \sum_{n=0}^{p}(-1)^n B_n(X^{c+\epsilon}, X^{c-\epsilon}). \tag{20}$$

In particular, $\dim C_n(\varphi, v)$ is finite for every n and equal to zero for $n \geq p+1$.

4) By definition of the topological index and of the topological degree, we have

$$i(\nabla\tilde{\varphi}, u_i) = (-1)^{k_i}.$$

It follows from (19) and from the additivity of the topological degree that

$$\sum_{n=0}^{p}(-1)^n \tilde{M}_n(X^{c+\epsilon}, X^{c-\epsilon}) = \sum_{n=0}^{p}(-1)^n \left(\sum_{i=1}^{j}\delta_{n,k_i}\right) = \sum_{i=1}^{j}(-1)^{k_i}$$

$$= \sum_{i=1}^{j} i(\nabla\tilde{\varphi}, u_i) = d(\nabla\tilde{\varphi}, B(v, 2\rho)). \tag{21}$$

By continuity of the topological degree, we have

$$d(\nabla\tilde{\varphi}, B(v, 2\rho)) = d(\nabla\varphi, B(v, 2\rho)) = i(\nabla\varphi, v). \tag{22}$$

Theorem 8.5 then follows from (18), (20), (21), and (22). □

Theorem 8.6. *Let U be an open subset of $\mathbf{R}^p$ and let v be an isolated critical point $\varphi \in C^2(U, \mathbf{R})$. If v is neither a local minimum nor a local maximum, then*

$$C_0(\varphi, v) = C_p(\varphi, v) = 0.$$

Proof. 1) By diminishing U if necessary, we can assume that v is the only critical point of φ located in U. Moreover, the Palais–Smale condition is satisfied over any closed ball $B[v, r] \subset U$. Let $\epsilon > 0$ and $X \subset U$ be given by Lemma 8.5. Then, by Lemma 8.3, X^c is a deformation retract of $X^{c+\epsilon}$, so that

$$C_n(\varphi, v) = H_n(X^c, X^c \setminus \{0\}) = H_n(X^{c+\epsilon}, X^c \setminus \{v\}).$$

Let $\eta \in C([0,1] \times X^{c+\epsilon}, X^{c+\epsilon})$ be the corresponding deformation.

2) In order to prove that $H_0(X^{c+\epsilon}, X^c \setminus \{0\}) = \{0\}$, it suffices to show that every point $u \in X^{c+\epsilon}$ is connected to a point in $X^c \setminus \{v\}$ by a continuous path contained in $X^{c+\epsilon}$. Let $\rho > 0$ be such that $B[v, \rho] \subset X^{c+\epsilon}$. Since v is not a local minimum, there exists $w \in B[v, \rho]$ such that $\varphi(w) < c$. Thus

v is connected to the point $w \in X^c \setminus \{v\}$ by a continuous path contained in $X^{c+\epsilon}$. Now, every point $u \in X^{c+\epsilon}$ is connected by a continuous path contained in $X^{c+\epsilon}$ to $\eta(1,u)$, which either is v or belongs to $X^c \setminus \{v\}$.

3) Any continuous map

$$f : S^{p-1} \to B[v,\rho] \cap \varphi^c \setminus \{v\}$$

has a continuous extension $g_1 : B^p \to B[v,\rho]$. It follows from Lemmas 8.3 and 8.5 that f has a continuous extension $g_2 : B^p \to \varphi^c$. Since v is not a local maximum, v is not an interior point of $g_2(B^p)$. Thus f has a continuous extension $g_3 : B^p \to \varphi^c \cup S_\delta \setminus B_\delta$ where $\delta > 0$ is small, $S_\delta = S(v,\delta)$, and $B_\delta = B(v,\delta)$. Using the argument of Lemma 6.5, we obtain a continuous extension $g_4 : B^p \to \varphi^c \setminus \{v\}$ of f. Thus $H_{p-1}(\varphi^c \cap B[v,\rho] \setminus \{v\}) \approx 0$. Since $H_p(\varphi^c \cap B[v,\rho]) \approx 0$, we obtain by exactness $C_p(\varphi, v) \approx 0$. □

Corollary 8.4. *Under the assumptions of Theorem 8.3, if 0 is an isolated critical point of φ with finite Morse index k and nullity ν, then the following are true.*

i) dim $C_n(\varphi,0)$ *is finite for every n and is equal to zero if* $n \notin \{k, k+1, \ldots, k+\nu\}$;

ii) *if 0 is a local minimum of $\tilde{\varphi}$, then*

$$C_n(\varphi,0) = \delta_{n,k} F;$$

iii) *if 0 is a local maximum of $\tilde{\varphi}$, then*

$$C_n(\varphi,0) = \delta_{n,k+\nu} F;$$

iv) *if 0 is neither a local minimum nor a local maximum of $\tilde{\varphi}$, then*

$$C_k(\varphi,0) = C_{k+\nu}(\varphi,0) = 0;$$

v) *if there exist integers $n_1 \neq n_2$ such that $C_{n_1}(\varphi,0) \neq 0$ and $C_{n_2}(\varphi,0) \neq 0$, then*

$$|n_1 - n_2| \leq \nu - 2.$$

Proof. By Theorem 8.4, $C_n(\varphi,0) \approx C_{n-k}(\tilde{\varphi},0)$, so that $C_n(\varphi,0) = 0$ if $n \leq k-1$. It follows from Theorem 8.5 and $\dim C_n(\varphi,0)$ is finite for every n and is equal to zero for $n \geq k+\nu+1$. It is easy to obtain ii) and iii) by a direct calculation. Theorem 8.6 implies iv). Finally, v) follows from i) to iv). □

8.8 Critical Groups at a Point of Mountain Pass Type

Interesting multiplicity results can be obtained by combining the minimax theorems and the Morse theory. Let us illustrate this fact by the mountain pass theorem situation.

Theorem 8.7. *Let X be a Hilbert space and let $\varphi \in C^2(X, \mathbf{R})$. Assume that there exists $u_0 \in X$, $u_1 \in X$ and a bounded open neighborhood Ω of u_0 such that $u_1 \in X \setminus \overline{\Omega}$ and*

$$\inf_{\partial\Omega} \varphi > \max(\varphi(u_0), \varphi(u_1)).$$

Let $\Gamma = \{g \in C([0,1], X) \,:\, g(0) = u_0, g(1) = u_1\}$ and

$$c = \inf_{g\in\Gamma} \max_{s\in[0,1]} \varphi(g(s)).$$

If φ satisfies the Palais–Smale condition over X, and if each critical point of φ in K_c is isolated in X, then there exists $u \in K_c$ such that $\dim C_1(\varphi, u) \geq 1$.

Proof. Let $\epsilon > 0$ be such that $c - \epsilon > \max(\varphi(u_0), \varphi(u_1))$ and c is the only critical value of φ in $[c-\epsilon, c+\epsilon]$. Consider the exact sequence

$$\ldots \to H_1(\varphi^{c+\epsilon}, \varphi^{c-\epsilon}) \xrightarrow{\partial} H_0(\varphi^{c-\epsilon}, \phi) \xrightarrow{i_*} H_0(\varphi^{c+\epsilon}, \phi) \to \ldots$$

where i_* is induced by the inclusion mapping $i \,:\, (\varphi^{c-\epsilon}, \phi) \to (\varphi^{c+\epsilon}, \phi)$. The definition of c implies that u_0 and u_1 are path connected in $\varphi^{c+\epsilon}$ but not in $\varphi^{c-\epsilon}$. Thus, $\ker i_* \neq \{0\}$ and, by exactness, $H_1(\varphi^{c+\epsilon}, \varphi^{c-\epsilon}) \neq \{0\}$. It follows from Theorem 8.1 that

$$M_1(\varphi^{c+\epsilon}, \varphi^{c-\epsilon}) = B_1(\varphi^{c+\epsilon}, \varphi^{c-\epsilon}) = \dim H_1(\varphi^{c+\epsilon}, \varphi^{c-\epsilon}) \geq 1.$$

Thus $\varphi^{-1}([c-\epsilon, c+\epsilon])$ contains a critical point u such that $\dim C_1(\varphi, u) \geq 1$ and, necessarily, $u \in K_c$. □

Corollary 8.5. *Besides the above assumptions, assume moreover that each $u \in K_c$ satisfies the following conditions:*

a) *$\varphi''(u)$ is a Fredholm operator;*

b) *the nullity of u is less than 2 provided the Morse index of u is equal to 0.*

Then there exists $u \in K_c$ such that

$$\dim C_n(\varphi, u) = \delta_{n,1}, \quad n \in \mathbf{N}.$$

Proof. 1) Let $u \in K_c$, with Morse index k and nullity ν, be such that dim $C_1(\varphi, u) \geq 1$. We can assume that $u = 0$. By Corollary 8.4, $k \leq 1$ and $\nu \geq 1$ if $k = 0$.

2) If $k = 0$, assumption b) implies that $\nu = 1$. It then follows from Corollary 8.4 that 0 is a local maximum of $\hat{\varphi}$ and

$$C_n(\varphi, 0) = \delta_{n,k+\nu} F = \delta_{n,1} F.$$

3) If $k = 1$, then, by Corollary 8.4, either $\nu = 0$ or 0 is a local minimum of $\hat{\varphi}$. In both cases,

$$C_n(\varphi, 0) = \delta_{n,k} F = \delta_{n,1} F. \quad \square$$

Corollary 8.6. *Under the assumptions of Corollary 8.5, if $X = \mathbf{R}^p$, there exists $u \in K_c$ such that*

$$i(\nabla\varphi, u) = -1.$$

Proof. By Corollary 8.5, there exists $u \in K_c$ such that $\dim C_n(\varphi, u) = \delta_{n,1}$. Theorem 8.5 implies that

$$i(\nabla\varphi, u) = \sum_{n=0}^{p} (-1)^n \dim C_n(\varphi, u) = -1. \quad \square$$

8.9 Continuity of the Critical Groups and Bifurcation Theory

The critical groups are continuous with respect to the C^1 topology.

Theorem 8.8. *Let U be an open neighborhood of v in a Hilbert space V and let $\varphi, \psi \in C^{2-0}(U, \mathbf{R})$. Assume that φ and ψ have v as the only critical point and satisfy the Palais-Smale condition over a closed ball $B[v, r] \subset U$. Then there exists $\eta > 0$, depending only upon φ, such that the condition*

$$\sup_{u \in U} (|\psi(u) - \varphi(u)| + |\nabla\psi(u) - \nabla\varphi(u)|) \leq \eta \tag{23}$$

implies

$$\dim C_n(\psi, v) = \dim C_n(\varphi, v), \quad n \in \mathbf{N}. \tag{24}$$

Proof. 1) Let $\epsilon > 0$ and $X \subset U$ be given by Lemma 8.5 applied to φ. The definition of the Morse numbers and Theorem 8.1 imply that

$$\dim C_n(\varphi, v) = M_n(X^{c+\epsilon}, X^{c-\epsilon}) = B_n(X^{c+\epsilon}, X^{c-\epsilon}) \tag{25}$$

where $c = \varphi(v)$.

2) Let $\rho > 0$ be such that

$$B[v, 2\rho] \subset \varphi^{-1}\left(\left[c - \frac{\epsilon}{3}, c + \frac{\epsilon}{3}\right]\right) \cap X. \tag{26}$$

By the Palais–Smale condition,

$$\delta = \inf\{|\nabla\varphi(u)| \, : \, \rho/2 \le |u-v| \le \rho\} > 0. \tag{27}$$

Let $\omega \in C^2(U, \mathbf{R})$ be such that

$$\begin{gathered} \omega(u) = 1 \quad \text{if} \quad |u-v| \le \rho/2 \\ \omega(u) = 0 \quad \text{if} \quad |u-v| \ge \rho \\ 0 \le \omega(u) \le 1 \\ \gamma = \sup_{u\in U} |\nabla\omega(u)| < \infty, \end{gathered} \tag{28}$$

and let

$$\eta = \min(\epsilon/3, \delta/2(1+\gamma)).$$

Assume that ψ satisfies the assumptions of the theorem and define $\tilde{\psi}$ on U by

$$\tilde{\psi}(u) = \varphi(u) + \omega(u)(\psi(u) - \varphi(u)).$$

It follows from (23), (27), and (28) that, for $\rho/2 \le |u-v| \le \rho$,

$$\begin{aligned} |\nabla\tilde{\psi}(u)| &\ge |\nabla\varphi(u)| - \omega(u)|\nabla\psi(u) - \nabla\varphi(u)| - |\nabla\omega(u)|\,|\psi(u) - \varphi(u)| \\ &\ge \delta - (1+\gamma)\eta \ge \delta/2. \end{aligned} \tag{29}$$

We obtain from (23) and (28) that, for $u \in U$,

$$|\tilde{\psi}(u) - \varphi(u)| = \omega(u)\,|\psi(u) - \varphi(u)| \le \eta \le \epsilon/3. \tag{30}$$

3) Since $\tilde{\psi}(u) = \varphi(u)$ if $|u-v| \ge \rho$, relations (26) and (30) imply that $\tilde{\psi}^{c\pm\epsilon} = \varphi^{c\pm\epsilon}$. Thus $\tilde{\psi}^{-1}([c-\epsilon, c+\epsilon]) \cap X = \varphi^{-1}([c-\epsilon, c+\epsilon]) \cap X$ is complete. It follows easily from (29) that $\tilde{\psi}$ satisfies the Palais–Smale condition over $\tilde{\psi}^{-1}([c-\epsilon, c+\epsilon]) \cap X$. Moreover, $B[v,\rho]$ is contained in the interior of X, so that X is positively invariant for the flow $\tilde{\sigma}$ defined by

$$\dot{\tilde{\sigma}}(t) = -\nabla\tilde{\psi}(\tilde{\sigma}(t))$$

$$\tilde{\sigma}(0) = u.$$

Finally, the definition of $\tilde{\psi}$ and (29) imply that v is the only critical point of $\tilde{\psi}$. If we denote by $\tilde{M}_n(X^{c+\epsilon}, X^{c-\epsilon})$ the Morse numbers corresponding to $\tilde{\psi}$, we have

$$\dim C_n(\tilde{\psi}, v) = \tilde{M}_n(X^{c+\epsilon}, X^{c-\epsilon}) = B_n(X^{c+\epsilon}, X^{c-\epsilon}). \tag{31}$$

But, by the definition of $\tilde{\psi}$,

$$C_n(\tilde{\psi}, v) = C_n(\psi, v), \tag{32}$$

and (24) follows from (25), (31), and (32). □

The preceding theorem is useful in *bifurcation theory*.

Let V, W be Banach spaces, let U be an open neighborhood of 0 in V, and let Λ be an open interval. Consider a mapping $f \in C(\Lambda \times U, W)$ such that $f(\lambda, 0) = 0$ for every $\lambda \in \Lambda$. A point $(\lambda_0, 0) \in \Lambda \times U$ is a *bifurcation point* for the equation

$$f(\lambda, u) = 0 \tag{33}$$

if every neighborhood of $(\lambda_0, 0)$ in $\Lambda \times U$ contains at least one solution (λ, u) of (32) such that $u \neq 0$.

If f is a C^1 mapping, the implicit function theorem implies that a necessary condition for $(\lambda_0, 0)$ to be a bifurcation point is the non-invertibility of $D_u f(\lambda_0, 0)$. However, this condition is not sufficient in general, as shown by the simple example with $V = W = \mathbf{R}^2$ and

$$f(\lambda, u_1, u_2) = (u_1 - \lambda u_1 + u_2^3, u_2 - \lambda u_2 - u_1^3)$$

for which

$$D_u f(1, 0, 0) = 0$$

is not invertible and which, however, has no bifurcation point, as $f(\lambda, u_1, u_2) = 0$ implies

$$0 = u_2(u_1 - \lambda u_1 + u_2^3) - u_1(u_2 - \lambda u_2 - u_1^3) = u_2^4 + u_1^4 = 0$$

and hence $(u_1, u_2) = 0$. Notice however that f is not a gradient mapping with respect to u. We shall describe a rather wide class of gradient mappings for which the necessary condition above is sufficient.

The proof of the following simple lemma is left to the reader.

Lemma 8.7. *Let $K \subset \Lambda$ be a non-empty compact interval such that $K \times \{0\}$ contains no bifurcation point for* (33). *Then there exists $\rho > 0$ such that $B[0, \rho] \subset U$ and each solution (λ, u) of* (33) *in $K \times B[0, \rho]$ satisfies $u = 0$.*

Theorem 8.9. *Let U be an open neighborhood of 0 in a Hilbert space V, let Λ be an open interval and let $f(\lambda, u)$ be the gradient with respect to u of $\varphi \in C^2(\Lambda \times U, \mathbf{R})$. Assume that the following conditions are satisfied:*

α) 0 is a critical point of $\varphi_\lambda = \varphi(\lambda, .)$ for every $\lambda \in \Lambda$ and 0 is an isolated critical point of φ_a and φ_b for some reals $a < b$ in Λ.

β) φ_λ satisfies the Palais–Smale condition over a closed ball $B[0, r] \subset U$ for every $\lambda \in [a, b]$.

γ) There exists $n \in \mathbf{N}$ such that

$$\dim C_n(\varphi_a, 0) \neq \dim C_n(\varphi_b, 0).$$

Then there exists a bifurcation point $(\lambda_0, 0) \in [a,b] \times \{0\}$ *for* (33).

Proof. If $[a,b] \times \{0\}$ contains no bifurcation point for (33), then, by Lemma 8.7, there exists $\rho > 0$ such that $B[0,\rho] \subset U$ and each solution (λ, u) of (33) in $[a,b] \times B[0,\rho]$ satisfies $u = 0$. We can assume, without loss of generality, that $\varphi(\lambda, 0) = 0$. Since φ is of class C^2, we can choose ρ small enough so that

$$|D_\lambda \varphi(\lambda, u)| + |D_{\lambda u} \varphi(\lambda, u)| \leq 1$$

whenever $\lambda \in [a,b]$ and $u \in B[0,\rho]$. Thus φ_λ and $\nabla \varphi_\lambda$ depend continuously on $\lambda \in [a,b]$, uniformly on $B[0,\rho]$. By Theorem 8.8, $\dim C_n(\varphi_\lambda, 0)$ is locally constant, and hence constant, on $[a,b]$, for every $n \in \mathbf{N}$. In particular,

$$\dim C_n(\varphi_a, 0) = \dim C_n(\varphi_b, 0), \quad n \in \mathbf{N},$$

a contradiction with assumption γ). □

8.10 Lower Semi-Continuity of the Betti Numbers

We shall prove in this section a lower semi-continuity property for the Betti numbers $B_n(\varphi^b, \varphi^a)$ with respect to the C^0 topology. It is interesting to notice that this property is weaker than the corresponding continuity property of the topological degree whenever both concepts are defined.

Lemma 8.8. *Let* $B \subset F \subset B' \subset A \subset E \subset A'$ *be topological spaces. Suppose that*

$$H_n(B', B) \approx H_n(A', A) \approx \{0\}, \quad n = 0, 1, \ldots. \tag{34}$$

Then

$$B_n(A, B) \leq B_n(E, F), \quad n = 0, 1, \ldots.$$

Proof. Let us consider the following diagrams:

$$\begin{array}{ccccccc} H_{n+1}(A', A) \to & H_n(A,B) & \xrightarrow{i_*} & H_n(A', B) & \to H_n(A', A) \\ & f_* \searrow & & \nearrow g_* & \\ & & H_n(E, B) & & \end{array}$$

$$\begin{array}{ccccccc} H_n(B', B) \to & H_n(E,B) & \xrightarrow{j_*} & H_n(E, B') & \to H_{n-1}(B', B) \\ & f'_* \searrow & & \nearrow g'_* & \\ & & H_n(E, F) & & \end{array}$$

By exactness, assumption (34) implies that i_* and j_* are isomorphisms. But $i_* = g_* \circ f_*$ and $j_* = g'_* \circ f'_*$, so that f_* and f'_* are injections. Thus

$$h_* = f'_* \circ f_* : H_n(A, B) \to H_n(E, F)$$

is an injection. □

Theorem 8.10. *Let M be a complete Riemannian manifold of class C^2 and let $\varphi \in C^{2-0}(M, \mathbf{R})$. Suppose that there exists $\hat{\varphi} : M \to \mathbf{R}$, $c \in \mathbf{R}$ and $\epsilon > 0$ such that*

i) $\sup_{u \in M} |\varphi(u) - \hat{\varphi}(u)| \leq \epsilon/3$;

ii) *c is the only critical value of φ in $[c-\epsilon, c+\epsilon]$;*

iii) *φ satisfies the* (PS) *condition over $\varphi^{-1}([c-\epsilon, c+\epsilon])$.*

Then

$$B_n(\varphi^{c+\epsilon}, \varphi^{c-\epsilon}) \leq B_n(\hat{\varphi}^{c+\epsilon/2}, \hat{\varphi}^{c-\epsilon/2}), \quad (n = 0, 1, \ldots).$$

Proof. By assumption i), we have

$$\varphi^{c-\epsilon} \subset \hat{\varphi}^{c-\epsilon/2} \subset \varphi^{c-\epsilon/6} \subset \varphi^{c+\epsilon/6)} \subset \hat{\varphi}^{c+\epsilon/2} \subset \varphi^{c+\epsilon}.$$

It follows from Lemma 8.3 and assumptions ii) and iii) that $\varphi^{c-\epsilon}$ (resp. $\varphi^{c+\epsilon/6}$ is a strong deformation retract of $\varphi^{c-\epsilon/6}$ (resp. $\varphi^{c+\epsilon}$). Hence

$$H_n(\varphi^{c-\epsilon/6}, \varphi^{c-\epsilon}) \approx H_n(\varphi^{c+\epsilon}, \varphi^{c+\epsilon/6}) \approx= 0 \quad (n = 0, 1, \ldots).$$

Applying Lemma 8.8, we obtain

$$B_n(\varphi^{c+\epsilon/6}, \varphi^{c-\epsilon}) \leq B_n(\hat{\varphi}^{c+\epsilon/2}, \hat{\varphi}^{c-\epsilon/2}) \quad (n = 0, 1, \ldots).$$

Since

$$B_n(\varphi^{c+\epsilon/6}, \varphi^{c-\epsilon}) = B_n(\varphi^{c+\epsilon}, \varphi^{c-\epsilon}),$$

the proof is complete. □

8.11 Critical Groups at a Saddle Point

Let X be a Hilbert space and assume that $\varphi \in C^2(X, \mathbf{R})$ satisfies the Palais–Smale condition over X. Assume also, as in the saddle point theorem, that X splits into a direct sum of closed subspaces X^- and X^+ with

$$2 \leq m := \dim X^- < +\infty,$$

$$b := \sup_{S_R^-} < d := \inf_{X^+} \varphi$$

where $S_R^- = \{u \in X^- : |u| = R\}$. Regard the identity mappings $\sigma : S_R^- \to S_R^-$ as the generator of the homology $H_{m-1}(S_R^-, \phi)$ and define

$$c = \inf_{\partial\tau = \sigma} \sup_{u \in |\tau|} \varphi(u),$$

where τ is any chain of m dimensional singular simplices on X such that $\partial\tau = \sigma$. In this way we obtain a natural variant of the saddle point theorem.

Theorem 8.11. *Under the above assumptions, c is a critical value of φ. Moreover, if each critical point of φ in K_c is isolated in X, then there exists $u \in K_c$ such that*

$$\dim C_m(\varphi, u) \geq 1.$$

Proof. 1) Let us show that $|\tau| \cap X^+$ is nonempty for any chain τ with $\partial\tau = \sigma$. Consider the exact sequence

$$\ldots \rightarrow H_m(X, X \setminus X^+) \xrightarrow{\partial} H_{m-1}(X \setminus X^+, \phi) \approx H_{m-1}(S_R^-, \phi) \rightarrow \ldots.$$

Since $\partial[\tau] = [\sigma] \neq 0$, necessarily $[\tau] \neq 0$. Thus $|\tau| \not\subset X \setminus X^+$. In particular, we have that

$$\sup_{u \in |\tau|} \varphi(u) \geq d$$

for any chain τ with $\partial\tau = \sigma$. Hence $c \geq d$.

2) Using Lemma 6.5, one can easily prove, by contradiction, that K_c is nonempty.

3) Assume now that each critical point of φ in K_c is isolated in X. Let $\epsilon > 0$ be such that $c-\epsilon > b$ and c is the only critical value of φ in $[c-\epsilon, c+\epsilon]$. Consider the exact sequence

$$\ldots \rightarrow H_m(\varphi^{c+\epsilon}, \varphi^{c-\epsilon}) \xrightarrow{\partial} H_{m-1}(\varphi^{c-\epsilon}, \phi) \xrightarrow{i_*} H_{m-1}(\varphi^{c+\epsilon}, \phi) \rightarrow \ldots.$$

There exists a chain τ such that $\sigma = \partial\tau$ and $|\tau| \subset \varphi^{c+\epsilon}$. Thus $[\sigma] = 0$ in $H_{m-1}(\varphi^{c+\epsilon}, \phi)$. On the other hand, if $[\sigma] = 0$ in $H_{m-1}(\varphi^{c-\epsilon}, \phi)$, there exists a chain τ such that $\sigma = \partial\tau$ and $|\tau| \subset \varphi^{c-\epsilon}$. But this contradicts the definition of c. Thus $[\sigma]$ is a nonzero element of Ker i_*. By exactness, $H_m(\varphi^{c+\epsilon}, \varphi^{c-\epsilon}) \neq \{0\}$. The conclusion then follows from Theorem 8.1. □

Using Corollary 8.3, we obtain the following result.

Corollary 8.7. *Under the assumptions of Theorem* 8.11, *if each critical point of φ in K_c is nondegenerate, then there exists $u \in K_c$ such that*

$$\dim C_n(\varphi, u) = \delta_{n,m}, \quad n \in \mathbf{N}.$$

Historical and Bibliographical Notes

The reader can consult [Wal₁] for a brief and lucid exposition of singular homology. Surveys of the mathematical work of Morse are given by [Bot₁], [Cai₁], and [Tho₂], and surveys of Morse theory are given by [Bot₂], [Cha₁], and [Rot₅].

Morse's first paper on this theory [Mrs₁] already includes such essential ingredients as the Morse lemma, gradient deformations, and Morse inequalities for a nondegenerate function on a smooth domain in $\mathbf{R}^N$. It was aimed as a generalization of the Birkhoff minimax theory [Bir₁]. The theory is extended to compact smooth manifolds in [Mrs₂], which also contains the

Morse index theorem and applications to the calculus of variations by the method of "broken extremals."

The Morse theory was extended to Hilbert spaces by Rothe [Rot_6] and to infinite-dimensional Riemannian manifolds by Palais and Smale ([Pal_1], [PaS_1], [Sma_1]). As in the Leray–Schauder theory, the compactness of the domain is replaced by a compactness condition on the function (the PS condition).

The classical Morse lemma for nondegenerate critical points was extended by Palais [Pal_1] to Hilbert spaces. Because of the loss of two orders of differentiability, the Palais method is only applicable to functions of class C^3. Using the Lyapunov–Schmidt method and the Palais approach, Gromoll–Meyer [GrM_1] succeeded in treating the case of degenerate critical points when the second differential of the function is a compact perturbation of the identity. On the other hand, Kuiper [Kui_1] and Cambini [Cam_1] independently gave a proof of the Morse lemma for a nondegenerate critical point of a C^2 function. This result was extended to the degenerate case by Hofer [Hof_3] when the second differential is a compact perturbation of identity. Hofer's proof uses deformations by a gradient flow (see [GoM_1] for extensions). Theorem 8.3 generalizes the previous results. We follow the proof of Hofer [Hof_3] (see [MaW_4] for another proof).

The shifting theorem is due to Gromoll–Meyer [GrM_1]. A new proof is given here. Theorem 8.5 was first proved by Rothe [Rot_7], to whom we also owe the first results on the continuity of the critical groups and the lower semicontinuity of the Betti numbers in the Hilbert space case [Rot_8].

Lemma 8.6, Lemma 8.8 and Theorem 8.10 are contained in the important paper of Marino–Prodi [MaP_1] on perturbation methods in Morse theory.

Of course the genericity of the non-degenerate case is known and has been used since Morse. Theorem 8.6 is due to Dancer [Dan_2] and Theorem 8.7 to Ambrosetti [$\text{Amb}_{2,3}$] in the nondegenerate case and to Hofer [Hof_3] in the general case.

Minimax methods were introduced in bifurcation theory by Krasnosel'skii [Kra_2] and Morse theory by Marino–Prodi [MaP_2] (see the surveys [Cha_1], [Rab_6]).

The results of Section 8.11 are due to Liu [Liu_1].

For Morse theory on Banach manifolds, the reader can consult [Sk_1], [Tr_1], and [U_1]. The completeness of the Morse inequalities is studied in [Joh_1] and [Sma_2]. Degenerate critical points are considered in [CGR_1], [Dan_9], and [Ro_7]. More results on bifurcation through variational methods can be found in [$\text{Boh}_{1,2}$], [Ch_7], [Clk_4], [Dan_6], [Mar_1], [Prod_1], [BenP_1], and [Wil_{12}].

Exercises

1. Let M be a complete C^2-Riemannian manifold and assume that $\varphi \in C^2(M, \mathbf{R})$ satisfies the Palais–Smale condition on M. If φ has a global minimum and $\chi(M)$ nondegenerate critical points with finite Morse index, then φ has at least $\chi(M) + 2$ critical points.

2. Let U be an open subset of $\mathbf{R}^2$ and let v be an isolated critical point of $\varphi \in C^2(U, \mathbf{R})$. Then $i(\nabla\varphi, v) \leq 1$.

3. Let U be an open subset of $\mathbf{R}^p$ and let v be an isolated local minimum point of $\varphi \in C^2(U, \mathbf{R})$. Then $i(\nabla\varphi, v) = 1$.

4. Let M be a complete C^2-Riemannian manifold. Assume that $\varphi \in C^1([0,1] \times M, \mathbf{R})$ and $a < b$ are such that the following conditions hold:

   i) $\varphi(\lambda, u)$ is continuous with respect to λ uniformly in $u \in M$.
   ii) For every $\lambda \in [0,1]$, $\varphi_\lambda \in C^{2-0}(M, \mathbf{R})$ and $\nabla\varphi_\lambda(u) \neq 0$ whenever $\varphi_\lambda(u) \in \{a, b\}$.
   iii) Every sequence (λ_j, u_j) such that $(\varphi(\lambda_j, u_j))$ is bounded and $\nabla\varphi_{\lambda_j}(u_j) \to 0$ contains a convergent subsequence.

   Then
   $$P(t, \varphi_1^b, \varphi_1^a) = P(t, \varphi_0^b, \varphi_0^a).$$

   Hint. Find $0 < c < (b-a)/2$ and, for $\lambda_0 \in [0,1]$, find $\eta > 0$ such that for $|\lambda - \lambda_0| \leq \eta$ one has
   $$\varphi_{\lambda_0}^{a-c} \subset \varphi_\lambda^{a+c/\epsilon} \subset \varphi_{\lambda_0}^{a+c/\epsilon} \subset \varphi_\lambda^{a+c} \subset \varphi_\lambda^{b-c} \subset \varphi_{\lambda_0}^{b-c/\epsilon} \subset \varphi_\lambda^{b+c/\epsilon} \subset \varphi_\lambda^{b+c}.$$
   Use Lemma 8.3 and compactness of $[0,1]$.

5. (Principle of symmetric criticality, [Pal$_4$]). Let $\{T(g)\}_{g \in G}$ be an isometric representation of the topological group G over a Hilbert space X. Let $\psi \in C^2(X, \mathbf{R})$ be an invariant functional. If $u \in V = \mathrm{Fix}(G)$ is a critical point of $\psi|_V$, then u is a critical point of ψ.

   Hint. Prove that $\nabla\psi(u) \in V$.

6. ([Wil$_{12}$]). Let $\{T(g)\}_{g \in G}$ and X be as in Exercise 5. Let $\psi \in C^2(\Lambda \times X, \mathbf{R})$ where Λ is an open interval. Assume that

   i) $\psi_\lambda = \psi(\lambda, .)$ is invariant for every $\lambda \in \Lambda$,
   ii) ψ restricted to $\Lambda \times \mathrm{Fix}(G)$ satisfies the assumptions of Theorem 8.8, where $V = \mathrm{Fix}\,G$.

Then there exists $[\lambda_0, 0] \in [a, b] \times \{0\}$ such that every neighborhood of $[\lambda_0, 0]$ in $\Lambda \times \mathrm{Fix}(G)$ contains at least one solution (λ, u) of

$$\nabla \psi_\lambda(u) = 0$$

such that $u \neq 0$.

9

Applications of Morse Theory to Second Order Systems

Introduction

The *Liapunov center theorem* is the classical result which follows easily from the equivariant Crandall–Rabinowitz bifurcation theorem. Consider the second order autonomous system

$$\ddot{u} + f(u) = 0$$

and assume that 0 is a solution and that

$$\beta_1^2 < \beta_2^2 < \ldots < \beta_s^2$$

($\beta_r \geq 0, 0 \leq r \leq s$) are the non-negative eigenvalues of $f'(0)$. The Liapunov theorem insures that if the geometric multiplicity of β_i^2 is one and if $\beta_r/\beta_i \notin \mathbf{N}$ for $r \neq i$, then this system has a family of periodic solutions with minimal period tending to $2\pi/\beta_i$ and with amplitude tending to 0.

In many applications, the multiplicity of β_i^2 is bigger than one. We prove in Section 9.2 that, in the variational case ($f = \nabla F$), it suffices to assume that $\beta_r/\beta_i \notin \mathbf{N}$ for $r \neq i$ in order to obtain a sequence (u_k) of solutions with minimal period tending to $2\pi/\beta_i$ and with amplitude tending to zero when $k \to \infty$.

The application given in Section 9.3 concerns *asymptotically linear non-autonomous systems* of the form

$$\ddot{u}(t) + \nabla F(t, u(t)) = 0.$$

Since the problem is no more S^1-invariant as in the autonomous case, the results of Chapter 7 are no more applicable. Using the Morse inequalities, one can prove the existence of one or two nontrivial solutions when $\nabla F(t, 0) = 0$. The basic condition, namely a distinct behavior of ∇F at the origin and at infinity, is an extension of the "twist" condition of the famous Poincaré–Birkhoff geometric fixed point theorem.

Finally, in Section 9.4, a strong multiplicity result for non-autonomous second order systems with periodic potential and non-degenerate periodic solutions is given, which corresponds, in the more difficult case of Hamiltonian systems with periodic Hamiltonian, to a famous solution by Conley–Zehnder of a *conjecture of Arnold*.

For some of the above results, it was necessary to analyze in more detail the Morse index of the action associated to non-autonomous linear second order system (this index is finite here because the system has order two), and it is the object of Section 9.1.

9.1 The Index of a Linear Second Order Differential System

Let A be a continuous mapping from $\mathbf{R}$ into the space of symmetric matrices of order N. We consider the periodic boundary value problem

$$\begin{gathered}\ddot{u}(t) + A(t)u(t) = 0\\ u(0) - u(T) = \dot{u}(0) - \dot{u}(t) = 0\end{gathered} \tag{1}$$

where $T > 0$ is fixed. The corresponding action is defined on H_T^1 by

$$\chi_T(u) = \int_0^T (1/2)[|\dot{u}(t)|^2 - (A(t)u(t), u(t))]\, dt.$$

Definition 9.1. The *index* $j(A,T)$ is the Morse index of χ_T.

Let us define the linear operator K on H_T^1 (with its usual norm and inner product) by the formula

$$((Ku, v)) = \int_0^T (u(t) + A(t)\, u(t), v(t))\, dt.$$

It is easy to check that K is self-adjoint and compact, and that

$$2\chi_T(u) = ((u - Ku, u)).$$

The space H_T^1 can be written as the orthogonal direct sum of $\ker(I-K)$, H^+ and H^- with $I - K$ positive (resp. negative) definite on H^+ (resp. H^-). Since K has at most finitely many eigenvalues (having, moreover, finite multiplicity) greater than one,

$$j(A,T) = \dim H^- < \infty,$$

i.e. *the index* $j(A,T)$ *is finite.*

Definition 9.2. The *nullity* $\nu(A,T)$ is the dimension of $\ker(I-K)$.

It is easy to verify that the nullity is equal to the number of linearly independent solutions of (1), so that *the nullity* $\nu(A,T)$ *is less or equal to* $2N$. The linear operator $I - K$ is a Fredholm operator of index zero and hence is invertible if and only if $\nu(A,T) = 0$.

In the autonomous case, it is easy to compute the index and the nullity after diagonalization of A.

Proposition 9.1. *Let $\alpha_1 \leq \alpha_2 \leq \ldots \leq \alpha_N$ be the eigenvalues of the constant matrix A. Then*

$$j(A,T) = \#\{k : \alpha_k > 0\} + 2\sum_{k=1}^{N} \#\left\{j \in \mathbf{N}^* : \frac{4\pi^2 j^2}{T^2} < \alpha_k\right\}$$

$$\nu(A,T) = \#\{k : \alpha_k = 0\} + 2\sum_{k=1}^{N} \#\left\{j \in \mathbf{N}^* : \frac{4\pi^2 j^2}{T^2} = \alpha_k\right\}.$$

Let us now consider the functional φ defined on H_T^1 by

$$\varphi(u) = \int_0^T [(1/2)|\dot{u}(t)|^2 - F(t, u(t))]\, dt$$

where $F \in \mathbf{C}^2([0,T] \times \mathbf{R}^N, \mathbf{R})$. Since

$$\varphi''(u_0)(u,v) = \int_0^T [(\dot{u}(t), \dot{v}(t)) - (D_u^2 F(t, u_0(t))u(t), v(t))]\, dt,$$

every critical point u_0 of φ satisfies the following properties:

i) The Morse index of u_0 is equal to $j(A,T)$ where $A(t) = D_u^2 F(t, u_0(t))$.

ii) The nullity of u_0 is equal to $\nu(A,T)$.

iii) $\varphi''(u_0)$ is a Fredholm operator.

iv) u_0 is non-degenerate if and only if $\nu(A,T) = 0$.

By Corollary 8.4, if u_0 is an isolated critical point of φ, then $\dim C_n(\varphi, u_0)$ is finite for every n and equal to zero except if $n \in \{j(A,T), j(A,T)+1, \ldots, j(A,T)+\nu(A,T)\}$.

9.2 Periodic Solutions of Autonomous Second Order Systems Near an Equilibrium

We consider the existence of small non-trivial periodic solutions for the autonomous system

$$\ddot{u}(t) + \nabla F(u(t)) = 0 \tag{2}$$

where $F \in \mathbf{C}^2(\mathbf{R}^N, \mathbf{R})$ is such that

$$\nabla F(u) = Au + o(|u|)$$

as $|u| \to 0$. Let

$$\beta_1^2 < \beta_2^2 < \ldots < \beta_s^2$$

($\beta_r \geq 0, 0 \leq r \leq s$) be the non-negative eigenvalues of the symmetric matrix A.

Theorem 9.1. *If β_i is such that $\beta_r/\beta_i \notin \mathbf{N}$ for all $r \neq i$, then there exists a sequence (u_k) of periodic solutions of (2), with minimal period T_k such that $|u_k|_\infty \to 0$ and $T_k \to 2\pi/\beta_i$ as $k \to \infty$.*

It is easy to verify that u is a periodic solution of (2) with minimal period $2\pi\lambda$ if and only if $u(t) = v(t/\lambda)$ where v is a solution of

$$\begin{aligned} \ddot{v}(t) + \lambda^2 \nabla F(v(t)) = 0 \\ v(0) - v(2\pi) = \dot{v}(0) - \dot{v}(2\pi) = 0 \end{aligned} \tag{3}$$

with minimal period 2π.

Let us define φ on $\mathbf{R} \times H^1_{2\pi}$ by

$$\varphi(\lambda, u) = \varphi_\lambda(u) = \int_0^{2\pi} [(1/2)|\dot{u}(t)|^2 - \lambda^2 F(u(t))]\, dt$$

so that the solutions of (3) are the critical points of φ_λ. For each $\lambda \in \mathbf{R}$, 0 is a critical point of φ_λ. Let us also define the operators K and N on $H^1_{2\pi}$ by the formulas

$$((Ku, v)) = \int_0^{2\pi} (u(t), v(t))\, dt$$

$$((Nu, v)) = \int_0^{2\pi} (\nabla F(u(t)), v(t))\, dt,$$

so that

$$\langle \varphi'_\lambda(u), v \rangle = ((u - Ku - \lambda^2 Nu, v)).$$

The proof of Theorem 9.1 requires the following lemma.

Lemma 9.1. *Let $\lambda \in \mathbf{R}$ and $r > 0$. The functional φ_λ satisfies the Palais–Smale condition over $B[0, r]$.*

Proof. Let (u_j) be a sequence in $B[0, r]$ such that $\nabla\varphi_\lambda(u_j) \to 0$, i.e.

$$u_j - Ku_j - \lambda^2 Nu_j = f_j, \quad j \in \mathbf{N}^*,$$

with $f_j \to 0$ in $H^1_{2\pi}$. Going if necessary to a subsequence, we can assume that $u_j \rightharpoonup u$ in $H^1_{2\pi}$ and that $u_j \to u$ uniformly on $[0, 2\pi]$. This implies that $Ku_j \to Ku$ and $Nu_j \to Nu$. Therefore, $u_j \to Ku + \lambda^2 Nu$.

Proof of Theorem 1. 1) Let us first prove that $(1 \mid \beta_i, 0)$ is a bifurcation point for the equation

$$\nabla\varphi_\lambda(u) = 0. \tag{4}$$

By assumption, if $\epsilon \in]0, \beta_i/2[$ is sufficiently small, we have $\lambda\beta_r \notin \mathbf{N}$ for $r \neq i$ whenever

$$\lambda \in [1/(\beta_i + \epsilon), 1/(\beta_i - \epsilon)].$$

We shall apply Theorem 8.8 with $a = 1/(\beta_i + \epsilon)$ and $b = 1/(\beta_i - \epsilon)$. Proposition 9.1 implies that

$$\nu(a^2A, 2\pi) = \nu(b^2A, 2\pi) = 0$$

and

$$j(b^2A, 2\pi) - j(a^2A, 2\pi) = 2m$$

where m is the multiplicity of β_i^2 as an eigenvalue of A. Thus, 0 is an isolated critical point of φ_a and φ_b and, if $n = j(a^2A, 2\pi)$, Corollary 8.3 implies that

$$\dim C_n(\varphi_a, 0) = 1 \neq 0 = \dim C_n(\varphi_b, 0).$$

By Lemma 9.1 and Theorem 8.8, there exists a bifurcation point $(\lambda_0, 0) \in [a, b] \times \{0\}$ for (4). Letting $\epsilon \to 0$, we obtain the desired conclusion.

2) By the first part of the proof, there exists a sequence (λ_k, v_k) of solutions of (3) such that $\lambda_k \to 1/\beta_i$, $v_k \neq 0$ and $v_k \to 0$ in $H^1_{2\pi}$. Since $v_k \to 0$ uniformly on $[0, 2\pi]$, we have

$$\frac{\|\nabla F(v_k) - Av_k\|_\infty}{\|v_k\|_\infty} \to 0 \quad \text{as} \quad k \to \infty. \tag{5}$$

In particular, there exists $C > 0$ such that

$$\frac{\|\nabla F(v_k)\|_\infty}{\|v_k\|_\infty} \leq C, \quad k \in \mathbf{N}^*. \tag{6}$$

Let $w_k = v_k/\|v_k\|_\infty$. It follows from (3) and (6) that $(\|\ddot{w}_k\|_\infty)$ is bounded, and so is $(\|\dot{w}_k\|_\infty)$ by the Sobolev inequality. Using the Ascoli–Arzela theorem, we can assume, going if necessary to a subsequence, that $w_k \to w$ and $\dot{w}_k \to \dot{w}$ uniformly on $[0, 2\pi]$. It follows then from (5) that

$$\left\| \frac{\nabla F(v_k)}{\|v_k\|_\infty} - Aw \right\|_\infty \to 0 \quad \text{as} \quad k \to \infty. \tag{7}$$

By (3) and (7), we obtain

$$\ddot{w}(t) + \frac{1}{\beta_i^2} Aw = 0$$

$$w(0) - w(2\pi) = \dot{w}(0) - \dot{w}(2\pi) = 0.$$

Since $\|w\|_\infty = 1$ and, by assumption, $\beta_r/\beta_i \notin \mathbf{N}$ for $r \neq i$, 2π is the minimal period of w. Thus, for k sufficiently large, 2π is also the minimal period of w_k and, hence, of v_k, which completes the proof. □

Remarks. 1) If the multiplicity of β_i^2 as an eigenvalue of A is equal to one, then Theorem 9.1 follows from the classical Liapunov Center Theorem.

2) By Theorem 9.1, the small periodic solutions of system (2) are related to the periodic solutions of the linearized system $\ddot{u} + Au = 0$. This is not

the case in general, as shown by the following examples where, respectively, the nonlinearity is not a gradient and the differential operator is not of the second order.

Example 9.1. Assume that $u = (u_1, u_2)$ is a T-periodic solution of the system

$$\ddot{u}_1 + u_1 + u_2^3 = 0$$
$$\ddot{u}_2 + u_2 - u_2^3 = 0.$$

After multiplying the first equation by u_2, the second by u_1, integrating from 0 to T and subtracting, we obtain

$$\int_0^T [u_2^4(t) + u_1^4(t)]\, dt = 0,$$

i.e. $u = 0$. On the other hand, all the solutions of the linearized system are 2π-periodic.

Example 9.2. Consider the Hamiltonian

$$\begin{aligned} H(u) &= H(u_1, u_2, u_3, u_4) \\ &= (1/2)(u_1^2 + u_3^2 - u_2^2 - u_4^2) + (u_1^2 + u_2^2 + u_3^2 + u_4^2)(u_3u_4 - u_1u_2). \end{aligned}$$

If u is a solution of the corresponding system

$$J\dot{u} + \nabla H(u) = 0,$$

then

$$\frac{d}{dt}(u_1u_4 + u_2u_3) = 4(u_3u_4 - u_1u_2)^2 + 2u_1^2u_2^2 + 2u_3^2u_4^2.$$

Since the right-hand side is positive for $u \neq 0$, we conclude that $u = 0$ is the unique periodic solution of the system. But, in this case also, all the solutions of the linearized system are 2π-periodic.

9.3 Periodic Solutions of Asymptotically Linear Non-Autonomous Second Order Systems

We consider the existence of multiple solutions of the periodic boundary value problem

$$\begin{gathered} \ddot{u}(t) + \nabla F(t, u(t)) = 0 \\ u(0) - u(T) = \dot{u}(0) - \dot{u}(T) = 0 \end{gathered} \tag{8}$$

where $F \in \mathbf{C}^2([0,T] \times \mathbf{R}^N, \mathbf{R})$ satisfies the conditions

$$\nabla F(t,u) = A_0(t)u + o(|u|) \quad \text{as} \quad |u| \to 0 \tag{9}$$

and

$$\nabla F(t,u) = A_\infty(t)u + o(|u|) \quad \text{as} \quad |u| \to \infty, \tag{10}$$

uniformly in $t \in [0,T]$. We shall write $j_0 = j(A_0,T)$ and $j_\infty = j(A_\infty,T)$.

Theorem 9.2. *Assume that $T > 0$ is such that the following conditions hold.*

A$_1$. $\nu(A_0,T) = 0$

A$_2$. $\nu(A_\infty,T) = 0$

A$_3$. $j_0 \neq j_\infty$.

Then the problem (8) *has at least one non-zero solution. Assume, moreover, that*

A$_4$. $|j_0 - j_\infty| \geq 2N$.

Then the problem (8) *has at least two non-zero solutions.*

The solution of (8) are the critical points of the functional φ defined on H_T^1 by

$$\varphi(u) = \int_0^T [(1/2)|\dot{u}(t)|^2 - F(t,u(t))]\,dt.$$

Let us also define the operator L and the functional ψ on H_T^1 by

$$((Lu,v)) = \int_0^T (\dot{u}(t) - A_\infty(t)u(t), v(t))\,dt$$

$$\psi(u) = \varphi(u) - (1/2)((Lu,u)).$$

Assumption A$_2$ implies that L is invertible. Since

$$\begin{aligned} |((\nabla\psi(u),v))| &= \left|\int_0^T (A_\infty(t)u(t) - \nabla F(t,u(t)), v(t))\,dt\right| \\ &\leq \|A_\infty u - \nabla F(.,u)\|_{L^2}\|v\|_{L^2} \\ &\leq \|A_\infty u - \nabla F(.,u)\|_{L^2}\|v\|, \end{aligned}$$

it follows from (10) that for every $\epsilon > 0$, there exists $c(\epsilon) \geq 0$ such that

$$\|\nabla\psi(u)\| \leq \epsilon\|u\| + c(\epsilon) \tag{11}$$

for every $u \in H_T^1$.

Proposition 9.2. *Under assumptions* (10) *and* A$_2$, *there exists $\rho > 0$ and $\hat{\sigma} \in C^\infty(H_T^1, \mathbf{R})$ satisfying the following conditions:*

a) $\nabla\varphi(u) = 0$ *implies* $\|u\| < \rho$.

b) $\hat{\sigma}(u) = 1$ *if* $0 \leq \|u\| \leq \rho$ *and* $\hat{\sigma}(u) = 0$ *if* $\|u\| \geq 2\rho$.

c) *the functional* $\hat{\varphi}(u) = (1/2)((Lu, u)) + \hat{\sigma}(u)\psi(u)$ *is such that* $\|\nabla\hat{\varphi}(u)\| \geq 1$ *if* $\rho \leq \|u\| \leq 2\rho$.

Proof. Taking $\epsilon = \|L^{-1}\|^{-1}/9$, there will exist by (11) $c_1(\epsilon)$ such that

$$\|\nabla\psi(u)\| \leq \epsilon\|u\| + c_1 \tag{12}$$

on H_T^1. Therefore, by the mean value theorem, we have

$$\begin{aligned}|\psi(u)| &\leq \int_0^T ((\nabla\psi(su), u))\, ds + |\psi(0)| \\ &\leq (\epsilon/2)\|u\|^2 + c_1\|u\| + |\psi(0)|.\end{aligned}$$

Thus, there exists $c_2 > 0$ such that

$$|\psi(u)| \leq \epsilon\|u\|^2 + c_2. \tag{13}$$

Let

$$\rho = 1 + \left(1 + c_1 + \frac{3c_2}{2}\right)/\epsilon.$$

It follows from (12) that

$$\|\nabla\varphi(u)\| \geq 9\epsilon\|u\| - \epsilon\|u\| - c_1.$$

Thus, the critical points of φ satisfy the a priori estimate

$$\|u\| \leq c_1/8\epsilon < \rho$$

and (a) is verified.

Let $\sigma \in C^\infty(\mathbf{R}, \mathbf{R})$ be such that

$$\begin{aligned}\sigma(s) &= 1 \quad \text{for } s \leq 0 \\ &= 0 \quad \text{for } s \geq 1 \\ -3/2 \leq \sigma'(s) &\leq 0 \quad \text{for } s \in \mathbf{R}.\end{aligned}$$

The function $\hat{\sigma}$ defined on H_T^1 by

$$\hat{\sigma}(u) = \sigma\left(\frac{\|u\| - \rho}{\rho}\right)$$

satisfies (b). If $\rho \leq \|u\| \leq 2\rho$, we deduce from (12) and (13) that

$$\begin{aligned}\|\nabla\hat{\varphi}(u)\| &= \left\|Lu + \sigma\left(\frac{\|u\| - \rho}{\rho}\right)\nabla\psi(u) + \sigma'\left(\frac{\|u\| - \rho}{\rho}\right)\frac{\psi(u)}{\rho\|u\|}u\right\| \\ &\geq 9\epsilon\rho - 2\epsilon\rho - c_1 - (3/2\rho)(4\epsilon\rho^2 + c_2) \geq \epsilon\rho - c_1 - (3/2)c_2 \geq 1,\end{aligned}$$

and the proof is complete. □

By Proposition 9.2, $\nabla\varphi(u) = 0$ if and only if $\nabla\hat{\varphi}(u) = 0$. Thus, in order to solve problem (8), it suffices to find the critical points of $\hat{\varphi}$.

Lemma 9.2. *Under the assumptions* (20) *and* A_2, *every sequence* (u_j) *in* H_T^1 *such that* $\nabla\hat{\varphi}(u_j) \to 0$ *contains a convergent subsequence.*

Proof. Assumption A_2 and Proposition 9.2 imply that $\|u_j\| < \rho$ $(j \in \mathbf{N})$. Thus,

$$\nabla\hat{\varphi}(u_j) = \nabla\varphi(u_j), \quad (j \in \mathbf{N}).$$

Arguing as in Lemma 9.1, we can conclude that (u_j) contains a convergent subsequence. □

Lemma 9.3. *Under the Assumption* (10) *and* A_2, *there exist* $a < b$ *such that the critical points of* $\hat{\varphi}$ *belong to* $\hat{\varphi}^{-1}(]a, b[)$ *and*

$$P(t, \hat{\varphi}^b, \hat{\varphi}^a) = t^{j_\infty}.$$

Proof. Define

$$a = \inf_{B[0,2\rho]} \hat{\varphi} - 1, \quad b = \sup_{B[0,2\rho]} \hat{\varphi} - 1,$$

and $\varphi_\infty(u) = (1/2)((Lu, u))$. Proposition 9.2 implies that $\hat{\varphi}^{-1}(]a, b[)$ contains the critical points of $\hat{\varphi}$ and that $\hat{\varphi}^a = \varphi_\infty^a$, $\hat{\varphi}^b = \varphi_\infty^b$. Hence

$$P(t, \hat{\varphi}^b, \hat{\varphi}^a) = P(t, \varphi_\infty^b, \varphi_\infty^a).$$

Since, by assumption A_2, 0 is the only critical point of the quadratic functional φ_∞, it follows from Theorem 8.1 and Corollary 8.3 that

$$P(t, \varphi_\infty^b, \varphi_\infty^a) = t^{j_\infty}. \quad \square$$

Proof of Theorem 9.2. We can assume that problem (8) has only a finite number of solutions, i.e. that $\hat{\varphi}$ has only a finite number of critical points. By Lemma 9.2, $\hat{\varphi}$ satisfies the Palais–Smale condition over H_T^1. Theorem 8.2 and Lemma 9.3 imply the existence of a polynomial $Q(t)$ with non-negative integer coefficients such that

$$M(t, \hat{\varphi}^a, \hat{\varphi}^b) = t^{j_\infty} + (1 + t)\,Q(t). \tag{14}$$

Assumptions (9) and A_1 and Corollary 8.3 imply that

$$\dim C_n(\hat{\varphi}, 0) = \delta_{n,j_0}. \tag{15}$$

Since $j_0 \neq j_\infty$, we obtain from (14) and (15) the existence of at least one non-zero critical point.

Now assume that $|j_0 - j_\infty| \geq 2N$ and that u is the only non-zero critical point of φ_1. Since, by (14),

$$t^{j_0} + \sum_{n=0}^{\infty} \dim C_n(\varphi_1, u) t^n = t^{j_\infty} + (1 + t)Q(t),$$

we necessarily have

$$\dim C_{j_\infty}(\hat{\varphi}, u) \geq 1$$

and either $\dim C_{j_0-1}(\hat{\varphi}, u) \geq 1$ or $\dim C_{j_0+1}(\hat{\varphi}, u) \geq 1$. Let us consider the case where $\dim C_{j_0-1}(\hat{\varphi}, u) \geq 1$, the other one being similar. By assumption, $j_0 - 1 \neq j_\infty$. Since the nullity of u is less or equal to $2N$, Corollary 8.4 implies that

$$|j_0 - 1 - j_\infty| \leq 2N - 2.$$

Hence, we obtain $|j_0 - j_\infty| \leq 2N - 1$, which is impossible since $|j_0 - j_\infty| \geq 2N$. □

9.4 Multiple Solutions of Lagrangian Systems

We consider the periodic boundary value problem

$$\begin{gathered}\frac{d}{dt} D_y L(t, u(t), \dot{u}(t)) = D_x L(t, u(t), \dot{u}(t)) \\ u(0) - u(T) = \dot{u}(0) - \dot{u}(T) = 0\end{gathered} \tag{16}$$

where $L = L(t, x, y)$ satisfies the assumptions (L_1) to (L_4) of Section 4.2.

Theorem 9.3. *Under the above assumptions, if all the weak solutions of* (16) *are non-degenerate, then* (16) *has at least* 2^N *geometrically distinct weak solutions.*

Proof. The weak solutions of (16) are the critical points of the functional φ defined on H_T^1 by

$$\varphi(u) = \int_0^T L(t, u(t), \dot{u}(t))\, dt.$$

By Proposition 4.1, φ is bounded from below and continuously differentiable. Since, by assumption (L_3),

$$\varphi(u + T_i e_i) = \varphi(u), \quad 1 \leq i \leq N,$$

it is natural to define φ on the manifold $M = T^N \times \tilde{H}_T^1$, where T^N is the N-dimensional torus and

$$\tilde{H}_T^1 = \left\{ u \in H_T^1 \,:\, \int_0^T u(t)\, dt = 0 \right\}.$$

By Proposition 4.1, φ satisfies the Palais–Smale condition over M. Without loss of generality, we can assume that φ has only a finite number of critical points $u_1, \ldots, u_j$. By a classical result of algebraic topology,

$$P(t, M, \phi) = P(t, T^N, \phi) = \sum_{n=0}^{N} \binom{N}{n} t^n.$$

By Corollary 8.2, there exists a polynomial $Q(t)$, with non-negative integer coefficients, such that

$$\sum_{n=0}^{\infty}\sum_{i=1}^{j}\dim C_n(\varphi,u_i)t^n=\sum_{n=0}^{N}\binom{N}{n}t^n+(1+t)Q(t). \tag{17}$$

Since the critical points of φ are non-degenerate, Corollary 8.3 implies that

$$\dim C_n(\varphi,u_i)=\delta_{n,k_i}, \tag{18}$$

where k_i is the Morse index of u_i. It follows from (17) and (18) that φ has at least

$$\sum_{n=0}^{N}\binom{N}{n}=2^N$$

critical points in M, so that (16) has at least 2^N geometrically distinct weak solutions. □

Historical and Bibliographical Notes

Theorem 9.1 was proved by Berger [Ber$_1$] using the Lyapunov–Schmidt method. Example 9.2 is due to Moser. The existence of a non-zero solution in Theorem 9.2 follows from Amann–Zehnder [AmZ$_1$] and the existence of two non-zero solutions from Dancer [Dan$_2$]. These authors use a finite-dimensional reduction which is in fact Cesari's method [Ces$_2$] and a wide generalization of Morse theory, the Conley index (see [Con$_1$, CoZ$_1$]). By the same method, Conley and Zehnder were able to solve the Arnold's conjecture for a torus ([CoZ$_2$]) and to consider general first order asymptotically linear Hamiltonian systems [CoZ$_1$]. But, since those problems are variational, it suffices to apply Morse theory or minimax methods. Chang [Ch$_1$] uses Proposition 9.2 and some deformation arguments which are in fact superfluous. The simple approach of Section 9.3 is also applicable to first order Hamiltonian systems and to semi-linear wave equations after a finite-dimensional reduction.

An earlier application of homology to obtain multiple critical points was made by Castro–Lazer [CaL$_1$]. See also [Cha$_2$] and [Cot$_{1,2}$] for applications of Morse theory to related problems.

Other applications of Morse theory are given in [Amb$_8$], [AmbL$_1$], [Ch$_{6,8}$], [TsW$_1$], [MerP$_1$], [Ts$_1$].

We have not considered here the important concept of Conley's index and its generalizations. See [Ben$_{7,8}$], [Ryb$_{2,3,6,8}$], [RybZ$_1$] and, for applications to Hamiltonian systems [Bart$_1$], [BentZ$_1$], [CZ$_3$], to boundary value problems [Dan$_{3,4,5,8}$], [Ryb$_{4,5,7}$].

The Arnold conjecture is considered in [Chap$_1$], [Hof$_6$].

Concerning singular dynamical systems, the reader can consult [AmC$_2$], [Cot$_{3,4,5}$] for a treatment by Morse theory and [Ben$_4$], [CGS$_1$], [CaS$_3$], [Gor$_{3,4}$], [Gre$_{2,3,4}$], [PiT$_3$], [DGM$_1$] for a treatment by minimax methods.

Exercises

1. Assume that $F \in C^2([0,T] \times \mathbf{R}^N, \mathbf{R})$ satisfies conditions (9) and (10) of Section 9.3. If

   A1. $\nu(A_0, T) = 0$

   A2. $\nu(A_\infty, T) = 0$

   A3. $j_\infty = 0 \neq j_0$,

   then the problem (8) has at least two non-zero solutions.

2. Assume that $F \in C^2([0,T] \times \mathbf{R}^N, \mathbf{R})$ satisfies condition (9) and (10) of Section 9.3. If

   A1. $\nu(A_0, T) = 0$

   A2. $\nu(A_\infty, T) = 0$

   A3. $j_0 = 0, \quad j_\infty = 1$,

   then problem (8) has at least two non-zero solutions.

3. Assume that $F \in C^2([0,T] \times \mathbf{R}^N, \mathbf{R})$ satisfies conditions (9) and (10) of Section 9.3. If

   A1. $\nu(A_\infty, T) = 0$

   A2. $j_\infty \notin [j_0, j_0 + \nu(A_0, T)]$,

   then problem (8) has at least one non-zero solution.

10

Nondegenerate Critical Manifolds

Introduction

After recalling some preliminary notions from differential geometry, this chapter presents the local and global aspects of the theory of *nondegenerate critical manifolds*. These manifolds are a natural extension of the notion of non-degenerate critical point.

The theory is applied to proving the existence of infinitely many periodic solutions of the *forced superlinear second order equation*

$$\ddot{u} + |u|^{p-2}u = f(t), \quad p \in]2, \infty[.$$

The periodic solutions of the forced equation are obtained from the periodic orbits of the corresponding autonomous equation

$$\ddot{u} + |u|^{p-2}u = 0$$

by a global perturbation argument. This approach depends upon a precise description of the solutions of the autonomous equation.

The last section is devoted to the existence of T-periodic solutions of the *perturbed second order equation*

$$\ddot{u}(t) + g(u(t)) = \epsilon f(t)$$

near a T-periodic orbit of the autonomous equation

$$\ddot{u}(t) + g(u(t)) = 0.$$

Since this equation is conservative, 1 is a Floquet multiplier with multiplicity 2 of its variational equation, so that classical perturbation arguments are not applicable. The periodic solutions are obtained here by combining the *Liapunov–Schmidt method* with an elementary variational argument. An application is given to the *subharmonics of the forced pendulum equation.*

10.1 Submanifolds

We define a class of sets locally diffeomorphic to a subspace of a Banach space.

Definition 10.1. *Let X be a Banach space and $Y \subset X$ be a closed subspace with a closed complement. A subset Z of X is a C^k-submanifold of X modelled on Y if, for every $z \in Z$, there exists an open neighborhood A of 0 in X, an open neighborhood B of z in X and a C^k-diffeomorphism $\Phi : A \to B$ such that*

$$\Phi(A \cap Y) = B \cap Z.$$

The tangent space of Z at z is defined by

$$T_z Z = \{\Phi'(\Phi^{-1}(z))v : v \in Y\}.$$

Remark 10.1. It is clear that

$$\Phi^{-1}|_{B \cap Z} : B \cap Z \to A \cap Y$$

is a chart at z and that Z is a C^k-manifold modelled on Y.

Remark 10.2. The space $T_z Z$ is independent of Φ and isomorphic to the tangent space at z of Z considered as a manifold.

Remark 10.3. The restriction to $A \cap Y$ of Φ is called a parametrization of $B \cap Z$.

The following theorem describes an interesting class of submanifolds.

Theorem 10.1 (Embedding theorem). *Let M be a C^k-manifold, X be a Hilbert space and let $f \in C^k(M, X)$. Assume that*

a) *M is compact.*

b) *f is injective.*

c) *$df|_{T_u M}$ is injective for every $u \in M$.*

Then $f(M)$ is a compact submanifold of X.

Proof. Let $x : D(x) \subset M \to R(x) \subset \mathbf{R}^N$ be a chart at $u \in M$. (Since M is compact we can assume that M is modelled on $\mathbf{R}^N$.) The set $D(x)$ being open, assumptions a) and b) imply the existence of an open set C of X such that $f(M) \cap C = f(D(x))$. By assumption, $(f \circ x^{-1}) \in C^k(R(x), X)$ and $(f \circ x^{-1})'(x(u))$ is injective. Define

$$V = (f \circ x^{-1})'(x(u))(\mathbf{R}^N)$$

and consider the mapping $\Phi : R(x) \times V^{\perp} \to X$ given by

$$\Phi(v, w) = (f \circ x^{-1})(v) + w.$$

It is easy to verify that $\Phi'(x(u), 0)$ is invertible. By the inverse function theorem, the mapping Φ is a C^k-diffeomorphism from an open neighborhood A of $(x(u), 0)$ in $R(x) \times V^{\perp}$ to an open neighborhood B of $f(u)$ in X. Moreover we can assume that $B \subset C$. Since f is injective, we have

$$\Phi(A \cap (R(x) \times \{0\})) = B \cap f(D(x)) = B \cap f(M).$$

It is then easy to verify the definition of submanifold by a formal manipulation. □

10.2 Normal Bundle

In this section, we assume that Z is a C^k-submanifold of a Hilbert space X modelled on a subspace Y.

Definition 10.2. *The normal space of Z at z, N_z, is the orthogonal complement of T_zZ in X. The normal bundle of Z is defined by*

$$NZ = \{(z,n) \in Z \times X \,:\, n \in N_zZ\}.$$

We also define, for $\rho > 0$, $N_\rho Z$ by

$$N_\rho Z = \{(z,n) \in NZ \,:\, |n| \leq \rho\}.$$

Lemma 10.1. *If Z is a compact C^2-submanifold of X modelled on Y, then there exists $\rho_0 > 0$, an open neighborhood $\mathcal{N}$ of Z and mappings $F \in C^1(\mathcal{N}, X)$, $G \in C^1(\mathcal{N}, X)$ such that, for every $u \in \mathcal{N}$,*

i) $u = F(u) + G(u)$;

ii) $(F(u), G(u)) \in N_{\rho_0}Z$;

iii) *the above decomposition is unique.*

Proof. 1) Let z, A, B, and Φ be like in Definition 10.1. Consider the decomposition

$$u = m + n, \tag{1}$$

where $m = \Phi(y)$, $y \in A \cap Y$, $n \in N_mZ$. Equation (1) is equivalent to

$$(\Phi(y) - u, \Phi'(y)v) = 0 \quad \text{for all} \quad v \in Y. \tag{2}$$

Let us denote by $L(y)$ the map $\Phi'(y)$ restricted to Y, so that equation (2) is equivalent to

$$L(y)^*(\Phi(y) - u) = 0. \tag{3}$$

Since Z is compact, the dimension of Y is finite and it is easy to verify that $L^*(.) : A \cap Y \to \mathcal{L}(X, Y)$ is continuously differentiable on $A \cap Y$. By the implicit function theorem, near $u_0 = z$ and $y_0 = \Phi^{-1}(z)$, equation (3) defines a C^1-function $y = g(u)$ such that $y_0 = g(u_0)$. It is then possible to locally define F and G near z by

$$F(u) = \Phi(g(u)), \quad G(u) = u - \Phi(g(u)).$$

In addition we can assume that, for $\rho > 0$ fixed,

$$|G(u)| = |G(u) - G(z)| \leq \rho.$$

By the local uniqueness of the decomposition (1), the mappings F and G are well defined on an open neighborhood of Z.

2) Let us prove that (iii) is satisfied for ρ_0 small enough. If it is not the case we can find sequences $(m_j, n_j) \in NZ$, $(m'_j, n'_j) \in NZ$ such that

$$m_j + n_j = m'_j + n'_j$$

$$n_j \to 0, n'_j \to 0 \quad \text{as} \quad j \to \infty$$

and $n_j \neq n'_j$, $j \in \mathbf{N}$. Going if necessary to a subsequence, we can assume that $m_j \to z \in Z$. Necessarily, $m'_j \to z$ and this contradicts the local uniqueness of the decomposition (1). □

A proof by contradiction of the following lemma is straightforward.

Lemma 10.2. *Let f be a continuous mapping between two metric spaces A and B. If $Z \subset A$ is compact, then, for each $\epsilon > 0$, there exists $\delta > 0$ such that*

$$d(f(a), f(z)) \leq \epsilon$$

whenever $a \in A$, $z \in Z$ and $d(a, z) \leq \delta$.

Let us recall that, if Z is a subspace of a metric space X, we use the notation

$$Z_\delta = \{u \in X \, : \, d(u, Z) \leq \delta\}.$$

Theorem 10.2. *Under the assumptions of Lemma 10.1, there exists $\epsilon_0 > 0$ such that for each $\epsilon \in]0, \epsilon_0[$, the map*

$$\Psi : NZ \to X, \quad (z, u) \to z + u$$

induces a homeomorphism from $N_\epsilon Z$ onto a closed neighborhood of Z.

Proof. Let $\epsilon_0 \in]0, \rho_0]$ be such that $Z_\epsilon \subset \mathcal{N}$. By Lemma 10.2, for each $\epsilon \in]0, \epsilon_0]$, there exists $\delta \in]0, \epsilon_0]$ such that $u \in Z_\delta$ implies $|G(u)| \leq \epsilon$. Hence, $Z_\delta \subset \Psi(N_\epsilon Z)$ and $\Psi(N_\epsilon Z)$ is a neighborhood of Z. Using the compactness of Z and the definition of NZ, it is easy to verify that $\Psi(N_\epsilon Z)$ is closed. Finally, since

$$\Psi(N_\epsilon Z) \subset \Psi(N_{\epsilon_0} Z) \subset Z_{\epsilon_0} \subset \mathcal{N},$$

the mapping Ψ is a homeomorphism from $N_\epsilon Z$ onto $\Psi(N_\epsilon Z)$. □

Proposition 10.1. *Let Z be a finite-dimensional C^2-submanifold of X modelled on Y and let P_m* (resp. Q_m) *be the orthogonal projectors onto $N_m Z$* (resp. $T_m Z$). *Then the mappings*

$$m \to P_m, \quad m \to Q_m$$

are continuously differentiable.

Proof. Since $P_m = I - Q_m$, it suffices to prove the result for Q_m. Let $(y_1, \ldots, y_j)$ be a basis of Y and let A, B, and Φ be as in Definition 10.1. For $m \in B \cap Z$ and $1 \leq i \leq j$, define $e_i(m)$ by

$$e_i(m) = \Phi'(\Phi^{-1}(m)) y_i,$$

so that $(e_1(m), \ldots, e_j(m))$ is a basis of $T_m Z$. If $(f_1(m), \ldots, f_j(m))$ is the basis obtained from $(e_1(m), \ldots, e_j(m))$ by orthonormalization, then

$$Q_m u = \sum_{i=1}^{j} (u, f_i(m))\, f_i(m).$$

By construction, the mapping $m \to Q_m$ is continuously differentiable. □

10.3 Critical Groups of a Nondegenerate Critical Manifold

The purpose of this section is to extend to compact manifolds of critical points the notion of nondegenerate critical point and to compute the corresponding critical groups.

Definition 10.3. *Let Z be a compact connected C^2-submanifold of a Hilbert space X and let $\varphi \in C^2(X, \mathbf{R})$. We say that Z is a nondegenerate critical manifold of φ if*

a) *all points of Z are critical points of φ;*

b) *the nullity of each $z \in Z$ is equal to the dimension of Z;*

c) *$\varphi''(z)$ is a Fredholm operator for each $z \in Z$.*

Remarks. 1) Under Assumption a) the nullity of each $z \in Z$ is greater or equal to the dimension of Z. Indeed, using the notations of Definition 10.1, we have

$$\nabla\varphi(\Phi(y)) = 0 \quad \text{for all} \quad y \in A \cap Y.$$

Hence,

$$\varphi''(\Phi(y))\Phi'(y)v = 0 \quad \text{for all} \quad y \in A \cap Y \text{ and } v \in Y.$$

In particular,

$$\varphi''(z)\Phi'(\Phi^{-1}(z))v = 0 \quad \text{for all} \quad v \in Y$$

so that $T_z Z \subset \ker \varphi''(z)$ and

$$\dim \ker \varphi''(z) \geq \dim T_z Z = \dim Z.$$

2. By the preceding Remark and Assumption b), $T_z Z = \ker \varphi''(z)$ for each $z \in Z$. Since, by Assumption c), $\varphi''(z)$ is a Fredholm operator, the spectral theorem implies that $N_z Z$ is the orthogonal sum of $N_z^+ Z$ and $N_z^- Z$ with $\varphi''(z)$ positive (resp. negative) definite on $N_z^+ Z$ (resp. $N_z^- Z$). Let us

denote by P_z^+ and P_z^- the corresponding orthogonal projectors. After the complexification of X and $\varphi''(z)$, P_z^+ is given by

$$P_z^+ = \frac{1}{2i\pi}\int_{\partial D}(\varphi''(z) - \lambda I)^{-1} d\lambda,$$

where D is an open disk in $\mathbf{C}$ containing the positive part of $\sigma(\varphi''(z))$ and such that $\partial D \cap \sigma(\varphi''(z)) = \phi$. Thus P_z^+ depends continuously upon z for the operator norm, and the same is true for P_z^-.

Definition 10.4. *Let X be a Hilbert space and let Z be a nondegenerate critical manifold of $\varphi \in C^2(X, \mathbf{R})$. The Morse index of Z is defined as the Morse index of $\varphi''(z)$. (By continuity this last number is independent of z.) The critical groups of Z are defined by*

$$C_n(\varphi, Z) = H_n(\varphi^c \cap U, \varphi^c \cap U \setminus Z), \quad n = 0, 1, \dots,$$

where c is the (constant) value of φ on Z, where U is a closed neighborhood of Z and where one takes $F = \mathbf{Z}_2$. (By excision, the critical groups are independent of U.)

Theorem 10.3. *Let X be a Hilbert space and let Z be a nondegenerate critical manifold with finite Morse index k of $\varphi \in C^2(X, \mathbf{R})$. Then the critical groups of Z are given by*

$$C_n(\varphi, Z) \approx H_{n-k}(Z, \phi), \quad n = 0, 1, \dots.$$

The proof of Theorem 10.3 depends upon the following lemma. Without loss of generality, we can assume that $\varphi(z) = 0$ for all $z \in Z$.

Lemma 10.3. *Under the assumptions of Theorem 10.3, for every $\epsilon > 0$, there exists $\delta > 0$ such that, if $z \in Z$, $u \in X$, and $|u - z| \leq \delta$, one has*

$$\|\varphi''(z) - \varphi''(u)\| \leq \epsilon$$

$$\left|\varphi(u) - \frac{1}{2}(\varphi''(z)(u - z), u - z)\right| \leq \epsilon |u - z|^2.$$

Proof. It suffices to apply Lemma 10.2 to obtain the first conclusion. Let $\epsilon > 0$ and assume that $h = u - z$ is such that $|h| \leq \delta$ with $\delta > 0$ given by the first part of the lemma. For every $z \in Z$, we have

$$\varphi(z) = 0, \quad \nabla\varphi(z) = 0.$$

Consequently,

$$\varphi(z + h) = \int_0^1 (1 - s)(\varphi''(z + sh)h, h)\, ds$$

and

$$\left|\varphi(z+h) - \frac{1}{2}(\varphi''(z)h,h)\right| = \left|\int_0^1 (1-s)[(\varphi''(z+sh)h,h) - (\varphi''(z)h,h)]\,ds\right| \le \epsilon|h|^2. \qquad \square$$

Proof of Theorem 10.3. 1) By continuity and compactness,

$$\mu_+ = \min_{z\in Z} \inf_{\substack{|n|=1\\ n\in N_z^+Z}} (\varphi''(z)n,n) > 0,$$

$$\mu_- = \max_{z\in Z} \inf_{\substack{|n|=1\\ n\in N_z^-Z}} (\varphi''(z)n,n) > 0.$$

Let $\epsilon = \min\{(\mu_+/2), (-\mu_-/2)\}$ and let $\delta > 0$ be given by Lemma 10.3. (In particular $z \in Z$ and $|n| \le \delta$ imply

$$\left|\varphi(z+n) - \frac{1}{2}(\varphi''(z)n,n)\right| \le \epsilon|n|^2.)$$

Using the notations of Theorem 10.2, we can assume that $\delta < \epsilon_0$ and we can identify $N_\delta Z$ and $\Psi(N_\delta Z)$. Define

$$N_\delta^+ Z = \{(z,n) \in N_\delta Z \,:\, n \in N_z^+ Z\}$$

and similarly for $N_\delta^- Z$. Lemma 10.3 and the definition of ϵ imply that

$$(\varphi''(u)n,n) \ge 0 \text{ if } u \in X, n \in N_\delta^+ Z \text{ and } |u-z| \le \delta, \tag{4}$$

$$N_\delta^- Z \subset \varphi^0, \tag{5}$$

$$N_\delta^+ Z \cap \varphi^0 = Z. \tag{6}$$

By Theorem 10.2, $U = N_\delta Z$ is a closed neighborhood of Z. Let us define η by

$$\eta \,:\, [0,1] \times U \to U,\ (t,z,n) \to (z, n - tP_z^+ n).$$

This deformation η is continuous as P_z^+ depends continuously upon z. For $(z,n) \in U \cap \varphi^0$, let us define by $f(t) = \varphi(\eta(t,z,n))$. It follows from (4) that

$$f''(t) = (\varphi''(z + n - tP_z^+ n)\, P_z^+ n, P_z^+ n) \ge 0$$

for all $t \in [0,1]$, so that f is convex on $[0,1]$. But $f(0) = \varphi(z,n) \le 0$ since $(z,n) \in \varphi^0$ and $f(1) = \varphi(z, P_z^- n) \le 0$ by (5). Thus, $\varphi(\eta(t,z,n)) \le 0$ for all $t \in [0,1]$. Moreover, if $\eta(t,z,n) = (z,0)$ for some $t \in [0,1]$ and some $(z,n) \in U \cap \varphi^0$, then $n = 0$. Indeed, $\eta(t,z,n) = (z,0)$ implies $P_z^- n = 0$ so that, by (6), $n = 0$. Finally, $N_\delta^- Z \setminus Z$ is a deformation retract of $\varphi^0 \cap U \setminus Z$ and $N_\delta^- Z$ is a deformation retract of $\varphi^0 \cap U$. Hence,

$$H_n(\varphi^c \cap U, \varphi^c \cap U \setminus Z) \approx (H_n(N_\delta^- Z, N_\delta^- Z \setminus Z), \quad n = 0,1,\ldots.$$

2) Since $F = \mathbf{Z}_2$, by Thom's isomorphism theorem we have

$$H_n(N_\delta^- Z, N_\delta^- Z \setminus Z) \approx H_{n-k}(Z,\phi),$$

and the proof is complete. $\square$

10.4 Global Theory

In this section, the relative homology of the pair (φ^b, φ^a), with $b > a$, will be related to the critical groups of the nondegenerate critical manifolds contained in $\varphi^{-1}([a,b])$.

Lemma 10.4. *Let X be a Hilbert space and let Z be a nondegenerate critical manifold of $\varphi \in C^2(X, \mathbf{R})$. Let P_z be the orthogonal projector from X onto $N_z Z$ ($z \in Z$). Then there exists $\rho > 0$ and $\eta > 0$ such that if $z \in Z$, $u \in X$, $u - z \in N_z Z$, and $|u - z| \leq \rho$, one has*

$$|P_z \nabla\varphi(u)| \geq \eta |u - z|.$$

Proof. Since Z is a nondegenerate critical manifold, the mapping $L(z)$: $N_z Z \to N_z Z$ defined by $L(z)h = \varphi''(z)h$ is invertible for each $z \in Z$. By continuity and compactness we have

$$\mu = \min_{z \in Z} \|L(z)^{-1}\|^{-1} > 0. \tag{7}$$

Let $\epsilon = \mu/2$ and assume that $h = u - z$ is such that $|h| \leq \delta$ where $\delta > 0$ is given by Lemma 10.3. The mean value theorem implies that

$$|P_z \nabla\varphi(z+h) - \varphi''(z)h| = |P_z(\nabla\varphi(z+h) - \varphi''(z)h)|$$

$$\leq |\nabla\varphi(z+h) - \varphi''(z)h| \leq |h| \sup_{s \in]0,1[} \|\varphi''(z+sh) - \varphi''(z)\| \leq \epsilon |h|. \tag{8}$$

Assume now that $h \in N_z Z$. It follows from (7) and (8) that

$$|P_z \nabla\varphi(z+h)| \geq |\varphi''(z)h| - |P_z \nabla\varphi(z+h) - \varphi''(z)h| \geq \mu|h| - \epsilon|h| = \frac{\mu}{2}|h|,$$

and the proof is complete with $\rho = \delta$ and $\eta = \mu/2$. □

Let us now introduce the following framework.

(A)

i) X is a Hilbert space and $\varphi \in C^2(X, \mathbf{R})$;

ii) $a < b$ are real numbers such that the Palais–Smale condition is satisfied over $\varphi^{-1}([a,b])$;

iii) $c \in]a,b[$ is the only critical value of φ contained in $[a,b]$;

iv) K_c consists of a finite number of isolated critical points and nondegenerate critical manifolds.

For simplicity, we shall replace assumption iv) in the proofs by

iv′) K_c is a nondegenerate critical manifold Z.

The general case can be obtained by an easy adaptation of the proof.

Using the notations of Theorem 10.2 and Lemma 10.4, we can assume that $\rho < \epsilon_0$ and that $\Psi(N_\rho Z) \subset \varphi^{-1}(]a, b[)$, and we can identify $N_\rho Z$ and $\Psi(N_\rho Z)$. Let $\omega \in C^1(X, \mathbf{R})$ be such that $\omega(u) \in [0,1]$ and

$$\begin{aligned} \omega(u) &= 1 \quad \text{if} \quad u \in N_{\rho/2}Z \\ &= 0 \quad \text{if} \quad u \notin N_\rho Z. \end{aligned}$$

Consider the vector field $f : X \to X$ defined by

$$\begin{aligned} f(u) &= -\nabla\varphi(u) \quad \text{if } u \notin N_\rho Z \\ &= -(1-\omega(u))\nabla\varphi(u) - \omega(u)P_{F(u)}\nabla\varphi(u) \quad \text{if } u \in N_\rho Z. \end{aligned}$$

Lemma 10.1 and Proposition 10.1 imply that f is continuously differentiable. The Cauchy problem

$$\begin{aligned} \dot{\sigma}(t) &= f(\sigma(t)) \\ \sigma(0) &= u \end{aligned}$$

has, therefore, a unique maximal solution $\sigma(.) = \sigma(., u)$ defined on $]\omega_-(u), \omega_+(u)[$. Since

$$\frac{d}{dt}(\varphi \circ \sigma)(t) = (\nabla\varphi(\sigma(t)), f(\sigma(t))), \tag{9}$$

the definition of f and Lemma 10.4 imply that either $\varphi(\sigma(t)) = \varphi(u)$ for all $t \in \mathbf{R}$ or that $\varphi \circ \sigma$ is decreasing.

Lemma 10.5. *If $\omega_+(u) = +\infty$, then $\varphi(\sigma(t)) \to -\infty$ as $t \to \omega_+(u)$.*

Proof. For $0 \le s \le t < \omega_+(u)$, we have

$$|\sigma(t) - \sigma(s)| \le \int_s^t |f(\sigma(r))|\, dr \le \int_s^t |\nabla\varphi(\sigma(r))|\, dr.$$

It then suffices to use the argument of Proposition 8.4. □

Lemma 10.6. *For every neighborhood A of $z \in Z$, there exists a neighborhood B of z such that, if $u \in B$, either $\sigma(t, u)$ stays in A for $0 < t < \omega_+(u)$ or $\sigma(t, u)$ stays in A until $\varphi(\sigma(t, u))$ becomes less than $c = \varphi(z)$.*

Proof. 1) Let $(m, n) \in N_{\rho/2}Z$. By the uniqueness of the solution of the Cauchy problem, $\sigma(t, m+n) = m + \sigma_m(t, n)$, where $\sigma_m(., n)$ is the solution of

$$\dot{\sigma}_m(t) = -P_m\nabla\varphi(m + \sigma_m(t))$$

$$\sigma_m(0) = n \in N_m Z,$$

provided $|\sigma_m(t, n)| \le \rho/2$.

2) Let $r \in]0, \rho/2[$ be such that

$$\{(m, n) \in NZ : |m - z| \le r, |n| \le r\} \subset A$$

and let

$$C = \{(m,n) \in NZ \ : \ |m - z| \leq r, r/2 \leq |n| \leq r\}.$$

By Lemma 10.4,

$$\delta = \inf\{|P_m \nabla \varphi(m+n)| \ : \ (m,n) \in C\} \geq \eta r/2 > 0.$$

Let us define

$$B = \{(m,n) \in NZ \ : \ |m - z| \leq r, |u| \leq r/2\} \cap \varphi^{c+\delta r/2}.$$

If $(m,n) \in B$ is such that $\sigma(t, m+n)$ doesn't stay in A for all $0 < t < \omega_+(m+n)$, then there exists $0 \leq t_1 < t_2 < \omega_+(m+n)$ such that $\sigma_m(t,n) \in B[0,r]$ for $0 \leq t \leq t_1$, $|\sigma_m(t_1,n)| = r/2$, $\sigma(t, m+n) \in C$ for $t_1 \leq t \leq t_2$ and $|\sigma_m(t_2,n)| = r$. It follows from (9) that, if $u = m + n$,

$$\begin{aligned}
\varphi(\sigma(t_2,u)) &= \varphi(\sigma(t_1,u)) - \int_{t_1}^{t_2} |P_m \nabla \varphi(\sigma(r,u))|^2 dr \\
&\leq \varphi(u) - \delta \int_{t_1}^{t_2} |P_m \nabla \varphi(\sigma(r,u))|^2 dr \\
&\leq c + \frac{\delta r}{2} - \delta \int_{t_1}^{t_2} |\dot{\sigma}(r,u)|\, dr \\
&\leq c + \frac{\delta r}{2} - \delta |\sigma(t_2,u) - \sigma(t_1,u)| \leq c + \frac{\delta r}{2} - \frac{\delta r}{2} = c. \qquad \square
\end{aligned}$$

Lemma 10.7. *For every $u \in \varphi^{-1}([c,b])$, either there is a (unique) $t \geq 0$ such that $\varphi(\sigma(t,u)) = c$ or $\omega_+(u) = +\infty$ and there is a $v \in K_c$ such that $\sigma(t,u) \to v$ as $t \to +\infty$.*

Proof. If $\varphi(\sigma(t,u)) > c$ for all $t \in]0, \omega_+(u)[$, Lemma 10.5 implies that $\omega_+(u) = +\infty$. Let us prove that for each $j \geq 1/\rho$, there exists $t_j \geq j$ such that $\sigma(t_j, u) \in N_{1/j} Z$. If this is not the case, we can find $j \geq 1/\rho$ such that

$$\sigma([j, +\infty[, u) \cap N_{1/j} Z = \phi.$$

Lemma 10.4 and the Palais–Smale condition over $\varphi^{-1}([a,b])$ imply that

$$\alpha = \sup\{(\nabla \varphi(u), f(u)) \ : \ u \in \varphi^{-1}([a,b]) \setminus N_{1/j} Z\} < 0.$$

Setting $\sigma(t) = \sigma(t,u)$, we obtain, for $t \geq j$,

$$c < \varphi(\sigma(t)) = \varphi(\sigma(j)) + \int_j^t (\nabla \varphi(\sigma(r)), f(\sigma(r)))\, dr \leq \varphi(\sigma(j)) + (t - j)\alpha,$$

which is impossible. Now, going if necessary to a subsequence, we can assume that $\sigma(t_j, u) \to z \in Z$ and $t_j \to +\infty$ when $j \to \infty$. Since $\varphi(\sigma(t,u))$ is decreasing in t, $\varphi(\sigma(t,u)) \to \varphi(z) = c$ when $t \to +\infty$. It follows then from Lemma 10.6 that $\sigma(t,u) \to z$ as $t \to +\infty$. $\square$

Lemma 10.8. *Under assumption* (A), φ^c *is a strong deformation retract of* φ^b *and* φ^a *is a strong deformation retract on* $\varphi^c \setminus K_c$.

Proof. 1) The first part of the lemma follows from Lemma 10.6 and Lemma 10.7 by using the argument of Lemma 8.3.

2) It is easy to obtain the second part of the lemma by using the gradient flow η given by $\dot{\eta} = -\nabla\varphi(\eta)$. □

Under assumption (A), K_c consists of a finite number of isolated critical points $u_1, \ldots, u_j$ and a finite number of nondegenerate critical manifolds $Z_1, \ldots, Z_\ell$. The *Morse numbers* of the pair (φ^b, φ^a) are defined by

$$M_n(\varphi^b, \varphi^a) = \sum_{i=1}^{j} \dim C_n(\varphi, u_i) + \sum_{i=1}^{\ell} \dim C_n(\varphi, Z_i).$$

Theorem 10.4. *Under assumption* (A),

$$M_n(\varphi^b, \varphi^a) = B_n(\varphi^b, \varphi^a), \quad n = 0, 1, \ldots .$$

Proof. By Lemma 10.8,

$$H_n(\varphi^b, \varphi^a) \approx H_n(\varphi^c, \varphi^a) \approx H_n(\varphi^c, \varphi^c \setminus K_c).$$

It suffices then to use the proof of Theorem 8.1. □

10.5 Second Order Autonomous Superlinear Equations

This section is devoted to the study of the autonomous problem

$$\begin{gathered} \ddot{u}(t) + g(u(t)) = 0 \\ u(0) - u(T) = \dot{u}(0) - \dot{u}(T) = 0 \end{gathered} \tag{10}$$

where $g \in C^1(]-\ell, \ell[, \mathbf{R})$ for some $\ell \in]0, +\infty[$. We assume that the following condition holds:

$$(\mathrm{A}_0) \quad g(-u) = -g(u), \quad 0 < g(u)u \quad \text{for} \quad 0 < |u| < \ell$$

and that g satisfies one of the following growth assumptions

$$(\mathrm{A}_1) \quad u^{-1}g(u) < g'(u) \quad \text{for} \quad 0 < |u| < \ell$$

$$(\mathrm{A}_1') \quad g'(u) < g(u)u^{-1} \quad \text{for} \quad 0 < |u| < \ell,$$

each of which prevents g from being linear.

The energy $(1/2)\dot{u}^2 + G(u)$, with

$$G(u) = \int_0^u g(s)\, ds,$$

is a first integral of (10). It follows classically from this fact that the initial conditions $u(0) = a > 0$ and $\dot{u}(0) = 0$ provide a periodic solution of (10) with minimal period

$$P(a) = 2\sqrt{2} \int_0^a \frac{dx}{\sqrt{G(a) - G(x)}}.$$

On the other hand, the solutions of (10) are the critical points of the functional φ defined on H_T^1 by

$$\varphi(u) = \int_0^T [(1/2)\dot{u}^2(t) - G(u(t))]\, dt.$$

The following theorem gives the Morse index and the nullity of the variational equation relative to a periodic solution of (10).

Theorem 10.5. *Under assumptions* (A_0) *and* (A_1) (resp. (A_j')), *if* u *is a solution of* (10) *with minimal period* T/k *such that* $u(0) \in]0, \ell[$ *and* $\dot{u}(0) = 0$, *then*

$$j(g' \circ u, T) = 2k, \quad \nu(g' \circ u, T) = 1$$

$$(\text{resp.} j(g' \circ u, T) = 2k - 1, \quad \nu(g' \circ u, T) = 1).$$

The proof of Theorem 10.5 requires the following classical result.

Spectral Theorem for the Periodic Boundary Value Problem. Let q and ρ be positive real continuous functions defined on $[0, T]$. We consider the eigenvalue problem

$$\ddot{h}(t) - q(t)h(t) + \lambda\rho(t)h(t) = 0$$

$$h(0) - h(T) = \dot{h}(0) - \dot{h}(T) = 0.$$

Then

a) there exists an infinite sequence of eigenvalues

$$0 < \lambda_0 < \lambda_1 \leq \lambda_2 < \lambda_3 \leq \lambda_4 < \dots.$$

b) The eigenfunctions corresponding to λ_{2k-1} or λ_{2k} have exactly $2k$ zeros in $[0, T[$.

c) If ρ is replaced by $\hat{\rho}$ such that $0 < \hat{\rho}(t) < \rho(t)$ on $[0, T]$, the corresponding eigenvalues $(\hat{\lambda}_j)$ are such that $\hat{\lambda}_j > \lambda_j$ $(j \in \mathbf{N})$.

Proof of Theorem 10.5. 1) By conservation of energy, $u(0) = \max_{t \in \mathbf{R}} |u(t)|$. Equation (10) implies that u is a solution of

$$\ddot{h}(t) + \frac{g(u(t))}{u(t)} h(t) = 0 \tag{11}$$

$$h(0) - h(T) = \dot{h}(0) - \dot{h}(T) = 0 \tag{12}$$

(with $g(u)/u = g'(0)$ when $u = 0$) and that $\dot{u}$ is a solution of

$$\ddot{h}(t) + g'(u(t))h(t) = 0 \tag{13}$$

satisfying (12). Let $c > 0$ be such that $g'(u(t)) + c > 0$ on $\mathbf{R}$ and denote by (λ_k) the eigenvalues of the problem

$$\ddot{h}(t) - ch(t) + \lambda(g'(u(t)) + c)h(t) = 0 \tag{14}$$

with the boundary conditions (12), and by $(\hat{\lambda}_k)$ those of

$$\ddot{h}(t) - ch(t) + \lambda\left(\frac{g(u(t))}{u(t)} + c\right)h(t) = 0$$

with the boundary conditions (12). Assume for definiteness that (A_1) holds (the case of (A_1') is similar and left to the reader). It follows from (A_1) that $\hat{\lambda}_j > \lambda_j$ for all $j \in \mathbf{N}$. Since T/k is the minimal period of u, u and $\dot{u}$ have exactly $2k$ zeros on $[0, T[$ as it follows from a direct phase plane analysis. Since u (resp. $\dot{u}$) is a solution of (11)–(12) (resp. (13)–(12)) we necessarily have

$$\lambda_{2k-1} = 1 \text{ or } \lambda_{2k} = 1, \quad \hat{\lambda}_{2k-1} = 1 \text{ or } \hat{\lambda}_{2k} = 1.$$

If $\lambda_{2k-1} = 1$, then $\hat{\lambda}_{2k} \geq \hat{\lambda}_{2k-1} > \lambda_{2k-1} = 1$, a contradiction. Thus,

$$\lambda_{2k-1} < 1 = \lambda_{2k} < \lambda_{2k+1}. \tag{15}$$

In particular, 1 is a simple eigenvalue of (14)–(12) so that $\nu(g' \circ u, T) = 1$.

2) Taking on H_T^1 the inner product

$$((h, v)) = \int_0^T [\dot{h}(t)\dot{v}(t) + ch(t)v(t)]\, dt,$$

let us define on H_T^1 the operator K by the formula

$$((Kh, v)) = \int_0^T [g'(u(t))h(t) + ch(t)v(t)]\, dt.$$

Then,

$$\chi_T(h) = (1/2)\int_0^T \dot{h}^2(t) - g'(u(t))h^2(t))\, dt = (1/2)((h - Kh, h)).$$

It is then easy to verify that the Morse index $j(g' \circ u, T)$ or χ_T is equal to the number of characteristic values of K contained in $]0, 1[$. But λ is a characteristic value of K if and only if λ is an eigenvalue of (14)–(12). Formula (15) then implies that $j(g' \circ u, T) = 2k$. □

Corollory 10.1. *Under the assumptions of Theorem* 10.5,

$$Z = \{u(.+\theta) : \theta \in \mathbf{R}\}$$

is a nondegenerate critical manifold of φ. *Moreover, the Morse index of* Z *is equal to* $2k$ *when* (A_1) *holds and to* $2k-1$ *when* (A_2) *holds.*

Proof. Equation (10) implies that u is a C^3 T-periodic function. Let us define $f : S^1 \to H_T^1$, where $S^1 \approx \mathbf{R}/(Tk^{-1}\mathbf{Z})$, by $f(\theta)(t) = u(t+\theta)$, so that $Z = f(S^1)$. It is easy to verify that $f \in C^2(S^1, H_T^1)$, that f is injective and that $df|T_\theta S^1$ is injective for every $\theta \in S^1$.

Since S^1 is compact, it follows from Theorem 10.1 that Z is a compact submanifold of H_T^1. Moreover, Z is connected and $\varphi''(z)$ is a Fredholm operator for each $z \in Z$. Since problem (10) is autonomous, all points of Z are critical points of φ. By Theorem 10.5, the nullity of all $z \in Z$ equals one, the dimension of Z. The value of the Morse index is also given by Theorem 10.5. □

Remark 10.4. If $u \in H_T^1$ is not smooth, then the set $\{u(\cdot+\theta) : \theta \in \mathbf{R}\}$ is not a smooth submanifold of H_T^1.

We finally study the solutions of (10) for an important special class of functions g satisfying (A_0) and (A_1).

Proposition 10.2. *Let* $p \in]1,\infty[$. *If* $g(u) = |u|^{p-2}u$, *then*

$$P(a) = a^{1-p/2}P(1).$$

Proof. Since $G(u) = |u|^p/p$, we obtain

$$\begin{aligned} P(a) &= 2\sqrt{2}\int_0^a \frac{dx}{\sqrt{G(a)-G(x)}} = 2\sqrt{2}\int_0^1 \frac{dx}{\sqrt{G(a)-G(xa)}} \\ &= 2\sqrt{2}a^{1-p/2}p^{1/2}\int_0^1 \frac{dx}{\sqrt{1-xP}}. \end{aligned}$$ □

For the above g, if $p \neq 2$, there exists a unique a_k such that $P(a_k) = T/k$. We shall denote by u_k the solution of (10) with the initial conditions $u(0) = a_k$, $\dot{u}(0) = 0$.

Proposition 10.3. *If* $p > 2$, *then, for each* $k \geq 2$,

$$\varphi(u_k) = k^{2p/(p-2)}\varphi(u_1) > 0.$$

Proof. It is easy to verify that the u_k can be deduced from u_1 by the formula $u_k(t) = k^{2/(p-2)}u_1(kt)$. By a direct calculation, we get $\varphi(u_k) = k^{2p/(p-2)}\varphi(u_1)$. Finally, equation (10) implies that

$$\int_0^T |\dot{u}_1(t)|^2 dt = \int_0^T |u_1(t)|^p dt.$$

Thus,

$$\varphi(u_1) = \left(\frac{1}{2} - \frac{1}{p}\right)\int_0^T |u_1(t)|^p dt > 0.$$ □

10.6 Periodic Solutions of Forced Superlinear Second Order Equations

Let us consider the problem

$$\begin{aligned} \ddot{u}(t) + |u(t)|^{p-2}u(t) &= f(t) \\ u(0) - u(T) = \dot{u}(0) - \dot{u}(T) &= 0 \end{aligned} \tag{16}$$

where $p > 2$, $f \in L^q(0,T;\mathbf{R})$, and $\frac{1}{p} + \frac{1}{q} = 1$. We shall prove that each T-periodic orbit of

$$\ddot{u}(t) + |u(t)|^{p-2}u(t) = 0, \tag{17}$$

but a finite number of them, generates two distinct solutions of (16).

Let $g(u) = |u|^{p-2}u$, $G(u) = |u|^p/p$ and let $\sigma \in C^\infty(\mathbf{R},\mathbf{R})$ be such that

$$\begin{aligned} \sigma(s) &= 1 \quad \text{for } s \le 0 \\ &= 0 \quad \text{for } s \ge 1 \\ \sigma'(s) &\le 0 \quad \text{for } s \in \mathbf{R}. \end{aligned}$$

For $\rho > 0$, define φ_ρ on H_T^1 by

$$\varphi_\rho(u) = \int_0^T \left[(1/2)\dot{u}^2(t) - G(u(t)) + \sigma\left(\frac{|u|_p^p - \rho}{\rho}\right) f(t)u(t)\right] dt,$$

where $|u|_p = \|u\|_{L^p}$. When $|u|_p^p \le \rho$, φ_ρ is equal to the action integral

$$\psi(u) = \int_0^T [(1/2)\dot{u}^2(t) - G(u(t)) + f(t)u(t)]\, dt$$

corresponding to the forced problem (16). When $|u|_p^p \ge 2\rho$, φ_ρ is equal to the action integral

$$\varphi(u) = \int_0^T [(1/2)\dot{u}^2(t) - G(u(t))]\, dt$$

corresponding to the autonomous problem (17).

Lemma 10.9. *There exists $\mu > 0$ such that, for every $u \in H_T^1$, one has*

$$|\varphi(u) - \varphi_\rho(u)| \le \mu\rho^{1/p}.$$

Proof. Hölder's inequality implies that

$$|\varphi(u) - \varphi_\rho(u)| \le \sigma\left(\frac{|u|_p^p - \rho}{\rho}\right) |f|_q |u|_p$$

and

$$\sigma\left(\frac{|u|_p^p - \rho}{\rho}\right) = 0 \quad \text{whenever} \quad |u|_p \ge (2\rho)^{1/p}. \qquad \square$$

Lemma 10.10. *There exists $\alpha > 0$ and $\beta > 0$ such that, for every $\rho > 0$, $\nabla\varphi_\rho(u) = 0$ and $\varphi_\rho(u) \leq \alpha\rho - \beta$ imply that*

$$\psi(u) = \varphi_\rho(u), \ \nabla\psi(u) = 0 \quad \text{and} \quad |u|_p^p < \rho.$$

Proof. Let us assume that $\nabla\varphi_\rho(u) = 0$, i.e.

$$\ddot{u} + g(u) = \sigma\left(\frac{|u|_p^p - \rho}{\rho}\right) f + \frac{p}{\rho}\sigma'\left(\frac{|u|_p^p - \rho}{\rho}\right)\left(\int_0^T fu\right)|u|^{p-2}u.$$

After multiplication by $u(t)$ and integration from 0 to T, we obtain

$$-\int_0^T \dot{u}^2 + \int_0^T g(u)u = \sigma\left(\frac{|u|_p^p - \rho}{\rho}\right)\left(\int_0^T fu\right)$$

$$+ \frac{p}{\rho}\sigma'\left(\frac{|u|_p^p - \rho}{\rho}\right)\left(\int_0^T fu\right)|u|_p^p. \tag{18}$$

Since $ug(u) = pG(u)$, we have, using (18),

$$\varphi_\rho(u) = \frac{1}{2}\int_0^T ug(u) - \int_0^T G(u) + \frac{p}{2}\left[\sigma'\left(\frac{|u|_p^p - \rho}{\rho}\right)\frac{|u|_p^p}{\rho}\right]\int_0^T fu$$

$$= \left(\frac{p}{2} - 1\right)\int_0^T G(u) - c_1\int_0^T fu \geq \left(\frac{p}{2} - 1\right)\int_0^T G(u) - c_1|f|_q|u|_p,$$

where c_1 is independent of ρ. Thus, there exists $\alpha > 0$ and $\beta > 0$, independent of ρ, such that

$$\varphi_\rho(u) \geq \alpha|u|_p^p - \beta. \tag{19}$$

If we assume, moreover, that $\varphi_\rho(u) \leq \alpha\rho - \beta$, inequality (19) implies that $|u|_p^p < \rho$. Thus, φ_ρ and ψ are equal on a neighborhood of u in H_T^1. Since $\nabla\varphi_\rho(u) = 0$, necessarily $\nabla\psi(u) = 0$. □

Lemma 10.11. *The functional φ and, for any $\rho > 0$, the functional φ_ρ satisfy the Palais-Smale condition over H_T^1.*

Proof. 1) Let us define $N : H_T^1 \to H_T^1$ by the formula

$$((Nu, v)) = \int_0^T (g(u(t)) + u(t), v(t))\,dt,$$

with $((.,.))$ the usual inner product in H_T^1. Let (u_j) be a sequence in H_T^1 such that $(\varphi(u_j))$ is bounded and $\nabla\varphi(u_j) \to 0$, i.e.

$$u_j - Nu_j = f_j$$

with $f_j \to 0$ in H_T^1 as $j \to \infty$. After taking the inner product with u_j, we obtain

$$\int_0^T \dot{u}_j^2 - \int_0^T g(u_j)u_j = ((f_j, u_j)).$$

Hence,

$$\varphi(u_j) = \frac{1}{2}\int_0^T \dot{u}_j^2 - \frac{1}{p}\int_0^T g(u_j)u_j = \left(\frac{1}{2} - \frac{1}{p}\right)\int_0^T |\dot{u}_j|^2 + \frac{1}{p}((f_j, u_j)).$$

Since $\varphi(u_j)$ is bounded, we obtain

$$\int_0^T \dot{u}_j^2 \le c\|f_j\|\,\|u_j\|$$

for some $c > 0$. We also have

$$\begin{aligned}\int_0^T u_j^2 &= \int_{|u_j(t)|<1} u_j^2 + \int_{|u_j(t)|<1} u_j^2\\ &\le T + \int_0^T |u_j|^p = T + \int_0^T g(u_j)u_j = T + \int_0^T \dot{u}_j^2 - ((f_j, u_j))\\ &\le T + (c+1)\|f_j\|\,\|u_j\|.\end{aligned}$$

Thus,

$$\|u_j\|^2 \le T + (2c+1)\|f_j\|\,\|u_j\|.$$

Since $\|f_j\| \to 0$, the sequence (u_j) is bounded in H_T^1. Going if necessary to a subsequence, we can assume that $u_j \rightharpoonup u$. It is easy to verify that $Nu_j \to Nu$, so that $u_j \to Nu$.

2) The proof that φ_ρ satisfies the Palais–Smale condition is similar and left to the reader. □

Theorem 10.6. *There exists $k_0 \in \mathbf{N}^*$ such that if $k \ge k_0$, each orbit of* (17) *with minimal period T/k generates two solutions of* (16). *In particular, problem* (16) *has infinitely many solutions.*

Proof. 1) Let $c_k = \varphi(u_k)$ be the sequence given in Proposition 10.3. Since (c_k/k^2) is increasing, we obtain

$$c_{k+1} - c_k > \left(\frac{2}{k} + \frac{1}{k^2}\right) c_k$$

$$c_k - c_{k-1} > \left(\frac{2}{k} + \frac{1}{k^2}\right) c_k,$$

and hence

$$\min(c_{k+1} - c_k, c_k - c_{k-1}) > \frac{c_k}{k}. \tag{20}$$

Let $\epsilon_k = c_k/2k$ and $\rho_k = (c_k/6k\mu)^p$. Assume that $\nabla\varphi_{\rho_k}(u) = 0$ and $\varphi_{\rho_k}(u) \in [c_k - \epsilon, c_k + \epsilon_k]$. Then, by Lemma 10.10, $\psi(u) = \varphi_{\rho_k}(u)$ and $\nabla\psi(u) = 0$ if

$$\alpha\rho_k - \beta \geq c_k + \epsilon_k = c_k\left(1 + \frac{1}{2k}\right). \tag{21}$$

Let $k_1 \in \mathbf{N}^*$ be such that $c_k \geq \beta + \frac{c_k}{2k}$ for $k \geq k_1$. If $k \geq k_1$, inequality (21) follows from $\alpha\rho_k \geq 2c_k$, i.e. $\alpha(c_k/6k\mu)^p \geq 2c_k$. This inequality is equivalent to

$$\frac{c_k^{p-1}}{k^p} \geq \frac{2}{\alpha}(6\mu)^p.$$

Since $p > 2$ and $c_k/k^2 \to +\infty$, there exists $k_0 \geq k_1$ such that the preceding inequality is satisfied for $k \geq k_0$.

2) By the first part of the proof, it suffices to show that for $k \geq k_0$, $\varphi_{\rho_k}^{-1}([c_k - \epsilon_k, c_k + \epsilon_k])$ contains at least two critical points of φ_{ρ_k}. Let $k \geq k_0$ and let us apply Theorem 8.10 with $\hat{\varphi} = \varphi_{\rho_k}$, $c = c_k$ and $\epsilon = \epsilon_k$. It follows from (20) that c is the only critical value of φ in $[c - \epsilon, c + \epsilon]$. By Lemma 10.9 we have, for every $u \in H_T^1$,

$$|\varphi(u) - \hat{\varphi}(u)| \leq \mu\rho_k^{1/p} = \epsilon/3.$$

Corollary 10.1 implies that

$$Z = \{u_k(. + \theta) \,:\, \theta \in \mathbf{R}\}$$

is a nondegenerate critical manifold of φ with Morse index $2k$. It follows from Theorem 10.4 that

$$H_n(\varphi^{c+\epsilon}, \varphi^{c-\epsilon}) \approx C_n(\varphi, Z), \quad n = 0, 1, \ldots.$$

Using Theorem 10.3, we have

$$\begin{aligned} C_n(\varphi, Z) &\approx H_{n-2k}(Z, \phi) \\ &\approx H_{n-2k}(S^1, \phi) \\ &\approx F && \text{if } 2k \text{ or } n = 2k+1 \\ &\approx \{0\} && \text{if } n \leq 2k-1 \text{ or } n \geq 2k+2, \end{aligned}$$

so that

$$B_n(\varphi^{c+\epsilon}, \varphi^{c-\epsilon}) = 1 \quad \text{for} \quad n = 2k \text{ or } 2k+1.$$

Since, by Lemma 10.11, φ satisfies the Palais–Smale condition, Theorem 8.10 implies that

$$B_n(\hat{\varphi}^{c+\epsilon/2}, \hat{\varphi}^{c-\epsilon/2}) \geq 1 \quad \text{for} \quad n = 2k \text{ or } n = 2k+1. \tag{22}$$

If $[c - \frac{\epsilon}{2}, c + \frac{\epsilon}{2}]$ is free of critical values of $\hat{\varphi}$, Lemma 10.11 and Lemma 8.3 imply that $\hat{\varphi}^{c-\epsilon/2}$ is a strong deformation retract of $\hat{\varphi}^{c+\epsilon/2}$. But then

$$B_n(\hat{\varphi}^{c+\epsilon/2}, \hat{\varphi}^{c-\epsilon/2}) = 0, \quad n = 0, 1, \ldots,$$

a contradiction.

Let us now assume that $\hat{\varphi}^{-1}([c-\frac{\epsilon}{2}, c+\frac{\epsilon}{2}])$ contains only one critical point u of $\hat{\varphi}$. Without loss of generality, we can also assume that $\hat{\varphi}(u) \neq c \pm \frac{\epsilon}{2}$. It follows from (22) and Theorem 8.1 that

$$\dim C_n(\hat{\varphi}, u) \geq 1, \quad n = 2k \text{ or } n = 2k+1. \tag{23}$$

By Lemma 10.10, $|u|_p^p < \rho_k$. Thus, the nullity of u, which is equal to the number of linearly independent solutions of

$$\ddot{h} + g'(u(t))\, h = 0$$

$$h(0) - h(T) = \dot{h}(0) - \dot{h}(T) = 0$$

is at most 2. According to Corollary 8.4, $\dim C_n(\hat{\varphi}, u)$ is different from zero for at most one value of n. This fact contradicts (23). Thus, $\hat{\varphi}^{-1}([c-\frac{\epsilon}{2}, c+\frac{\epsilon}{2}])$ contains at least two critical points of $\hat{\varphi}$. □

Theorem 10.7. *If all the solutions of* (16) *are nondegenerate critical points of the corresponding action* ψ, *there exists* $k_0 \in \mathbf{N}^*$ *such that, for each* $k \geq k_0$, (16) *has at least one solution with Morse index* k.

Proof. Let $k \geq k_0$, with k_0 given in the proof of Theorem 10.6. By assumption, the critical points of $\hat{\varphi} = \varphi_{\rho_k}$ in $\varphi_{\rho_k}^{-1}([c_k - \frac{\epsilon_k}{2}, c_k + \frac{\epsilon_k}{2}])$ contains at least one critical point u_{2k} with Morse index $2k$ and one critical point u_{2k+1} with Morse index $2k+1$. Since, by Lemma 10.10,

$$|u_{2k}|_p^p < \rho_k, \quad |u_{2k+1}|_p^p < \rho_k,$$

the functions u_{2k} and u_{2k+1} are critical points of ψ with respective Morse indices $2k$ and $2k+1$. □

10.7 Local Perturbations of Nondegenerate Critical Manifolds

Let Z be a compact connected C^2-submanifold of a Hilbert space X and let $\varphi \in C^2(\mathbf{R} \times X, \mathbf{R})$. Assume that all points of Z are critical points of $\varphi_0 = \varphi(0,.)$ and consider the existence of critical points of $\varphi_\epsilon = \varphi(\epsilon,.)$ near Z for ϵ near 0.

Theorem 10.8. *If* Z *is a nondegenerate critical manifold of* φ_0, *then there exists* $\bar{\epsilon} > 0$ *such that, for all* $0 < |\epsilon| < \bar{\epsilon}$, φ_ϵ *has at least* $\operatorname{cat}_Z(Z)$ *critical points near* Z.

We shall use the Liapunov–Schmidt method on the manifold Z. We denote by P_m (resp. Q_m) the orthogonal projector onto $N_m Z$ (resp. $T_m Z$) and we define $M(\epsilon, u) = \nabla \varphi_\epsilon(u)$.

Lemma 10.12. *If Z is a nondegenerate critical manifold of φ_0, then equation*

$$P_m M(\epsilon, m+n) + Q_m n = 0 \tag{24}$$

defines on an open neighborhood of $\{0\} \times Z$ a C^1 mapping $n = n(\epsilon, m)$.

Proof. 1) Let $z \in Z$. Since Z is a nondegenerate critical manifold, the map

$$x \to P_z D_u M(0,z)x + Q_z x$$

is invertible on X. By the implicit function theorem, equation (24) defines, on an open neighborhood of $[0, z]$, a C^1 mapping $n = n(\epsilon, m)$. Let us recall that, by Proposition 10.1, P_m and Q_m are C^1.

2) Using the compactness of Z, it is then easy to verify that $u(\epsilon, m)$ is well defined on an open neighborhood of $\{0\} \times Z$. □

By Lemma 10.12, for ϵ sufficiently small, the function

$$\psi_\epsilon(m) = \varphi(\epsilon, m + n(\epsilon, m))$$

is well defined and continuously differentiable on Z.

Lemma 10.13. *If Z is a nondegenerate critical manifold of φ_0, there exists $\bar{\epsilon} > 0$ such that, if $0 < |\epsilon| < \bar{\epsilon}$, $m \in Z$ and $\nabla\psi_\epsilon(m) = 0$ then $\nabla\varphi_\epsilon(m + n(\epsilon, m)) = 0$.*

Proof. 1) Clearly $n(0, m) = 0$ for all $m \in Z$. Differentiating the identity $Q_m n(\epsilon, m) = 0$ with respect to m at $\epsilon = 0$, we obtain $Q_m D_m n(0, m) = 0$. By compactness, there exists $\bar{\epsilon} > 0$ such that, for $|\epsilon| < \bar{\epsilon}$, $\|Q_m D_m n(\epsilon, m)\| \leq 1/2$ on Z.

2) Let $0 < |\epsilon| < \bar{\epsilon}$. If $m \in Z$ is a critical point of ψ_ϵ then

$$(\nabla\varphi_\epsilon(m + n(\epsilon, m)), (id + D_m n(\epsilon, m))v) = 0$$

for all $v \in T_m Z$. Since $P_m \nabla\varphi_\epsilon(m + n(\epsilon, m)) = 0$, we have

$$(\nabla\varphi_\epsilon(m + n(\epsilon, m)), (id + Q_m D_m n(\epsilon, m))v) = 0 \tag{25}$$

for all $v \in T_m Z$. Because $\|Q_m D_m n(\epsilon, m)\| \leq 1/2$, the map $id + Q_m D_m n(\epsilon, m)$ is invertible on $T_m Z$. It follows then from (25) that $Q_m \nabla\varphi_\epsilon(m + n(\epsilon, m)) = 0$. Since $P_m \varphi_\epsilon(m + n(\epsilon, m)) = 0$, $m + n(\epsilon, m)$ is a critical point of φ_ϵ and the proof is complete. □

Proof of Theorem 10.8. Let $\bar{\epsilon} > 0$ be given by Lemma 10.13. For $0 < |\epsilon| < \bar{\epsilon}$, the C^1 function ψ_ϵ has at least $\operatorname{cat}_z(z)$ critical points on the compact C^2 manifold Z (see [Pal$_3$]). It suffices then to apply Lemma 10.13. □

By Lemma 10.12, problem

$$M(\epsilon, u) = 0 \tag{26}$$

is equivalent near $\{0\} \times Z$, to $N(\epsilon, m) = 0$ where $N(\epsilon, m) = Q_m M(\epsilon, m + n(\epsilon, m))$. When Z is a circle, $N(\epsilon, .)$ can be considered as a periodic function of the real variable m. Moreover, $N(0, m) = 0$ so that problem (26) is a bifurcation problem in $\mathbf{R}^2$.

Theorem 10.9. *Let Z be a nondegenerate critical circle of φ_0. If $m \in Z$ is such that*

$$Q_m D_\epsilon M(0, m) \neq 0,$$

then there is a neighborhood $\mathcal{N}$ of $[0, m]$ in $\mathbf{R} \times X$ such that $[\epsilon, u] \in \mathcal{N}$ and $M(\epsilon, u) = 0$ implies $\epsilon = 0$ and $u \in Z$.

Proof. Since $D_\epsilon N(0, m) = Q_m D_\epsilon M(0, m)$, it suffices to apply the implicit function theorem. □

Theorem 10.10. *Let $\varphi \in C^3(\mathbf{R} \times X, \mathbf{R})$ and let Z be a nondegenerate critical circle of φ_0. If $z \in Z$ is a simple zero of*

$$h(m) = Q_m D_\epsilon M(0, m),$$

then there exists $\bar{\epsilon} > 0$ and a differentiable function $u^* :]-\bar{\epsilon}, \bar{\epsilon}[\to X$ such that

$$u^*(0) = z \quad \textit{and} \quad M(\epsilon, u^*(\epsilon)) = 0.$$

Moreover, there is a neighborhood $\mathcal{N}$ of $[0, z]$ in $\mathbf{R} \times X$ such that $[\epsilon, u] \in \mathcal{N}$ and $M(\epsilon, u) = 0$ implies either $u = u^*(\epsilon)$ or $\epsilon = 0$ and $u \in Z$.

Proof. Since $N(0, m) = 0$, we have that

$$N(\epsilon, m) = \epsilon H(\epsilon, m)$$

where

$$H(\epsilon, m) = \int_0^1 D_\epsilon N(s\epsilon, m)\, ds.$$

Thus, for $\epsilon \neq 0$, equation $N(\epsilon, m) = 0$ is equivalent to $H(\epsilon, m) = 0$. Since, by assumption, H is C^1 and

$$H(0, z) = h(z) = 0$$

$$D_m H(0, z) = h'(z) \neq 0,$$

it suffices to apply the implicit function theorem. □

Example. Let $f \in C(\mathbf{R}, \mathbf{R})$ be a τ-periodic function and let us consider the problem

$$\begin{aligned} \ddot{u}(t) + \sin u(t) &= \epsilon f(t) \\ u(j\tau) - u(0) &= \dot{u}(j\tau) - \dot{u}(0) = 0 \end{aligned} \tag{27}$$

where j is a positive integer. Assume that $k^{-1} j\tau > 2\pi$ and that v is a solution of

$$\ddot{v}(t) + \sin v(t) = 0$$

with minimal period $k^{-1} j\tau$. By Theorems 10.5 and 10.8, problem (27) has, for $|\epsilon|$ sufficiently small, at least two solutions near $Z = \{v(. + \theta) : \theta \in \mathbf{R}\}$.

Historical and Bibliographical Notes

In finite dimension, the notion of nondegenerate critical manifold and Theorem 10.3 are due to Bott [Bot$_3$]. Section 10.4 seems to be the first complete treatment of the global theory in infinite dimension.

Ekeland [Eke$_7$] introduced critical manifolds in the study of the fixed energy periodic problem. Theorems 10.5, 10.6, and 10.7 are due to Willem [Wil$_4$]. Lemmas 10.9, 10.10, and 10.11 are taken from Bahri–Berestycki [BaB$_1$] who were the first ones to prove, by contradiction, the existence of infinitely many periodic solutions of forced superlinear second order equations and of corresponding systems. See [BaB$_2$] for an extension to first order Hamiltonian systems and [BaB$_3$], [Str$_1$] for elliptic superlinear problems. When the forcing term is even or odd, the problem is simpler (see [Maw$_2$]).

Theorem 10.8 appears in [Rec$_{1,2}$], [Dan$_7$], [AmbCE$_1$]. Theorems 10.9 and 10.10 generalize results of Albizatti [Alb$_1$] and of Lazer and McKenna [LMc$_2$]. See also [Wil$_{11}$].

Perturbations of Lusternik–Schnirelman theory are given in [Kra$_2$], [Amb$_{9,10}$], [Poh$_{1,2,3}$].

See also [Amb$_7$] for autonomous superquadratic problems and [Ba$_{1,2}$], [BaL$_1$], [DoL$_1$], [Gne$_1$], [Lon$_1$], [Ol$_{11}$], [PitT$_{1,2}$], [Rab$_{21}$], [Ta$_{1,2,3}$] for forced superquadratic problems.

Exercises

1. Let X be a Hilbert space and let $\varphi \in C^2(X, \mathbf{R})$ be such that

   i) φ is bounded from below.

   ii) φ satisfies the PS-condition over X.

   iii) 0 is a non-degenerate critical point of φ with Morse index k_0.

   iv) the other critical points of φ are contained in j nondegenerate critical manifolds $Z_1, \ldots, Z_j$ with respective Morse index $k_1, \ldots, k_j$ and homeomorphic to S^1.

   Then there exists a polynomial $Q(t)$ with nonnegative integer coefficients such that

$$t^{k_0} + \left(t^{k_1} + \ldots + t^{k_j}\right)(1+t) = 1 + (1+t)Q(t).$$

2. Let $H \in C^2(\mathbf{R}^{2N}, \mathbf{R})$ and let v be a (non-constant) T-periodic solution of $J\dot{u} + \nabla H(u) = 0$ such that if h is a T-periodic solution of the linearized system

$$J\dot{h} + H''(v)h = 0$$

then h is proportional to $\dot{v}$. Consider the perturbed problem

$$J\dot{u} + \nabla H(u) = \epsilon f(t,u,\epsilon), \quad u(0) = u(T) \qquad (*)$$

where $f \in C^1(\mathbf{R} \times \mathbf{R}^{2N} \times \mathbf{R}, \mathbf{R})$. If

$$f(t,u,\epsilon) = \nabla_u F(t,u,\epsilon)$$

then, for $|\epsilon|$ small enough, problem $(*)$ has at least two solutions near

$$Z = \{v(.+\theta) : \theta \in \mathbf{R}\}.$$

3. Let H, v, and f be as in Exercise 2. Define

$$h(\lambda) = \int_0^T (f(t, v_\theta(t), 0), \dot{v}_\theta(t))\, dt$$

where $v_\theta(.) = v(.+\theta)$. If $h(\theta_0) \neq 0$, there exists a neighborhood $\mathcal{N}$ of $[(v_{\theta_0}, 0)]$ in $H_T^{\frac{1}{2}} \times \mathbf{R}$, such that, if $[u,\epsilon] \in \mathcal{N}$ is a solution of $(*)$, then $u \in Z$ and $\epsilon = 0$.

4. Now let $H \in C^3$ and $f \in C^2$. If θ_0 is a simple zero of h, there exists $\overline{\epsilon} > 0$ and a differentiable function $u^* \ :] - \overline{\epsilon}, \overline{\epsilon}[\to H_T^{\frac{1}{2}}$ such that $u^*(0) = v_{\theta_0}$, and $u^*(\epsilon)$ is a solution of $(*)$. Moreover there exists a neighborhood $\mathcal{N}$ of $[v_{\theta_0}, 0]$ is $X \times \mathbf{R}$ such that, if $[u,\epsilon] \in N$ is a solution of $(*)$ then either $u = u^*(\epsilon)$ or $\epsilon = 0$ and $u \in Z$.

Bibliography

[**ALL**$_1$] AHMAD, S. and A.C. LAZER, Critical point theory and a theorem of Amaral and Pera, *Boll. Un. Mat. Ital.* (6) 3B (1984) 583–598.

[**ALP**$_1$] AHMAD, S., and A.C. LAZER, and J. L. PAUL, Elementary critical point theory and perturbations of elliptic boundary value problems at resonance, *Indiana Univ. Math. J.* **25** (1976) 933–944.

[**AlG**$_1$] ALBIZZATI, A., Sélection de phase par un terme d'excitation pour les solutions périodiques de certaines éequations différentielles, *C. R. Acad. Sci. Paris I,* **296** (1983) 259–262.

[**AlY**$_1$] ALEXANDER, J.C. and J.A. YORKE, The homotopy continuation method: numerically implementable topological procedures, *Trans. Amer. Math. Soc.* **242** (1978) 271–284.

[**Ale**$_1$] ALEKSANDROV, P.S., In memory of Lazar Aranovich Lyusternik, *Russian Math. Surveys* **37, 1** (1982) 145–147.

[**AVD**$_1$] ALEXANDROV, P.S., M.I. VISHIK, V.A. DITKIW, A.N. KOLMOGOROV, M.A. LA VRENT'EV, and O.A. OLEINIK, Lazar' Aranovich Lyusternik (on the occasion of his eightieth birthday), *Russian Math. Surveys* **36, 6** (1980) 3–10.

[**AVS**$_1$] ALEXANDROV, P.S., M.I. VISHIK, V.K. SAUL'EV, and L.E. EL'GOL'TS, Lazar Aranovich Lyusternik (on the occasion of his 60th birthday), *Russian Math. Surveys* **15, 2** (1960) 153–168; (on the occasion of his 70th birthday), ibid. **25, 4** (1970) 2–10.

[**Ama**$_1$] AMANN, H., Saddle points and multiple solutions of differential equations, *Math. Z.* **169** (1979) 127–166.

[**Ama**$_2$] AMANN, H., Periodic solutions of Hamiltonian systems, *Delft Progress Report* **9** (1984) 4–18.

[**AmZ**$_1$] AMANN, H. and E. ZEHNDER, Nontrivial solutions for a class of nonresonance problems and applications to nonlinear differential equations, *Ann. Scuola Norm. Sup. Pisa* **4, 7** (1980) 539–603.

[**AmZ**$_2$] AMANN, A. and E. ZEHNDER, Periodic solutions of asymptotically linear Hamiltonian systems, *Manuscripta Math.* **32** (1980) 149–189.

[AmZ$_3$] AMANN, H. and E. ZEHNDER, Multiple periodic solutions for a class of nonlinear autonomous wave equations, *Houston J. Math.* **7** (1981) 147–174.

[Amb$_1$] AMBROSETTI, A., On the existence of multiple solutions of a class of nonlinear boundary value problems, *Rend. Sem. Mat. Univ. Padova* **49** (1973) 195–204.

[AmB$_2$] AMBROSETTI, A., Differential equations with multiple solutions and nonlinear functional analysis, in Equadiff 82, Wurzburg, *Lect. Notes in Math.* No. 1017, 1983, Springer, Berlin, 10–37.

[Amb$_3$] AMBROSETTI, A., Elliptic equations with jumping nonlinearities, *J. Math. Phys. Sci.* (Madras) (1984) 1–10.

[Amb$_4$] AMBROSETTI, A., Recent advances in the study of the existence of periodic orbits of Hamiltonian systems, in *Advances in Hamiltonian Systems,* Birkhauser Basel, 1983, 1–22.

[Amb$_5$] AMBROSETTI, A., Nonlinear oscillations with minimal period, *Proc. Symp. Pure Math.* **45** (1985) 29–36.

[Amb$_6$] AMBROSETTI, A., Teoria di Lusternik-Schnirelman su varietá con bordo negli spazi di Hilbert, *Rend. Sem. Mat. Univ. Padova* **45** (1971) 337–353.

[Amb$_7$] AMBROSETTI, A., Esistenza di infinite soluzioni per problemi non lineari, *Atti Accad. Naz. Lincei, Mem. Cl. Sci. FG. Mat. Natur. Sez. I,* **93** (1973) 231–246.

[Amb$_8$] AMBROSETTI, A., Some remarks on the buckling problem for a thin clamped shell, *Ricerche di Mat.* **23** (1974) 161–170.

[Amb$_9$] AMBROSETTI, A., A perturbation theorem for superlinear boundary value problems, Math. Res. Center, Univ. Wisconsin–Madison, TSR # 1446, 1974.

[Amb$_{10}$] AMBROSETTI, A., A note on the Lusternik–Schnirelman theory for functionals which are not even, *Ric. di Mat.* **25** (1976) 179–186.

[AmC$_1$] AMBROSETTI, A. and V. COTI-ZELATI, Solutions with minimal period for Hamiltonian system in a potential well, *Ann. Inst. H. Poincaré, Analyse non linéaire,* **4** (1987) 275–296.

[AmC$_2$] AMBROSETTI, A. and V. COTI-ZELATI, Critical points with lack of compactness and singular dynamical systems, preprint.

[AmCE$_1$] AMBROSETTI, A., V. COTI-ZELATI, and I. EKELAND, Symmetry breaking in Hamiltonian system, *J. Differential Equations* **67** (1987) 165–184.

[AmL$_1$] AMBROSETTI, A. and D. LUPO, On a class of nonlinear Dirichlet problems with multiple solutions, *J. Nonlinear Analysis* **8** (1984) 1145–1150.

[AmM$_1$] AMBROSETTI, A. and G. MANCINI, Solutions of minimal period for a class of convex Hamiltonian systems, *Math. Ann.* **255** (1981) 405–422.

[AmM$_2$] AMBROSETTI, A. and G. MANCINI, On a theorem of Ekeland and Lasry concerning the member of periodic Hamiltonian trajectories, *J. Differential Equations* **43** (1982) 249–256.

[AmR$_1$] AMBROSETTI, A. and P.H. RABINOWITZ, Dual variational methods in critical point theory and applications, *J. Funct. Anal.* **14** (1973) 349–381.

[AmS$_1$] AMBROSETTI, A. and P.N. SRIKANTH, Superlinear elliptic problems and the dual principle in critical point theory, *J. Math. Phys. Sci.,* **18** (1984) 441–451.

[ASt$_1$] AMBROSETTI, A. and M. STRUWE, A note on the problem $-\Delta u = \lambda u + u|u|^{2^*-2}$, *Manuscripte Math.,* **54** (1986) 373–379.

[Ano$_1$] ANOSOV, V.I., Critical points of periodic functionals, *Soviet Math. Dokl.* **1** (1960) 208–210.

[Arz$_1$] ARZELA, C., Sul principio di Dirichlet, *Rend. Ac. Sci. Bologna (NS)* **1** (1986- 1987), 71–84.

[Aub$_1$] AUBIN, J.P., *Applied Abstract Analysis,* Wiley, New York, 1977.

[Aub$_2$] AUBIN, J.P., *L'analyse non-linéaire et ses motivations économiques,* Masson, Paris, 1984.

[AuE$_1$] AUBIN, J.P. and I. EKELAND, *Applied Nonlinear Analysis,* Wiley, Interscience, New York, 1984.

[AuE$_2$] AUBIN, J.P. and I. EKELAND, Second order evolution equations associated with convex Hamiltonians, *Canad. Math. Bull.* **23** (1980) 81–94.

[Ba$_1$] BAHRI, A., Topological results on a certain class of functionals and applications, *J. Funct. Anal.* **41** (1981) 397–427.

[Ba$_2$] BAHRI, A., La fibration de Milnor et l'effet non-linéaire en analyse, *C.R. Acad. Sci. Paris I* **303** (1986) 65–68.

[BaB$_1$] BAHRI, A. and H. BERESTYCKI, Existence of forced oscillations for some nonlinear differential equations, *Comm. Pure Appl. Math.* **37** (1984) 403–442.

[**BaB$_2$**] BAHRI, A. and H. BERESTYCKI, Forced vibrations of superquadratic Hamiltonian systems, *Acta Math.* **152** (1984) 143–197.

[**BaB$_3$**] BAHRI, A. and H. BERESTYCKI, A perturbation method in critical point theory and applications, *Trans. Amer. Math. Soc.* **267** (1981) 1–32.

[**BaL$_1$**] BAHRI, A. and P.L. LIONS, Remarks on the variational theory of critical points and applications, *C.R. Acad. Sci. Paris, Sci. I. Math.* **301** (1985) 145–148.

[**BaM$_1$**] BAHRI, A. and J.M. MOREL, Image de la somme de deux opérateurs, *C.R. Acad. Sci. Paris, A* **287** (1978) 719–722.

[**Bai$_1$**] BAIRE, R., Sur les fonctions de variables réelles, *Annali Mat. Pure Appl. 3,* **3** (1899) 1–123.

[**Bar$_1$**] BARTOLO, P., An extension of Krasnosel'skiigenus, *Boll. Un. Math. Ital.* **1-C** (1982) 347–356.

[**BBF$_1$**] BARTOLO, P., V. BENCI, and D. FORTUNATO, Abstract critical point theorems and applications to some nonlinear problems with "strong resonance" at infinity, *J. Nonlinear Anal.* **9** (1983) 981–1012.

[**BMi$_1$**] BASILE, N. and M. MININNI, Multiple periodic solutions for a semilinear wave equation with nonmonotone nonlinearity, *J. Nonlinear Anal.* **9** (1985) 837–848.

[**Bat$_1$**] BATES, P.W., A variational approach to solving semilinear equations at resonance, in *Nonlinear Phenomena in Mathematical Sciences,* Academic Press, New York, 1982, 103–112.

[**Bat$_2$**] BATES, P.W., Reduction theorem for a class of semilinear equations at resonance, *Proc. Amer. Math. Soc.* **84** (1982) 73–78.

[**BaC$_1$**] BATES, P.W. and A. CASTRO, Necessary and sufficient conditions for existence of solutions to equations with noninvertible linear part, *Rev. Colombiana Mat.* **15** (1981) 7–24.

[**BaE$_1$**] BATES, P.W. and I. EKELAND, A saddle-point theorem, in *Differential Equations,* Academic Press, New York, 1980, 123–126.

[**Ben$_1$**] BENCI, V., A geometrical index for the group S^1 and some applications to the study of periodic solutions of ordinary differential equations, *Comm. Pure Appl. Math.* **34** (1981) 393–432.

[**Ben$_2$**] BENCI, V., On critical point theory for indefinite functionals in the presence of symmetries, *Trans. Amer. Math. Soc.* **274** (1982) 533–572.

[**Ben**$_3$] BENCI, V., Some critical point theorems and applications, *Comm. Pure Appl. Math.* **33** (1980) 147–172.

[**Ben**$_4$] BENCI, V., Normal modes of a Lagrangian system constrained in a potential well, *Ann. Inst. H. Poincaré, Anal. non-linéaire* **1** (1984) 379–400.

[**Ben**$_5$] BENCI, V., Closed geodesics for the Jacobi metric and periodic solutions of prescribed energy of natural Hamiltonian systems, *Ann. Inst. H. Poincaré, Anal. non-linéaire,* **1** (1984) 401–412.

[**Ben**$_6$] BENCI, V., Periodic solutions of Lagrangian systems on a compact manifold, *J. Differential Equations* **63** (1986) 135–161.

[**Ben**$_7$] BENCI, V., Some applications of the generalized Morse–Conley index, *Confer. Semin. Mat. Univ. Bari,* **218** (1987) 1–32.

[**Ben**$_8$] BENCI, V., A new approach to the Morse–Conley theory, in *Recent Advances in Hamiltonian Systems,* G. Dell'Antonio ed., World Scient. Publ., Singapore, to appear.

[**BCF**$_1$] BENCI, V., A. CAPOZZI, and D. FORTUNATO, Periodic solutions of Hamiltonian systems of prescribed period, MRC Technical Summary Report No. 2508, University of Wisconsin, Madison, 1983.

[**BCF**$_2$] BENCI, V., A. CAPOZZI, and D. FORTUNATO, Periodic solutions of Hamiltonian systems with "superquadratic" potential, *Annali. Mat. Pura Appl. 4,* **143** (1986) 1–46.

[**BCF**$_3$] BENCI, V., A. CAPOZZI, and D. FORTUNATO, On asymptotically quadratic Hamiltonian systems, *J. Nonl. Anal.* **8** (1983) 929–931.

[**BeF**$_1$] BENCI, V. and D. FORTUNATO, The dual method in critical point theory. Multiplicity results for indefinite functionals, *Ann. Mat. Pure Appl. 4,* **32** (1982) 215–242.

[**BeF**$_2$] BENCI, V. and D. FORTUNATO, A "Birkhoff–Lewis" type result for a class of Hamiltonian systems, *Manuscripta Math.* **59** (1987) 441–456.

[**BenP**$_1$] BENCI, V. and F. PACELLA, Morse theory for symmetric functionals in the sphere and an application to a bifurcation problem, *J. Nonlinear Anal.* **9** (1985) 763–773.

[**BeR**$_1$] BENCI, V. and P.H. RABINOWITZ, Critical point theorems for indefinite functionals, *Invent. Math.* **52** (1979) 241–273.

[**Bere**$_1$] BERESTYCKI, H., Solutions périodiques de systèmes hamiltoniens, *Séminaire Bourbaki 1982/83*, exposé No. 603, *Astérisque* 105–106, 1983, 105–128.

[**Bere**$_2$] BERESTYCKI, H., Orbites périodiques de systèmes conservatifs, *Sémin. Goulaouic-Meyer-Schwartz 1981-1982*, exposé No. 24, 1982.

[**BL**$_1$] BERESTYCKI, H. and J.M. LASRY, A topological method for the existence of periodic orbits to conservative systems, preprint.

[**BLM**$_1$] BERESTYCKI, H., M. LASRY, G. MANCINI, and B. RUF, Existence of multiple periodic orbits on star-shaped Hamiltonian surfaces, *Comm. Pure Appl. Math.* **38** (1985) 253–289.

[**Ber**$_1$] BERGER, M.S., Bifurcation theory and the type numbers of M. Morse, *Proc. Nat. Acad. Sci.* **69** (1972) 1737–1738.

[**Ber**$_2$] BERGER, M.S., *Nonlinearity and Functional Analysis*, Academic Press, New York, 1977.

[**Ber**$_3$] BERGER, M.S., Periodic solutions of second order dynamical systems and isoperimetric variational problems, *Amer. J. Math.* **93** (1971) 1–10.

[**Ber**$_4$] BERGER, M.S., On non-absolute minima in the global calculus of variations, in *Méthodes topologiques en analyse non linéaire*, Granas ed., Sémin. Math. Sup. No. 95, Université de Montréal, Montréal, 1985, 20–40.

[**Ber**$_5$] BERGER, M., On a family of perioidic solutions of Hamiltonian systems, *J. Differential Equations* **10** (1971) 17–26.

[**Ber**$_6$] BERGER, M., On periodic solutions of second order Hamiltonian systems, *J. Math. Anal. Appl.* **29** (1970) 512–522.

[**BeB**$_1$] BERGER, M.S. and E. BOMBIERI, On Poincaré's isoperimetric problem for simple closed geodesics, *J. Funct. Anal.* **42** (1981) 274–298.

[**BeS**$_1$] BERGER, M.S. and M. SCHECHTER, On the solvability of semilinear gradient operator equations, *Advances in Math.* **25** (1977) 97–132.

[**Bert**$_1$] BERTOTTI, M.L., Forced oscillations of asymptotically linear Hamiltonian systems, *Boll. U.M.I.* (7)1-B (1987), 729–740.

[**BertZ**$_1$] BERTOTTI, M.L. and E. ZEHNDER, A Poincaré–Birkhoff type result in higher dimensions, preprint.

[$\mathbf{Bir_1}$] BIRKHOFF, G.D., Dynamical systems with two degree of freedom, *Trans. Amer. Math. Soc.* **18** (1917) 199–300.

[$\mathbf{BiH_1}$] BIRKHOFF, G.D. and M.R. HESTENES, Natural isoperimetric conditions in the calculus of variations, *Duke Math. J.* **1** (1935) 198–286.

[$\mathbf{BiH_2}$] BIRKHOFF, G.D. and M.R. HESTENES, Generalized minimax principles in the calculus of variations, *Duke Math. J.* **1** (1935) 413–432.

[$\mathbf{BiK_1}$] BIRKHOFF, G.D. and O. KELLOGG, Invariant points in function spaces, *Trans. Amer. Math. Soc.* **23** (1922) 96–115.

[$\mathbf{Blo_1}$] BLOT, J., Les systèmes hamiltoniens, leurs solutions périodiques, CEDIC-Nathan, Paris, 1982.

[$\mathbf{Boh_1}$] BÖHME, R., Nichtlineare störung der isolierten Eigenwerte selbs adjungierten Operatoren, *Math. Z.* **123** (1971) 61–92.

[$\mathbf{Boh_2}$] BÖHME, R., Dir Lösung der Verzweigungsgleichungen für nichtlineare Eigenwertprobleme, *Math. Z.* **127** (1972) 105–126.

[$\mathbf{Bol_1}$] BOLZA, O., *Vorlesungen über Variationsrechnung,* Leipzig, 1909, Taubner.

[$\mathbf{BZS_1}$] BORISOVICH, Yu.G., V.G. ZVYAGIN, and Yu.J. SAPRONOV, Nonlinear Fredholm maps and the Leray–Schauder theorem, *Russian Math. Surveys* **32** (1977) 1–54.

[$\mathbf{Bor_1}$] BORSUK, K., Drei Sätze über die n-dimensional Euklidische Spbäre, *Fund. Math.* **20** (1933) 177–190.

[$\mathbf{Bot_1}$] BOTT, R., Marston Morse and his mathematical works, *Bull. Amer. Math. Soc.* (NS) **3** (1980) 907–950.

[$\mathbf{Bot_2}$] BOTT, R., Lectures on Morse theory, old and new, *Bull. Amer. Math. Soc.* **7** (1982) 331–358.

[$\mathbf{Bot_3}$] BOTT, R., Nondegenerate critical manifolds, *Ann. of Math.* **60** (1954) 248–261.

[$\mathbf{Brd_1}$] BREDON, G.E., *Introduction to Compact Transformation Groups,* Academic Press, London, 1972.

[$\mathbf{Bre_1}$] BREZIS, H., *Analyse fonctionelle. Theorie et applications,* Masson, Paris, 1983.

[$\mathbf{Bre_2}$] BREZIS, H., Periodic solutions of nonlinear vibrating strings and duality principles, *Bull. Amer. Math. Soc.* (NS) **8** (1983) 409–426.

[**BrC$_1$**] BREZIS, H. and J.M. CORON, Periodic solutions of nonlinear wave equations and Hamiltonian systems, *Amer. J. Math.* **103** (1981) 559–570.

[**BCN$_1$**] BREZIS, H., J.M. CORON, and L. NIRENBERG, Free vibrations for a nonlinear wave equation and a theorem of Rabinowitz, *Comm. Pure Appl. Math.* **33** (1980) 667–689.

[**BrE$_1$**] BREZIS, H. and I. EKELAND, Un principe variationnel associé à certaines équations paraboliques. Le cas dépendant du temps. *C.R. Acad. Sciences Paris,* **282 A** (1976) 971–974.

[**BrE$_2$**] BREZIS, H. and I. EKELAND, Un principe variationnel associé à certaines équations paraboliques. Le cas dépendant du temps, *C.R. Acad. Sci. Paris,* **282 A** (1976) 1197–1198.

[**Bro$_1$**] BRONDSTED, A., Conjugate convex functions in topological vector spaces, *Mat. Fys. Medd. Dansk. Vid. Selsk.* **34, 2** (1964), 1–26.

[**Bro$_2$**] BROUSSEAU, V., L'index d'un système hamiltonien linéaire, *C.R. Acad. Sci. Paris I,* **303** (1986) 351–354.

[**BuM$_1$**] BUSENBERG, S. and M. MARTELLI, Better bounds for periodic solutions of differential equations in Banach spaces, HMC Math. Dept. Techn. Rep., March 1986.

[**Cai**] CAIRNS, S., Marston Morse 1892–1977, *Bull. Inst. Math. Acad. Sinica* **6** (1978) 1–9.

[**Cam$_1$**] CAMBINI, A., Sul lemma di Morse, *Boll. Un. Mat. Ital.* (4) **7** (1973) 87–93.

[**BaM$_1$**] CAMBINI, A. and A.M. MICHELETTI, Soluzioni periodiche di sistemi non lineari conservativi, *Boll. Un. Mat. Ital.* (4) **10** (1974) 713–723.

[**Cap$_1$**] CAPOZZI, A., On subquadratic Hamiltonian systems, *J. Nonlinear Anal.* **8** (1984) 553–562.

[**Cap$_2$**] CAPOZZI, A., Remarks on periodic solutions of subquadratic nonautonomous Hamiltonian systems, *Boll. Un. Mat. Ita.,* B (6) **4** (1985), 113–124.

[**CaF$_1$**] CAPOZZI, A. and D. FORTUNATO, An abstract critical point theorem for strongly indefinite functionals, *Proc. Symp. Pure Math.* **45**, Amer. Math. Soc., Providence, 1986, 237–241.

[**CFS$_1$**] CAPOZZI, A., D. FORTUNATO and A. SALVATORE, Periodic solutions of Lagrangian systems with bounded potential, *J. Math. Anal. Appl.* **124** (1987) 482–494.

[**CFS$_2$**] CAPOZZI, A., D. FORTUNATO and A. SALVATORE, Periodic solutions of dynamical systems, *Meccanica* **20** (1985) 281–284.

[**CGS$_1$**] CAPOZZI, A., C. GRECO and A. SALVATORE, Lagrangian systems in presence of singularities, *Proc. Amer. Math. Soc.* **102** (1988), 125–130.

[**CaS$_1$**] CAPOZZI, A. and A. SALVATORE, Periodic solutions for nonlinear problems with strong resonance at infinity, *Comm. Math. Univ. Carolinae* **23** (1982) 415–425.

[**CaS$_2$**] CAPOZZI, A. and A. SALVATORE, Nonlinear problems with strong resonance at infinity: an abstract theorem and applications, *Proc. Roy. Soc. Edinb.* **99** A (1985) 333–345.

[**CaS$_3$**] CAPOZZI, A. and A. SALVATORE, Periodic solutions of Hamiltonian systems: the case of the singular potential, Proc. NATO-ASI on Nonlinear Functional Anal., Singh ed., Reidel, Dordrecht, 1986, 207–216.

[**Car$_1$**] CARISTI, G., Monotone perturbations of linear operators having nullspace made of oscillating functions, *J. Nonlinear Analysis,* **11** (1987) 851-860.

[**Cas$_1$**] CASTRO, A., Periodic solutions of the forced pendulum equation, in *Differential Equations,* Ahmad and Lazer, ed., Academic Press, New York, 1980, 149–160.

[**Cas$_2$**] CASTRO, A., A two-point boundary value problem with jumping nonlinearities, *Proc. Amer. Math. Soc.* **79** (1980) 207–211.

[**Cas$_3$**] CASTRO, A., A semilinear Dirichlet problem, *Can. J. Math.* **31** (1979) 337–340.

[**Cas$_4$**] CASTRO, A., Hammerstein equations with indefinite kernel, *Int. J. Math. and Math. Sci.* **1** (1978) 187–201.

[**Cas$_5$**] CASTRO, A., Méthodes de réduction via minimax, in *Differential Equations,* de Figraeiredo-Honig ed., Lect. Notes Math. No. 957, Springer, Berlin, 1982.

[**CaL$_1$**] CASTRO, A. and A.C. LAZER, Critical point theory and the number of solutions of a nonlinear Dirichlet problem, *Ann. Mat. Pura Appl.* **4, 70** (1979) 113–137.

[**CaL$_2$**] CASTRO, A. and A.C. LAZER, Applications of a max-min principle, *Rev. Columbiana Mat.* **10** (1976) 141–149.

[**Ces$_1$**] CESARI, L., *Optimization Theory and Applications,* Springer, New York, 1983.

[$\mathbf{Ces_2}$] CESARI, L., Functional analysis and Galerkin's method, *Michigan Math. J.* **11** (1964) 385–414.

[$\mathbf{Cha_1}$] CHANG, K.C., Infinite Dimensional Morse Theory and its Applications, Sémin. Math. Sup. No. 97, Presses Univ. Montréal, Montréal, 1985.

[$\mathbf{Cha_2}$] CHANG, K.C., Solution of asymptotically linear operator equations via Morse theory, *Comm. in Pure and Appl. Math.* **34** (1981) 693–712.

[$\mathbf{Cha_3}$] CHANG, K.C., Variational methods for non-differentiable functionals and its applications to partial differential equations, *J. Math. Anal. Appl.* **80** (1981) 102–129.

[$\mathbf{Cha_4}$] CHANG, K.C., A variant mountain pass lemma, *Sci. Sinica Ser. A* **26** (1983) 1241–1255.

[$\mathbf{Cha_5}$] CHANG, K.C., On the mountain pass lemma, in *Equadiff* **6**, Brno 1985, Lect. Notes in Math. No. 1192, Springer, Berlin, 1986, 203–208.

[$\mathbf{Cha_6}$] CHANG, K.C., Morse theory on Banach spaces and its applications to partial differential equation, *Chinese Ann. Math.* **4B** (1983) 381–399.

[$\mathbf{Cha_7}$] CHANG, K.C., On a bifurcation theorem due to Rabinowitz, *J. Systems Sci. Math. Sci.* **4** (1984) 191–195.

[$\mathbf{Cha_8}$] CHANG, K.C., Applications to homology theory to some problems in differential equations, *Proc. Symp. Pure Math.* vol. 45 (1985) 253–262.

[$\mathbf{ChH_1}$] CHANG, K.C. and C.W. HONG, Periodic solutions for the semilinear spherical wave equation, *Acta Math. Sinica* (NS) **1** (1985) 87–96.

[$\mathbf{CLD_1}$] CHANG, K.C., S. LI and G.C. DONG, A new proof and an extension of a theorem of P. Rabinowitz concerning nonlinear wave equations, *J. Nonlinear Anal.* **6** (1982) 139–150.

[$\mathbf{CWL_1}$] CHANG, K.C., S.P. WU and S. LI, Multiple periodic solutions for an asymptotically linear wave equation, *Indiana Math. J.* **31** (1982) 721–731.

[$\mathbf{Chap_1}$] CHAPERON, M., Quelques questions de géométrie symplectique, Séminaire Bourbaki 1982/83 No. 610, Astérisque No. 105-106, 1983, 231–249.

[$\mathbf{ClH_1}$] CHOW, S.N. and J.K. HALE, *Methods of Bifurcation Theory,* Springer, Berlin, 1982.

[Clk_1] CLARK, D.C., A variant of the Ljusternik–Schnirelmann theory, *Indiana J. Math.* **22** (1973) 65–74.

[Clk_2] CLARK, D.C., Periodic solutions of variational systems of ordinary differential equations, *J. Differential Equations* **28** (1978) 354–358.

[Clk_3] CLARK, D.C., On periodic solutions of autonomous Hamiltonian systems of ordinary differential equations, *Proc. Amer. Math. Soc.* **39** (1973) 579–584.

[Clk_4] CLARK, D.C., Eigenvalue bifurcation for odd gradient operators, *Rocky Mt. J. Math.* **5** (1975) 317–336.

[Cla_1] CLARKE, F.H., A classical variational principle for periodic Hamiltonian trajectories, *Proc. Amer. Math. Soc.* **76** (1979) 186–188.

[Cla_2] CLARKE, F.H., Periodic solutions to Hamiltonian inclusions, *J. Differential Equations* **40** (1981) 1–6.

[Cla_3] CLARKE, F.H., *Optimization and Nonsmooth Analysis,* Wiley Interscience, New York, 1983.

[Cla_4] CLARKE, F.H., Tonelli's regularity theory in the calculus of variations: recent pgoress, in *Optimization and Related Fields,* Conti, De Giorgi, Giannessi ed., Lect. Notes Math. No. 1190, Springer, 1986.

[Cla_5] CLARKE, F.H., Régularité, existence et condition nécessaires pour le problème fondamental en calcul des variations, Centre de Rech. Math., Univ. Montréal, CRM-1354, 1986.

[Cla_6] CLARKE, F.H., Periodic solutions of Hamilton's equations and local minima of the dual action, *Trans. Amer. Math. Soc.* **287** (1985) 239–251.

[Cla_7] CLARKE, F.H., On hamiltonian flows and symplectic transformations, *SIAM J. Control and Optimization* **20** (1982) 355–359.

[Cla_8] CLARKE, F.H., Action principles and periodic orbits, in Séminaire Brézis-Lions, vol. 8, Pitmann, 1988.

[Cla_9] CLARKE, F.H., Optimization and periodic trajectories, in *Periodic Solutions of Hamiltonian Systems and Related Topics,* Rabinowitz, Ambrosetti, Ekeland, Zehnder, eds., Reidel, Dordrecht, 1987, 99–110.

[ClE_1] CLARKE, F.H. and I. EKELAND, Hamiltonian trajectories with prescribed minimal period, *Comm. Pure Appl. Math.* **33** (1980) 103–116.

[CoP_1] COCLITE, M. and G. PALMIERE, Multiple solutions for variational problems and applications, *Boll. Un. Mat. Ital.* (7) 1-B (1987) 347–371.

[ClE$_2$] CLARKE, F.H. and I. EKELAND, Nonlinear oscillations and boundary value problems for Hamiltonian systems, *Arch. Rat. Mech. Anal.* **78** (1982) 315–333.

[Con$_1$] CONLEY, C.C., *Isolated Invariant Sets and the Morse Index*, CBMS 38, Amer. Math. Soc., Providence, R.I., 1978.

[CoZ$_1$] CONLEY, C. and E. ZEHNDER, Morse type index theory for flows and periodic solutions for Hamiltonian equations, *Comm. Pure Appl. Math.* **37** (1984) 207–253.

[CoZ$_2$] CONLEY, C. and E. ZEHNDER, The Birkhoff–Lewis fixed point theorem and a conjecture of V. Arnold, *Invent. Math.* **73** (1983) 33–49.

[CoZ$_3$] CONLEY, C. and E. ZEHNDER, Subharmonic solutions and Morse theory, *Physica* **124** A (1984) 649–658.

[CoF$_1$] CONNER, P.E. and E.E. FLOYD, Fixed point free involutions and equivariant maps, *Bull. Amer. Math. Soc.* **66** (1960) 416–441.

[CoF$_2$] CONNER, P.E. and E.E. FLOYD, Fixed point free involutions and equivariant maps, II, *Trans. Amer. Math. Soc.* **105** (1962) 222–228.

[Cor$_1$] CORON, J.M., Résolution de l'équation $Au + Bu = f$ où A est linéaire auto-adjoint et B dérive d'un potentiel convexe, *Ann. Fac. Sci. Toulouse* **1** (1979) 215–234.

[Cor$_2$] CORON, J.M., Solutions périodiques non triviales d'une équation des ondes, *Comm. Partial Differential Equations* **6** (1981) 829–848.

[Cor$_3$] CORON, J.M., Periodic solutions of a nonlinear wave equation without assumption of monotonicity, *Math. Ann.* **262** (1983) 273–285.

[Cos$_1$] COSTA, D.G., An application of the Lusternik–Schnirelmann theory, Proc. 15th Brazilian Seminar of Analysis, 1982, 211–233.

[CoW$_1$] COSTA, D.G. and M. WILLEM, Multiple critical points of invariant functionals and applications, *J. Nonlinear Anal.* **9** (1986) 843–852.

[CoW$_2$] COSTA, D.G. and M. WILLEM, Lusternik–Schnirelmann theory and asymptotically linear Hamiltonian systems, in *Colloquia Math. Soc. Janos Bolyai.* **47**, *Differential Equations: Qualitative Theory*, Szeged, Hungary, 1984, North Holland, 1986, 179–191.

[Cot$_1$] COTI-ZELATI, V., Perturbations of second order Hamiltonian systems via Morse theory, *Boll. Un. Mat. Ital.* **4-C** (1985) 307–322.

[Cot_2] COTI-ZELATI, V., Periodic solutions of dynamical systems with bounded potential, *J. Differential Equations* **67** (1987) 400–413.

[Cot_3] COTI-ZELATI, V., Remarks on dynamical systems with weak forces, *Manuscripta Math.* **57** (1987) 417–424.

[Cot_4] COTI-ZELATI, V., Dynamical systems with effective-like potentials, *J. Nonlinear Analysis* **12** (1988) 209–222.

[Cot_5] COTI-ZELATI, V., Morse theory and periodic solutions of Hamiltonian systems, Ph.D. Thesis, Trieste, 1987.

[Cou_1] COURANT, R., *Dirichlet Principle, Conformal Mappings and Minimal Surfaces,* Wiley Interscience, New York, 1950.

[CoH_1] COUNRANT, R. and D. HILBERT, *Methods of Mathematical Physics,* vol. 1, Wiley, New York, 1962.

[Cro_1] CROKE, B.C., Poincaré's problem and the length of the shortest closed geodesic on a convex hypersurface, *J. Diff. Geom.* **17** (1982) 595–634.

[CrW_1] CROKE, B.C. and A. WEINSTEIN, Closed curves on convex hypersurfaces and period of nonlinear oscillations, *Inv. Math.* **64** (1981) 199–202.

[CGR_1] CROUZEIX, M., G. GEYMONAT and G. RAUGEL, Some remarks about the Morse lemma in infinite dimension, preprint.

[Dan_1] DANCER, E.N., On the use of asymptotics in nonlinear boundary value problems, *Ann. Mat. Pura Appl.* (4) **131** (1982) 167–187.

[Dan_2] DANCER, E.N., Degenerate critical points, homotopy indices and Morse inequalities, *J. Reine Angew. Math.* **350** (1984) 1–22.

[Dan_3] DANCER, E.N., On the existence of solutions of certain asymptotically homogeneous problems, *Math. Z.* **177** (1981) 33–48.

[Dan_4] DANCER, E.N., Symmetries, degree, homotopy indices and asymptotically homogeneous problems, *J. Nonlinear Anal.* **6** (1982) 667–686.

[Dan_5] DANCER, E.N., Breaking of symmetries for forced equations, *Math. Ann.* **262** (1983) 473–486.

[Dan_6] DANCER, E.N., A new degree for S^1-invariant gradient mappings and applications, *Ann. Inst. H. Poincaré, Analyse non linéaire,* **2** (1985) 329–370.

[Dan_7] DANCER, E.N., The G-invariant implicit function theorem in infinite dimension, II, *Proc. Roy. Soc. Edinburgh* **102-A** (1986) 211–220.

[$\mathbf{Dan_8}$] DANCER, E.N., Multiple solutions of asymptotically homogeneous problems, preprint.

[$\mathbf{Dan_9}$] DANCER, E.N., Degenerate critical points, homotopy indices and Morse inequalities, II, preprint.

[$\mathbf{DCF_1}$] DE CANDIA, A. and D. FORTUNATO, Osservazioni su alcuni problemi ellittici non lineari, *Rendic. Istit. Mat. Univ. Trieste,* **17** (1985) 30–46.

[$\mathbf{DFS_1}$] DE FIGUEIREDO, D.G. and S. SOLIMINI, A variational approach to superlinear elliptic problems, *Comm. Partial Differential Equations* **9** (1984) 699–717.

[$\mathbf{Dei_1}$] DEIMLING, K., *Nonlinear Functional Analysis,* Springer, Berlin, 1985.

[$\mathbf{Den_1}$] DENG, S., Minimal periodic solutions for a class of Hamiltonian equations, preprint.

[$\mathbf{Des_1}$] DESOLNEUX-MOULIS, N., Orbites périodiques des systèmes hamiltoniens autonomes, Sem. Bourbaki 1979/80, No. 552, Lect. Notes in Math. No. 842, Springer, Berlin, 1983.

[$\mathbf{DiM_1}$] DING, S.H. and J. MAWHIN, The range of some monotone gradient perturbations of self-adjoint operators with finite-dimensional kernel having the unique continuation property, *Houston J. Math.,* to appear.

[$\mathbf{Dol_1}$] DOLPH, C.L., Nonlinear integral equations of the Hammerstein type, *Trans. Amer. Math. Soc.* **66** (1949) 289–307.

[$\mathbf{DoL_1}$] DONG, G.C. and LI, S., On the existence of infinitely many solutions of the Dirichlet problem for some nonlinear elliptic equations, *Sci. Sinica,* Ser. A **25** (1982) 468–475.

[$\mathbf{DuB_1}$] DU BOIS-REYMOND, P., Erläterungen zu den Anfangsgrunden der Variationsrechnung, *Math. Ann.* **15** (1879) 283–314.

[$\mathbf{Eel_1}$] EELLS, J., A setting for global analysis, *Bull. Amer. Math. Soc.* **72** (1966) 751–807.

[$\mathbf{Eke_1}$] EKELAND, I., Periodic solutions of Hamiltonian equation and a theorem of P. Rabinowitz, *J. Differential Equations* **34** (1979) 523–534.

[$\mathbf{Eke_2}$] EKELAND, I., Forced oscillations of nonlinear Hamiltonian systems, II, *Advances in Math.* **7A** (1981) 345–360.

[Eke$_3$] EKELAND, I., Oscillations de systèmes hamiltoniens non-linéaires, III, *Bull. Soc. Math. France* **109** (1981) 297–330.

[Eke$_4$] EKELAND, I., On the variational principle, *J. Math. Anal. Appl.* **47** (1974) 324–353.

[Eke$_5$] EKELAND, I., Nonconvex minimization problems, *Bull. Amer. Math. Soc.* (NS) **1** (1979) 443–474.

[Eke$_6$] EKELAND, I., Problèmes variationnels non convexes, in *Proceedings of the International Congress of Mathematicians,* Helsinki, 1978, 855–858.

[Eke$_7$] EKELAND, I., Une théorie de Morse pour les systèmes hamiltoniens convexes, *Ann. Inst. H. Poincaré, Anal. non linéaire* **1** (1984) 143–197.

[Eke$_8$] EKELAND, I., Index theory for periodic solutions of convex Hamiltonian systems, in *Proceedings AMS Summer Institute on Nonlinear Functional Analysis* (Berkeley, 1983). Proceedings of Symposia in Pure Mathematics, Vol. 45 (1986), I, 395–423.

[Eke$_9$] EKELAND, I., Legendre duality in non convex optimization and calculus of variations, *SIAM J. Control and Optimization* **15** (1977) 905.

[Eke$_{10}$] EKELAND, I., Two results in convex analysis, in *Optimization and Related Fields,* Conti, De Giorgi, Giannessi eds., Lect. Notes Math., No. 1190, Springer, Berlin, 1986, 215–228.

[Eke$_{11}$] EKELAND, I., A perturbation theory near convex Hamiltonian systems, *J. Differential Equations,* **50** (1983) 407–440.

[Eke$_{12}$] EKELAND, I., Ioffe's mean value theorem, in *Convex Analysis and Optimization,* Aubin, ed., Pitman, 1982, 35–42, Boston.

[EkH$_1$] EKELAND, I. and H. HOFER, Periodic solutions with prescribed minimal period for convex autonomous Hamiltonian system, *Invent. Math.* **81** (1985) 155–188.

[EkH$_2$] EKELAND, I. and H. HOFER, Subharmonics for convex nonautonomous Hamiltonian systems, *Comm. Pure Applied Math.* **40** (1987) 1–36.

[EkH$_3$] EKELAND, I. and H. HOFER, Convex Hamiltonian energy surfaces and their periodic trajectories, *Comm. Math. Phys.* **113** (1987) 419–469.

[EkL_1] EKELAND, I. and J.M. LASRY, On the number of periodic trajectories for a Hamiltonian flow on a convex energy surface, *Ann. of Math.* **112** (1980) 283–319.

[EkL_2] EKELAND, I. and J.M. LASRY, Duality in non convex variational problems, in *Advances in Hamiltonian Systems,* Aubin, Benssoussan, Ekeland ed., Birkhauser, Basel, 1983, 74–108.

[EkL_3] EKELAND, I. and J.M. LASRY, Sur le nombre de points critiques de fonctions invariantes par des groupes, *C.R. Acad. Sc. Paris,* **282** A (1976) 841–844.

[ELa_1] EKELAND, I. and L. LASSOUED, Multiplicité des trajectoires fermées de systèmes hamiltoniens convexes, *Ann. Inst. Poincaré, Anal. non linéaire,* **4** (1987) 307–336.

[ELa_2] EKELAND, I. and L. LASSOUED, Un flot hamiltonien a au moins deux trajectoires fermées sur toute surface d'énergie convexe et bornée, *C.R. Acad. Sc. Paris* **361** (1985) 161–164.

[ELe_1] EKELAND, I. and G. LEBOURG, Generic Frechet-differentiability and perturbed optimization problems in Banach spaces, *Trans. Amer. Math. Soc.* **224** (1976) 193–216.

[EkT_1] EKELAND, I. and R. TEMAM, *Convex Analysis and Variational Problems,* North Holland, Amsterdam, 1976.

[$EkTu_1$] EKELAND, I. and T. TURNBULL, *Infinite Dimensional Optimization and Convexity,* The University of Chicago Press, Chicago, 1983.

[Eul_1] EULER, L., Metodus nova et facilio calculum variationum tractandi, *Novi commentarii Academiae Scientiarum Petropolitanae* **16** (1771) 35–70 (= Oeuvres (1) **25**, 208–235).

[Fad_1] FADELL, E.R., The relationship between Ljusternik–Schnirelman category and the concept of genus, *Pacif. J. Math.* **89** (1980) 33–42.

[Fad_2] FADELL, E.R., The equivariant Ljusternik–Schnirelmann method for invariant functionals and relative cohomological index theories, in *Méth. topologiques en analyse non linéaires,* Granas ed., Sémin. Math. Sup. No. 95, Montreal, 1985, 41–70.

[Fad_3] FADELL, E.R., Cohomological methods in non-free G-spaces with applications to general Borsuk–Ulam theorems and critical point theorems for invariant functionals, in *Nonlinear Functional Anal. and Appl.,* Singh ed., NATO Asi series, Reidel, Dordrecht, 1986, 1–45.

[FaH_1] FADELL, E.R. and S. HUSSEINI, Relative cohomological index theories, *Advances in Math.* **64** (1987) 1–31.

[$\mathbf{FHR_1}$] FADELL, E.R., S.Y. HUSSEINI and P.H. RABINOWITZ, Borsuk-Ulam theorems for arbitrary S^1-actions and applications, *Trans. Amer. Math. Soc.* **274** (1982) 345–360.

[$\mathbf{FaR_1}$] FADELL, E.R. and P.H. RABINOWITZ, Bifurcation for odd potential operators and an alternative topological index, *J. Funct. Anal.* **26** (1977) 48–67.

[$\mathbf{FaR_2}$] FADELL, E.R. and P.H. RABINOWITZ, Generalized cohomological index theories for Lie group actions with an application to bifurcation questions for Hamiltonian systems, *Inventiones Math.* **45** (1978) 139–174.

[$\mathbf{Fen_1}$] FENCHEL, W., *Convex Cones, Sets and Functions,* Lecture Notes, Princeton Univ., 1951.

[$\mathbf{Fen_2}$] FENCHEL, W., On conjugate convex functions, *Canad. J. Math.* **1** (1949) 73–77.

[$\mathbf{Fen_3}$] FENCHEL, W., Convexity through the ages, in *Convexity and Its Applications,* P.M. Gruber and J.M. Wills ed., Birkhauser, Basel, 1983.

[$\mathbf{FoWi_1}$] FOURNIER, G. and M. WILLEM, Multiple solutions of the forced double pendulum equation, *Ann. de l'Inst. H. Poincaré, Analyse non linéaire,* to appear.

[$\mathbf{FuC_1}$] FUCIK, S., *Solvability of Nonlinear Equations and Boundary Value Problems,* Reidel, Dordrecht, 1980.

[$\mathbf{FuC_2}$] FUCIK, S., Nonlinear equations with linear part at resonance: variational approach, *Comment. Math. Univ. Carolinae* **18** (1977) 723–734.

[$\mathbf{FuC_3}$] FUCIK, S., Nonlinear potential equations with linear part at resonance, *Casopis Pest. Mat.* **103** (1978) 78–94.

[$\mathbf{FuC_4}$] FUCIK, S., Variational noncoercive nonlinear problems, in *Theory of Nonlinear Operators,* Akademie-Verlag, Berlin, 1978, 61–69.

[$\mathbf{Fun_1}$] FUNK, P., Variationsrechnung und ihre Anwendung in Physik un Technik, *Grundlehren Math. Wiss.* **94**, Springer, Berlin, 2d. ed., 1970.

[$\mathbf{GaM_1}$] GAINES, R.E. and J.L. MAWHIN, *Coincidence Degree and Nonlinear Differential Equations,* Springer, Berlin, 1977.

[$\mathbf{GeG_1}$] GEBA, K. and A. GRANAS, A proof of Borsuk's antipodal theorem for Fredholm maps, *J. Math. Anal. Appl.* **96** (1983) 196–202.

[Gia_1] GIANNONE, F., Soluzioni periodiche di sistemi Hamiltonian in presenza di vincoli, *Rapporti Dep. Mat. Pisa* **11** (1982).

[Gir_1] GIRARDI, M., Multiple orbits for starshaped Hamiltonian surfaces with symmetries, *Ann. Inst. H. Poincaré, Anal. non lin.*, **1** (1984) 285–294.

[GiM_1] GIRARDI, M. and M. MATZEU, Some results on solutions of minimal period to superquadratic Hamiltonian systems, *J. Nonlinear Anal.* **7** (1983) 475–482.

[GiM_2] GIRARDI, M. and M. MATZEU, Solutions of minimal period for a class of non convex Hamiltonian systems and applications to the fixed energy problem, *J. Nonlinear Anal.* **10** (1986) 371–382.

[GiM_3] GIRARDI, M. and M. MATZEU, Solutions of prescribed minimal period to convex and non-convex Hamiltonian systems, *Boll. Un. Mat. Ital.* B (6) **4** (1985) 951–967.

[GiM_4] GIRARDI, M. and M. MATZEU, Periodic solutions of convex Hamiltonian systems with a quadratic growth at the origin and superquadratic at infinity, *Ann. Mat. Pura Appl.* (4) **147** (1987) 21–72.

[GiM_5] GIRARDI, M. and M. MATZEU, A variational approach to periodic solutions of Hamiltonian systems with arbitrary long minimal period, preprint.

[Gol_1] GOLUMB, M., Zur Theorie der nichtlinearen Integral gleichungen, Integral gleichungssysteme und algemeinen Funktional gleichungen, *Math. Z.* **39** (1934) 45–75.

[GoM_1] GOLUBITSKY, M. and J. MARSDEN, The Morse lemma in infinite dimension via singularity theory, *SIAM J. Math. Anal.* **14** (1983) 1037–1044.

[Gor_1] GORDON, W., A theorem on the existence of periodic solutions to Hamiltonian systems with convex potential, *J. Differential Equations* **10** (1971) 324–335.

[Gor_2] GORDON, W.B., Physical variational principles which satisfy the Palais–Smale condition, *Bull. Amer. Math. Soc.* **78** (1972) 712–716.

[Gor_3] GORDON, W.B., A minimizing property of Keplerian orbits, *Amer. J. Math.* **99** (1977) 961–971.

[Gor_4] GORDON, W.B., Conservative dynamical systems involving strong forces, *Trans. Amer. Math. Soc.* **204** (1975) 113–135.

[Gos$_1$] GOSSEZ, J.P., Some nonlinear differential equations with resonance at the first eigenvalue, in Att. 3^e Semin. Analisi Funzionale ed Applic. SAFA III, Confer. Semin. Mat. Univ. Bari No. 167, 1979, 355–389.

[Gre$_1$] GRECO, C., On forced oscillations of Lagrangian systems, *Rend. Ist. Mat. Univ. Trieste* **18** (1986) 192–199.

[Gre$_2$] GRECO, C., Periodic solutions of some nonlinear ordinary differential equations with singular nonlinear part, *Boll. Un. Mat. Ital.*, B, to appear.

[Gre$_3$] GRECO, C., Periodic solutions of second order Hamiltonian systems in an unbounded potential well, *Proc. Roy. Soc. Edinburgh* A **105** (1987) 1–15.

[Gre$_4$] GRECO, C., Periodic solutions of a class of singular Hamiltonian systems, *J. Nonlinear Anal.* **12** (1988) 259–270.

[GrM$_1$] GROMOLL, D. and W. MEYER, On differentiable functions with isolated critical points, *Topology* **8** (1969) 361–369.

[GrS$_1$] GROSSINHO, M.R. and L. SANCHEZ, A note on periodic solutions of some nonautonomous differential equations, *Bull. Austral. Math. Soc.* **34** (1986) 253–265.

[Had$_1$] HADAMARD, J., *Lecons sur le Calcul des Variations,* Hermann, Paris, 1910.

[Ham$_1$] HAMEL, G., Uber erzwungene Schwingungen bei endlichen Amplituden, *Math. Ann.* **86** (1922) 1–13.

[Hes$_1$] HESTENES, M.R., The problem of Bolza in the calculus of variations, *Bull. Amer. Math. Soc.* **47** (1942) 57–75.

[Hil$_1$] HILBERT, D., Ueber das Dirhchlet'sche Prinzip Jahresber. *Deutsche Math. Verein.* **8** (1900) 184–188 (reprinted in *J. Reine Angew. Math.* **129** (1905) 63–67 french trans. by M.L. Langel in *Nouv. Ann. de Math.* (3) **19** (1900) 377–344).

[Hil$_2$] HILBERT, D., Ueber das Dirichlet'sche Prinzip, *Festchr. Feier* 150-Jähr. Best V. Ges. Wiss. Göttingen, 1901 (reprinted in *Math. Ann.* **59** (1904) 161–186).

[HiV$_1$] HIRIART-URRUTY, J.B., A short proof of the variational principle for approximate solutions of a minimization problem, *Amer. Math. Monthly* **90** (1983) 206–207.

[**Hof$_1$**] HOFER, H., A geometric description of the neighborhood of a critical point given by the mountain-pass theorem, *J. London Math. Soc.* (2) **31** (1985) 566–570.

[**Hof$_2$**] HOFER, H., A new proof of a result of Ekeland and Lasry concerning the number of periodic Hamiltonian trajectories on a prescribed energy surface, *Boll. Univ. Mat. Ital.* **16** B (1982) 931–942.

[**Hof$_3$**] HOFER, H., The topological degree at a critical point of mountain pass type, in *Proc. Symp. Pure Math.* **45**, part I, Amer. Math. Soc., 1986, 501–509.

[**Hof$_4$**] HOFER, H., Some theory of strongly indefinite functionals with applications, *Trans. Amer. Math. Soc.* **275** (1983) 185–214.

[**Hof$_5$**] HOFER, H., Variational and topological methods in partially ordered spaces, *Math. Ann.* **261** (1982) 493–514.

[**Hof$_6$**] HOFER, H., Lagrangian embeddings and critical point theory, *Ann. Inst. H. Poincaré, Analyse non linéaire* **2** (1985) 407–462.

[**HoS$_1$**] HOLM, P. and E.H. SPANIER, Involutions and Fredholm maps, *Topology* **10** (1971) 203–218.

[**Hor$_1$**] HORN, J., Beiträge zur Theorie der kleinen Schwingungen, *Z. Math. Phys.* **48** (1903) 400–434.

[**Jen$_1$**] JENSEN, J.L.W.V., Sur les fonctions convexes et les inégalités entre les valeurs moyennes, *Acta Math.* **30** (1908) 175–193.

[**Joh$_1$**] JOHN, F., Ueber die Volkständigkeit der Relationen von Morse für die Anzahlen kritischer Punkte, *Math. Ann.* **109** (1934) 381–394.

[**Kli$_1$**] KLINKENBERG, W., *Lectures on Closed Geodesics,* Springer, Berlin, 1978.

[**Koz$_1$**] KOZLOV, V.V., Calculus of variations in the large and classical mechanics, *Russian Math. Surveys* **40**, No. 2 (1985) 37–71.

[**Kra$_1$**] KRASNOSELSKI, M.A., On the estimation of the number of critical points of functionals, *Uspeki Math. Nauk* **7** (1952) No. 2 (48), 157–164.

[**Kra$_2$**] FRASNOSELSKI, M.A., *Topological Methods in the Theory of Nonlinear Integral Equations,* Pergamon, Oxford, 1965.

[**Kui$_1$**] KUIPER, C., C^1-equivalence of functions near isolated critical points, Symposium on Infinite-Dimensional Topology, *Ann. Math. Studies* No. 69, Princeton Univ. Press, Princeton, 1972, 199–218.

[$LaLa_1$] LANDESMAN, E. and A.C. LAZER, Nonlinear perturbations of linear eigenvalue problems at resonance, *J. Math. Mech.* **19** (1970) 609–623.

[$LasV_1$] LASSOUED, L. and C. VITERBO, La théorie de Morse pour les systèmes hamiltoniens.

[Laz_1] LAZER, A.C., Some resonance problems for elliptic boundary value problems, in *Nonlinear Functional Analysis and Differential Equations,* Cesari, Kannan, Schuur eds., Dekker, 1976, 269–289.

[LLM_1] LAZER, A.C., E.M. LANDESMAN and D.R. MEYERS, On saddle point problems in the calculus of variations, the Ritz algorithm and monotone convergence, *J. Math. Anal. Appl.* **52** (1975) 594–614.

[$LaLe_1$] LAZER, A.C. and D.E. LEACH, Bounded perturbations of forced harmonic oscillations at resonance, *Ann. Mat. Pura App.* (4) **82** (1969) 49–68.

[Leb_1] LEBESGUE, H., Integrale, longueur, aire, *Annali Mat. Pura Appl.* **3, 7** (1902) 231–358.

[Leb_2] LEBESGUE, H., Sur le problème de Dirichlet, *Rend. Circ. Mat. Palermo,* **24** (1907) 371–402.

[Leg_1] LEGENDRE, A.M., Mémoire sur l'intégration de quelques équations aux différences partielles, *Mém. Acad. Sciences,* 1787, 309–351.

[Lev_1] LEVY, P., *Lecons D'analyse Fonctionnelle,* Gauthier-Villars, Paris, 1922.

[Lic_1] LICHTENSTEIN, L., Uber einige Existenzprobleme der Variationsrechnung, *J. für Math.* **145** (1915) 24–85.

[Liu_1] LIU, G.Q., Morse index of a saddle point, to appear.

[LiP_1] LIOTARD, D. and J.P. PENOT, Critical paths and passes: applications to Quantum chemistry, in *Proc. Coll.* Carry-le-Rouet, Della Lora, Demongeot, Lacolle ed., Springer, Berlin, 1981, 213–221.

[Lju_1] LJUSTERNIK, L., *Topology of the Calculus of Variations in the Large,* Amer. Math. Soc., Providence, R.I., 1966.

[LJS_1] LJUSTERNIK, L. and L. SCHNIRELMANN, *Méthodes Topologiques dans les Problèmes Variationnels,* Hermann, Paris, 1934.

[Llo_1] LLOYD, N.G., *Degree Theory,* Cambridge Univ. Press, Cambridge, 1978.

[**Lon$_1$**] LONG, Y., Multiple solutions of perturbed superquadratic second order Hamiltonian systems, MRC Techn. Rep. No. 2963, Univ. of Wisconsin, Madison, 1987.

[**Lov$_1$**] LOVICAR, V., On infinitely many solutions to some nonlinear homogeneous equations, preprint.

[**LuS$_1$**] LUPO, D. and S. SOLOMINI, A note on a resonance problem, *Proc. Royal Soc. Edinburgh* **102** A (1986) 1–7.

[**Lya$_1$**] LYAPUNOV, A.M., Problème gégéral de la stabilité du mouvement, *Ann. Fac. Sci. Toulouse* **2** (1907) 203–474.

[**Mnc$_1$**] MANCINI, G., Esistenzia in grande di traiettoria periodiche per sistemi Hamiltoniani autonomi, in *Metodi asintotici e topologici in problemi differenziali non lineari, l'Aquila,* 1981, Boccardo-Micheletti ed., Pitagora, Bologna, 1981, 125–140.

[**Man$_1$**] MANCINI, G., Periodic solutions of Hamiltonian systems having prescribed minimal period, *Adv. in Hamiltonian Systems,* Birkhauser, Boston, 1983, 43–72.

[**Man$_1$**] MANDELBROJT, S., Sur les fonctions convexes, *C.R. Ac. Sci. Paris* **209** (1939) 977–978.

[**Mar$_1$**] MARINO, A., La bifurcazione sul caso variazionale, *Confer. Semin. Mat.,* Univ. Bari No. 132, 1973.

[**MaP$_1$**] MARINO, A. and G. PRODI, Metodi perturbativi nella teoria di Morse, *Boll. Un. Mat. Ital. Suppl. Facs.* **3** (1975) 1–32.

[**MaP$_2$**] MARINO, A. and G. PRODI, La teoria di Morse per spazi di Hilbert, *Rend. Sem. Mat. Univ. Padova* **41** (1968) 43–68.

[**Mar$_1$**] MARZANTOWICZ, W., On the nonlinear elliptic equations with symmetry, *J. Math. Anal. Appl.* **81** (1981) 156–181.

[**Maw$_1$**] MAWHIN, J., Semi-coercive monotone variational problems, *Acad. R. Belgique, Bull. Cl. Sciences* (5) **73** (1987) 118–130.

[**Maw$_2$**] MAWHIN, J., Problèmes de Dirichlet variationnels non-linéaires, *Sémin. Math. Supérieures* No. 104, Presses Univ. Montréal, Montréal, 1987.

[**Maw$_3$**] MAWHIN, J., Compacité monotonie et convexité dans l'étude de problèmes aux limites semi-linéaires, *Sém. d'Analyse Moderne* No. 19, Univ. de Sherbrooke, 1981.

[**Maw$_4$**] MAWHIN, J., Necessary and sufficient conditions for the solvability of nonlinear equations through the dual least action principle, in *Workshop on Applied Differential Equations,* Beijing, 1985, Xias, Pu ed., World Scientific, Singapore, 1986, 91–108.

[**Maw$_5$**] MAWHIN, J., Points fixes, points critiques et problèmes aux limites, *Sémin. Math. Sup.* No. 92, Presses Univ. Montréal, Montréal, 1985.

[**Maw$_6$**] MAWHIN, J., A Neumann boundary value problem with jumping monotone non linearity, *Delft Progress Report* **10** (1985) 44–52.

[**Maw$_7$**] MAWHIN, J., The dual least action principle and nonlinear differential equations, in *Intern. Conf. Qualitative Theory of Differential Equations,* Edmonton, 84, Allegreto, Butler ed., Math. Dept., Univ. Alberta, Edmonton, 1986, 262–274.

[**Maw$_8$**] MAWHIN, J., Critical point theory and nonlinear differential equations, in *Equadiff VI* Brno, 1985, Lect. Notes in Math. No. 1192, Springer, Berlin, 1986, 49–58.

[**Maw$_9$**] MAWHIN, J., Contractive mappings and periodically perturbed conservative systems, *Arch. Math.* (Brno) **12** (1976) 67–73.

[**Maw$_{10}$**] MAWHIN, J., Semilinear equations of gradient type in Hilbert space and applications to differential equations, in *Nonlinear Differential Equations: Stability, Invariance and Bifurcation,* Academic Press, New York, 1981, 269–282.

[**Maw$_{11}$**] MAWHIN, J., Periodic solutions of ordinary differential equations: the Poincaré's heritage, in *Differential Topology - Geometry and Related Fields,* Rassias, ed., Teubner, Leipzig, 1985, 287–307.

[**Maw$_{12}$**] MAWHIN, J., Forced second order conservation systems with periodic nonlinearity, *Ann. Inst. Poincaré, Anal. non linéaire,* to appear.

[**MWW$_1$**] MAWHIN, J., J.R. WARD and M. WILLEM, Variational methods and semi-linear elliptic equations, *Arch. Rat. Mech. Anal.* **95** (1986) 269–277.

[**MWW$_2$**] MAWHIN, J., J.R. WARD and M. WILLEM, Necessary and sufficient conditions for the solvability of a nonlinear two-point boundary value problem, *Proc. Amer. Math. Soc.* **93** (1985) 667–674.

[**MaW$_1$**] MAWHIN, J. and M. WILLEM, Critical points of convex perturbations of some indefinite quadratic forms and semi-linear boundary value problems at resonance, *Ann. Inst. Poincaré, Anal. non linéaire,* **3** (1986) 431–453.

[$\mathbf{MaW_2}$] MAWHIN, J. and M. WILLEM, Multiple solutions of the periodic boundary value problem for some forced pendulum-type equations, *J. Differential Equations* **52** (1984) 264–287.

[$\mathbf{MaW_3}$] MAWHIN, J. and M. WILLEM, Variational methods and boundary value problems for vector second order differential equations and applications to the pendulum equation, in *Nonlinear Analysis and Optimization,* Lect. Notes in Math. No. 1107, Springer, Berlin, 1984, 181–192.

[$\mathbf{MaW_4}$] MAWHIN, J. and M. WILLEM, On the generalized Morse Lemma, *Bull. Soc. Math.* Belgique (B) **37** (1985) 23–29.

[$\mathbf{Maz_1}$] MAZUR, S., Uber konvexe Mengen in linearen normierten Raümen, *Studia Math.* **4** (1933).

[$\mathbf{MaS_1}$] MAZUR, S. and J. SCHAUDER, Uber ein Prinzip in der Variationsrechnung, in *Comptes rendus du Congrès International des Mathématiciens,* Oslo, 1936, tome II, 65.

[$\mathbf{McL_1}$] McLINDEN, L., An application of Ekeland's theorem to minimax problems, *J. Nonlinear Anal.* **6** (1982) 189–196.

[$\mathbf{McS_1}$] McSHANE, E.J., Recent developments in the calculus of variations, *AMS Semicentennial Publications,* Providence, R.I., **19**, 69–97.

[$\mathbf{McS_2}$] McSHANE, E.J., The calculus of variations from the beginning through optimal control theory, in *Optimal Control and Differential Equations* (Proc. Conf. Univ. Oklahoma 1977), Academic Press, New York, 1978, 3–49.

[$\mathbf{MerP_1}$] MERCURI, F. and G. PALMIERI, Morse theory with low differentiability, preprint.

[$\mathbf{Mil_1}$] MILNOR, J., *Topology from the Differentiable Viewpoint,* Univ. Press of Virginia, Charlottesville, 1965.

[$\mathbf{Mil_2}$] MILNOR, J., *Morse Theory,* Princeton Univ. Press, Princeton, N.J., 1963.

[$\mathbf{Mor_3}$] MOREAU, J.J., Fonctionnelles conveses, *Séminaire Equations aux dérivées partielles,* Collège de France, 1967.

[$\mathbf{Moy_1}$] MORREY, C.B., *Multiple Integrals in the Calculus of Variations,* Springer, Berlin, 1966.

[$\mathbf{Mrs_1}$] MORSE, M., Relations between the critical points of a real function of n independent variables, *Trans. Amer. Math. Soc.* **27** (1925) 345–396.

[Mos$_1$] MOSCO, U., Dual variational inequalities, *J. Math. Anal. Appl.* **40** (1972) 202–206.

[Mos$_1$] MOSER, J., Periodic orbits near an equilibrium and a theorem of A. Weinstein, *Comm. Pure Appl. Math.* **29** (1976) 727–747.

[Nag$_1$] NAGUMO, M., A note on the theory of degree of mapping in Euclidean spaces, *Osaka Math. J.* **4** (1952) 1–10.

[Neh$_1$] NEHARI, Z., On a class of nonlinear integral equations, *Math. Z.* **72** (1959) 175–183.

[Ni$_1$] NI, Wei-Ming, Some minimax principles and their applications in nonlinear elliptic equations, *J. Anal. Math.* **37** (1980) 248–275.

[Nir$_1$] NIRENBERG, L., Variational and topological methods in nonlinear problems, *Bull. Amer. Math. Soc.* (NS) **4** (1981) 267–302.

[Oll$_1$] OLLIVRY, J.P., Vibrations forcées pour une équation d'onde non-linéaire, *C.R. Acad. Sci. Paris,* Ser. I, **297** (1983) 29–32.

[Pal$_1$] PALAIS, R., Morse theory on Hilbert manifolds, *Topology* **2** (1963) 299–340.

[Pal$_2$] PALAIS, R., Ljusternik–Schnirelmann theory on Banach manifolds, *Topology* **5** (1966) 115–132.

[Pal$_4$] PALAIS, R., The principle of symmetric criticality, *Comm. in Math. Phys.* **69** (1979) 19–30.

[PaS$_1$] PALAIS, R. and S. SMALE, A generalized Morse theory, *Bull. Amer. Math. Soc.* **70** (1964) 165–171.

[Pen$_1$] PENOT, J.P., Méthode de descente: point de vue topologique et géométrique, Notes de cours 3ème cycle, Univ. Pau, 1975.

[Pen$_2$] PENOT, J.P., The drop theorem, the petal theorem and Ekeland's variational principle, *J. Nonlinear Anal.* **10** (1986) 813–822.

[PiT$_1$] PISANI, R. and M. TUCCI, Existence of infinitely many periodic solutions for a perturbed Hamiltonian system, *J. Nonlinear Analysis* **8** (1984) 873–891.

[PiT$_2$] PISANI, R. and M. TUCCI, Soluzioni periodiche di sistemi Hamiltoniani perturbati di periodo fissato, preprint.

[PiT$_3$] PISANI, R. and M. TUCCI, Teoremi di molteplicità per equazioni differenziali con termine non lineare, singolare nell'origine, *le Matematiche* **36** (1981) 251–260.

[Pit$_1$] PITCHER, E., Inequalities of critical point theory, *Bull. Amer. Math. Soc.* **64** (1958) 1–30.

[Poh$_1$] POHOZAEV, S.I., The set of critical values of a functional, *Math. USSR* **4** (1968) 93–98.

[Poh$_2$] POHOZAEV, S.I., On an approach to nonlinear equations, *Soviet Math. Dokl.* **20** (1979) 912–916.

[Poh$_3$] POHOZAEV, S.I., Periodic solutions of certain nonlinear systems of ordinary differential equations, *J. Differential Equations* **16** (1980) 80–86.

[Poi$_1$] POINCARÉ, H., *Les Méthodes Nouvelles de la Mécanique Céleste,* Gauthier-Villars, Paris, 1892-1897.

[Poi$_2$] POINCARÉ, H., Sur les lignes géodésiques des surfaces convexes, *Trans. Amer. Math. Soc.* **6** (1905) 237–274.

[Prod$_1$] PRODI, G., Problemi di diramazione per equazioni funzionali, in *Atti Ottavo Congr. Un. Mat. Ital., Trieste,* 1967, Bologna, 1968, 118–137.

[PuS$_1$] PUCCI, P. and J. SERRIN, Extensions of the mountain pass theorem, *J. Funct. Anal.* **59** (1984) 185–210.

[PuS$_2$] PUCCI, P. and J. SERRIN, A mountain pass theorem, *J. Differential Equations* **60** (1985) 142–149.

[PuS$_3$] PUCCI, P. and J. SERRIN, The structure of the critical set in the mountain pass theorem, *Trans. Amer. Math. Soc.* **299** (1987) 115–132.

[Rab$_1$] RABINOWITZ, P.H., On subharmonic solutions of Hamiltonian systems, *Comm. Pure Applied Math.* **33** (1980) 609–633.

[Rab$_2$] RABINOWITZ, P.H., Periodic solutions of Hamiltonian systems: a survey, *SIAM J. Math. Anal.* **13** (1982) 343–352.

[Rab$_3$] RABINOWITZ, P.H., Periodic solutions of Hamiltonian systems, *Comm. Pure Appl. Math.* **31** (1978) 157–184.

[Rab$_4$] RABINOWITZ, P.H., Periodic solutions of a Hamiltonian system on a prescribed energy surface, *J. Differential Equations* **33** (1979) 336–352.

[Rab$_5$] RABINOWITZ, P.H., Some minimax theorems and applications to nonlinear partial differential equations, in *Nonlinear Analysis,* Academic Press, New York, 1978, 161–177.

[$\mathbf{Rab}_6$] RABINOWITZ, P.H., *Minimax Methods in Critical Point Theory with Applications to Differential Equations,* CBMS Reg. Conf. Ser. in Math. No. 65, Amer. Math. Soc., Providence, R.I., 1986.

[$\mathbf{Rab}_7$] RABINOWITZ, P.H., A variational method for finding periodic solutions of differential equations, in *Nonlinear Evolution Equations,* Crandall ed., Academic Press, New York, 1978, 225–251.

[$\mathbf{Rab}_8$] RABINOWITZ, P.H., Some aspects of nonlinear eigenvalue problems, *Rocky M. J. Math.* **3** (1973) 161–202.

[$\mathbf{Rab}_9$] RABINOWITZ, P.H., Some critical point theorems and applications to semilinear elliptic partial differential equations, *Ann. Scuola Norm. Sup. Pisa, Cl. Sci.* **4, 5** (1978) 215–223.

[$\mathbf{Rab}_{10}$] RABINOWITZ, P.H., A minimax principle and applications to elliptic partial differential equations, in *Nonlinear Partial Differential Equations and Applications,* Lect. Notes in Math. No. 648, Springer, Berlin, 1978, 97–115.

[$\mathbf{Rab}_{11}$] RABINOWITZ, P.H., Critical points of indefinite functionals and periodic solutions of differential equations, *Proceedings of the International Congress of Mathematicians,* Helsinki, 1978, 791–796.

[$\mathbf{Rab}_{12}$] RABINOWITZ, P.H., The mountain pass theorem: theme and variations, in *Differential Equations,* de Figueiredo, Hoing, ed., Lect. Notes Math. No. 957, Springer, Berlin, 1982, 237–271.

[$\mathbf{Rab}_{13}$] RABINOWITZ, P.H., Some aspects of critical point theory, MRC Techn. Summ. Rept. No. 2465, 1983.

[$\mathbf{Rab}_{14}$] RABINOWITZ, P.H., Minimax methods for indefinite functionals, MRC Techn. Summ. Rep. No. 2619, 1984.

[$\mathbf{Rab}_{15}$] RABINOWITZ, P.H., Free vibrations for a semilinear wave equation, *Comm. Pure Appl. Math.* **31** (1978) 31–68.

[$\mathbf{Rab}_{16}$] RABINOWITZ, P.H., Subharmonic solutions of a forced wave equation, in *Contr. to Anal. and Geom.,* John Hopkins Univ. Press, Baltimore, Md., 1981, 285–292.

[$\mathbf{Rab}_{17}$] RABINOWITZ, P.H., On a theorem of Weinstein, *J. Differential Equations* **68** (1987) 332–343.

[$\mathbf{Rab}_{18}$] RABINOWITZ, P.H., On the existence of periodic solutions for a class of symmetric Hamiltonian systems, *Nonlinear Analysis,* **11** (1987) 599–611.

[$\mathbf{Rab}_{19}$] RABINOWITZ, P.H., Variational methods for nonlinear eigenvalue problems, in *Eigenvalues of Nonlinear Problems,* Prodi ed., Cremonese, Roma, 1974, 141–195.

[$\mathbf{Rab}_{20}$] RABINOWITZ, P.H., A bifurcation theorem for potential operators, *J. Funct. Anal.* **25** (1977) 412–424.

[$\mathbf{Rab}_{21}$] RABINOWITZ, P.H., Multiple critical points of perturbed symmetric functionals, *Trans. Amer. Math. Soc.* **272** (1982) 753–770.

[$\mathbf{Rab}_{22}$] RABINOWITZ, P.H., Periodic solutions of large norm of Hamiltonian systems, *Differential Equations* **50** (1983) 33–48.

[$\mathbf{Rab}_{23}$] RABINOWITZ, P.H., On a class of functionals invariant under a $\mathbf{Z}^n$ action, C.M.S. Technical Summary Report, Univ. of Wisconsin, 1987.

[$\mathbf{Ree}_{1}$] REEKEN, M., Stability of critical points under small perturbations I and II, *Manuscripta Math.* **7** (1972) 387–411; **8** (1973) 69–92.

[$\mathbf{Ree}_{2}$] REEKEN, M., Stability of critical values and isolated critical continua, *Manuscripta Math.* **12** (1974) 163–193.

[$\mathbf{Roc}_{1}$] ROCKAFELLAR, R.T., *Convex Analysis,* Princeton Univ. Press, Princeton, N.J., 1970.

[$\mathbf{Rot}_{1}$] ROTHE, E.H., Gradient mappings, *Bull. Amer. Math. Soc.* **59** (1953) 5–19.

[$\mathbf{Rot}_{2}$] ROTHE, E.H., Some applications of functional analysis to the calculus of variations, in *Calculus of Variations and Its Applications,* Proc. Sympos. Appl. Math. No. 8, Amer. Math. Soc., Providence, R.I., 1958, 143–151.

[$\mathbf{Rot}_{3}$] ROTHE, E.H., Weak topology and calculus of variations, in *Calculus of Variations, Classical and Modern,* CIME 1966, Cremonese, Rome, 1967, 207–237.

[$\mathbf{Rot}_{4}$] ROTHE, E.H., *An Introduction to Various Aspects of Degree Theory in Banach Spaces,* American Math. Soc., Providence, R.I., 1986.

[$\mathbf{Rot}_{5}$] ROTHE, E.H., Morse theory in Hilbert space, *Rocky M. J. Math.* **3** (1973) 251–274.

[$\mathbf{Rot}_{6}$] ROTHE, E.H., Some remarks on critical point theory in Hilbert space, in *Proc. Symp. Nonlinear Problems,* Univ. Wisconsin Press, 1963, 233–256.

[$\mathbf{Rot}_{7}$] ROTHE, E.H., A relation between the type numbers of a critical point and the index of the corresponding field of gradient vectors, *Math. Nachr.* **4** (1950) 12–27.

[Rot_8] ROTHE, E.H., On continuity and approximation questions concerning critical Morse groups in Hilbert spaces, in *Symp. Infinite Dimensional Topology,* Baton Rouge, 1967, Anderson ed., Amer. Math. Stud., Princeton, vol. 69, 1972, 275–295.

[Rot_9] ROTHE, E.H., Some remarks on critical point theory in Hilbert space (continuation), *J. Math. Anal. Appl.* **20** (1967) 515–520.

[RuS_1] RUF, B. and S. SOLIMINI, On a class of superlinear Sturm–Liouville problems with arbitrarily many solutions, *SIAM J. Math. Anal.* **17** (1986) 761–771.

[Ryb_1] RYBAKOWSKI, K.P., *The Homotopy Index Theory on Metric Spaces with Applications to Partial Differential Equations,* Universitext, Springer, Berlin, 1987.

[Ryb_2] RYBAKOWSKI, K.P., On the homotopy index for infinite-dimensional semiflows, *Trans. Amer. Math. Soc.* **269** (1982) 351–382.

[Ryb_3] RYBAKOWSKI, K.P., The Morse index, repeller-attraction pairs and the connection index for semiflows on noncompact spaces, *J. Differential Equations* **47** (1983) 66–98.

[Ryb_4] RYBAKOWSKI, K.P., An index product-formula for the study of elliptic resonance problems, *J. Differential Equations* **56** (1985) 408–425.

[Ryb_5] RYBAKOWSKI, K.P., Nontrivial solutions of elliptic boundary value problem with resonance at zero, *Annali Mat. Pura Appl.* (4) **139** (1985) 237–278.

[Ryb_6] RYBAKOWSKI, K.P., On a relation between the Brouwer degree and the Conley index for gradient flows, *Bull. Soc. Math. Belgique* B **37** (1986) 87–96.

[Ryb_7] RYBAKOWSKI, K.P., A homotopy index continuation method and periodic solutions of second order gradient systems, *J. Differential Equations* **65** (1986) 203–218.

[Ryb_8] RYBAKOWSKI, K.P., On critical groups and the homotopy index in Morse theory on Hilbert manifolds, *Rend. Ist. Mat. Univ. Trieste* **18** (1986) 163–176.

[$RybZ_1$] RYBAKOWSKI, K.P. and E. ZEHNDER, A Morse equation in Conley's index theory for semiflows on metric spaces, *Ergodic Theory and Dynamical Systems* **5** (1985) 123–143.

[Sal_1] SALVATORE, A., Periodic solutions of Hamiltonian systems with a subquadratic potential, *Boll. Un. Mat. Ital.* **1-C** (1984) 393–406.

[$\mathbf{Snc_1}$] SANCHEZ, L., Solutions périodiques non triviales d'une équation d'évolution non linéaire du quatrième ordre, *C.R. Acad. Sci. Paris* **209** (1980) 305–307.

[$\mathbf{San_1}$] SANGER, R.G., Functions of lines and the calculus of variations, in *Contributions to the Calculus of Variations,* 1931-1932, Univ. Chicago Press, 1933, 190–293.

[$\mathbf{Sch_1}$] SCHWARTZ, J.T., *Nonlinear Functional Analysis,* Gordon and Breach, New York, 1969.

[$\mathbf{ScL_1}$] SCHWARTZ, L., *Analyse Hilbertienne,* Hermann, Paris, 1979.

[$\mathbf{Sei_1}$] SEIFERT, H., Periodische Bewegungen mechanischer systeme, *Math. Z.* **51** (1948) 197–216.

[$\mathbf{SeT_1}$] SEIFERT, H. and W. THRELFALL, *Variations-Rechnung im Grossem (Theorie von Marston Morse),* Teubner, Leipzig and Berlin, 1938.

[$\mathbf{Shi_1}$] SHI, S.Z., Ekeland's variational principle and the mountain pass lemma, *Acta Math. Sinica,* (NS) **1** (1985) 348–355.

[$\mathbf{ShC_1}$] SHI, S.Z. and K.C. CHANG, A local minimax theorem without compactness, in *Nonlinear and Convex Analysis,* Proc. Confer. in Honor of Ky Fan, Dekker, 1987, 211–233.

[$\mathbf{Sk_1}$] SKRYPNIK, I.V., On the application of Morse's method to nonlinear elliptic equations, *Soviet Math. Dokl.* **13** (1972) 202–205.

[$\mathbf{Sma_1}$] SMALE, S., Morse theory and a nonlinear generalization of the Dirichlet problem, *Ann. of Math.* **17** (1964) 307–315.

[$\mathbf{Sma_2}$] SMALE, S., On the structure of manifolds, *Amer. J. of Math.* **84** (1962) 387–399.

[$\mathbf{Smo_1}$] SMOLLER, J., *Shock Waves and Reaction-Diffusion Equations,* Springer-Verlag, New York, 1983.

[$\mathbf{Sol_1}$] SOLIMINI, S., Existence of a third solution for a class of B.V.D. with jumping nonlinearities, *J. Nonlinear Anal.* **7** (1983) 917–927.

[$\mathbf{Sol_2}$] SOLIMINI, S., On the solvability of some elliptic partial differential equations with the linear part at resonance, *J. Math. Anal. Appl.* **117** (1986) 138–152.

[$\mathbf{Ste_1}$] STEINLEIN, H., Borsuk's antipodal theorem and its generalizations and applications: a survey, in *Méth. Topol. en Analyse Non-Linéaire,* Granas, ed., Sémin. Math. Sup. No. 95, Montréal, 1985, 166–235.

[Str_1] STRUWE, M., Infinitely many critical points for functionals which are not even and applications to superlinear boundary value problems, *Manuscripta Math.* **32** (1980) 335–364.

[Str_2] STRUWE, M., Multiple solutions of differential equations without the Palais–Smale condition, *Math. Ann.* **261** (1982) 399–412.

[Str_3] STRUWE, M., Generalized Palais–Smale conditions and applications, *Vorlesungsreihe* SFB **72** No. 17, Bonn, 1983.

[Str_4] STRUWE, M., *Variationsmethoden der Nichtlinearen Funktional Analysis,* Univ. Bonn, 1985.

[Str_5] STRUWE, M., Plateau's Problem and the calculus of variations, *Vorlesungsreihe* SFB **72** No. 32, Bonn, 1986.

[Stu_1] STUART, C.A., An introduction to bifurcation theory based on differential calculus, in *Nonlinear Analysis and Mechanics, Heriot-Watt Symposium,* Vol. IV, p. 76–137, R.J. Knops ed., Pitman, San Francisco-London-Melbourne, 1979.

[Sul_1] SULLIVAN, F., A characterization of complete metric spaces, *Proc. Amer. Math. Soc.* **83** (1981) 345–346.

[Sva_1] SVARC, A.S., The genus of a fibre space, *Trudy Moskov. Mat. Obsc.* **10** (1961) 271–272; **11** (1962) 99–126 (Russian); English transl. in *Amer. Math. Soc. Transl.* (2) **55** (1966) 49–140.

[Szu_1] SZULKIN, A., Minimax principles for lower semicontinuous functions and applications to nonlinear boundary value problems, *Ann. Inst. Poincaré, Analyse non linéaire* **3** (1986) 77–109.

[Szu_2] SZULKIN, A., Ljusternik–Schnirelmann theory on C^1-manifolds, Reports No. 14, Dept. Math. Univ. Stockholm, 1987.

[Szu_3] SZULKIN, A., Critical Point Theory of Ljusternik–Schnirelmann Type and Applications to Partial Differential Equations, *Sémin. Math. Sup.,* Presses Univ. Montréal, to appear.

[Szu_4] SZULKIN, A., Morse theory and existence of periodic solutions of convex Hamiltonian systems, *Bull. Soc. Math.,* France, 1988, to appear.

[Ta_1] TANAKA, K., Infinitely many periodic solutions for the equation $u_{tt} - u_{xx} + |u|^{s-1}u = f(x,t)$, *Comm. Partial Differential Equations* **10** (1985) 1317–1345.

[Ta_2] TANAKA, K., Density of the range of a wave operator with non-monotonic superlinear nonlinearity, *Proc. Japan Acad.* **62** A (1986) 129–132.

[Ta_3] TANAKA, K., Infinitely many periodic solutions for a superlinear forced wave equation, *J. Nonlinear Anal.* **11** (1987) 85–104.

[Ter_1] TERSIAN, S.A., On a minimax theorem, *C.R. Acad. Bulgare Sci.* **38** (1985) 27–30.

[The_1] THEWS, K., A reduction method for some nonlinear Dirichlet problems, *J. Nonlinear Anal.* **3** (1979) 795–813.

[The_2] THEWS, K., T-periodic solutions of time dependent Hamiltonian systems with a potential vanishing at infinity, *Manuscripta Math.* **33** (1981) 327–338.

[Tho_1] THOM, R., Quelques propriétés des variétés différentielles, *Comm. Math. Helvetici* **28** (1954) 17–86.

[Tho_2] THOM, R., Marston Morse, *C.R. Acad. Sci. Paris,* Vie académique, **285** (1977) 148–149.

[Tho_3] THOM, R., Le degré Brouwerien en topologie différentielle moderne, *Nieuw Archief voor Wiskunde,* **19** (1971) 10–16.

[Tia_1] TIAN, G., On the mountain pass theorem, *Kexue Tongbao* **29** (1984) 1150–1154.

[Tia_2] TIAN, G., On the mountain pass lemma (in Chinese), *Kexue Tongbao* (Chinese) **28** (1983) 833–835.

[Tol_1] TOLAND, J.F., A duality principle for non-convex optimization and the calculus of variations, *Arch. Rat. Mech. Anal.* **71** (1979) 41–61.

[Tol_2] TOLAND, J.F., Duality in nonconvex optimization, *J. Math. Anal. Appl.* **66** (1978) 399–415.

[Ton_1] TONELLI, L., *Fondamenti di Calculo delle Variazioni,* Zanichelli, Bologna, 1921-1923.

[Tr_1] TROMBA, A.J., A general approach to Morse theory, *J. Diff. Geometry* **12** (1977) 47–85.

[TsW_1] TSHINANGA, S. and M. WILLEM, Morse theory, Cesari method and asymptotically linear Hamiltonian systems, *Boll. U.M.I.,* **VI-4** (1985) 297–305.

[U_1] UHLENBECK, K., Morse theory on Banach manifolds, *J. Funct. Anal.* **10** (1973) 430–445.

[Vai_1] VAINBERG, M.M., *Variational Methods for the Study of Nonlinear Operators,* Holden-Day, San Francisco, Cal., 1964.

[Vai$_2$] VAINBERG, M.M., *Variational Method and Method of Monotone Operators in the Theory of Nonlinear Equations,* Wiley, 1973.

[VGr$_1$] VAN GROESEN, E.W.C., On normal modes in classical Hamiltonian systems, *Int. J. Non-Linear Mechanics* **18** (1983) 55–70.

[VGr$_2$] VAN GROESEN, E.W.C., On an application of analytical minimax methods, *Delft Progress Report,* **10** (1985) 250–260.

[VGr$_3$] VAN GROESEN, E.W.C., Hamiltonian Flow on an Energy Surface: 240 Years after the Euler–Maupertuis Principle, Proceed. Sixth Scheveningen Conference, 1984, Lect. Notes in Physics no. 239, Springer, Berlin, 1985, 322–341.

[VGr$_4$] VAN GROESEN, E.W.C., Analytical mini-max methods for Hamiltonian brake orbits of prescribed energy, preprint.

[BGr$_5$] VAN GROESEN, E.W.C., A simplified proof for a result of Ekeland and Lasry concerning the number of periodic Hamiltonian trajectories, Dept. Math. Cath. Univ. Nijmegen, Report 8229, 1982.

[VGr$_6$] VAN GROESEN, E.W.C., Existence of multiple normal mode trajectories on convex energy surfaces of even, classical Hamiltonian systems, *J. Differential Equations* **57** (1985) 70–89.

[VGr$_7$] VAN GROESEN, E.W.C., Multiple normal modes in natural Hamiltonian systems, in *Méthodes Topologiques en Analyse Non-linéaire,* Sémin. Math. Super. No. 95, Presses Univ. Montréal, 1985, 136–155.

[VGr$_8$] VAN GROESEN, E.W.C., Applications of natural constraints in critical point theory to periodic solutions of natural Hamiltonian systems, MRC Techn. Summ. No. 2593, 1983.

[VGr$_9$] VAN GROESEN, E.W.C., On small period, large amplitude normal modes of natural Hamiltonian systems, *J. Nonlinear Anal.* **10** (1986) 41–54.

[VGr$_{10}$] VAN GROESEN, E.W.C., Applications of natural constraints in critical point theory to boundary value problems on domains with rotation symmetry, *Arch. Math.* (Basel) **44** (1985) 171–179.

[Vi$_1$] VITERBO, C., Une théorie de Morse pour les systèmes hamiltoniens étoilés, *C.R. Acad. Sci. Paris,* **301** (1985) 487–489.

[Vi$_2$] VITERBO, C., A proof of the Weinstein's conjecture in $\mathbf{R}^{2n}$, *Ann. Inst. Poincaré, Anal. non linéaire* **4** (1987) 337–356.

[Vol$_1$] VOLTERRA, V., *Lecons sur les fonctions de lignes,* Gauthier-Villars, Paris, 1913.

[Vol_2] VOLTERRA, V., Le calcul des variations, son évolution et ses progrès, son rôle dans la physique mathématique, Public. Fac. Sci. Univ. Charles et Univ. Masaryk, Praha-Brno, 1932 (*Opera Mat.* **5** 217–241).

[VoP_1] VOLTERRA, V. and J. PERES, *Théorie Générale des Fonctionnelles,* Gauthier–Villars, Paris, 1936.

[Wal_1] WALLACE, A.H., *Algebraic Topology,* Benjamin, New York, 1970.

[War_1] WARD, J.R., A boundary value problem with a periodic nonlinearity, *J. Nonlinear Anal.* **10** (1986) 207–213.

[Wei_1] WEINSTEIN, A., Periodic orbits for convex Hamiltonian systems, *Ann. Math.* **108** (1978) 507–518.

[Wei_2] WEINSTEIN, A., Normal modes for nonlinear Hamiltonian systems, *Invent. Math.* **20** (1973) 47–57.

[Wei_3] WEINSTEIN, A., On the hypotheses of Rabinowitz' periodic orbit theorems, *J. Differential Equations* **33** (1979) 353–358.

[Wil_1] WILLEM, M., Oscillations forcées de systèmes hamiltoniens, *Public. Sémin. Analyse non linéaire* Univ. Besancon, 1981.

[Wil_2] WILLEM, M., Subharmonic oscillations of convex Hamiltonian systems, *J. Nonlinear Anal.* **9** (1985) 1303–1311.

[Wil_3] WILLEM, M., *Lectures on Critical Point Theory,* Trabalho de Mat. No. 199, Fundacao Univ. Brasilia, Brasilia, 1983.

[Wil_4] WILLEM, M., Perturbation de variétés critiques non dégénérées et oscillations non-linéaires forcées, to appear.

[Wil_5] WILLEM, M., Analyse convexe et optimisation, CIACO, Louvain-la-Neuve, 1987.

[Wil_6] WILLEM, M., Remarks on the dual least action principle, *Z. Anal. Anw.* **1** (1982) 85–90.

[Wil_7] WILLEM, M., Density of the range of potential operators, *Proc. Amer. Math. Soc.* **83** (1981) 341–344.

[Wil_8] WILLEM, M., Periodic solutions of convex Hamiltonian systems, *Bull. Soc. Math. Belg.* **36** (A) (1984) 11–22.

[Wil_9] WILLEM, M., Subharmonic oscillations of a semilinear wave equation, *J. Nonlinear Anal.* **9** (1985) 503–514.

[Wil_{10}] WILLEM, M., Periodic oscillations of odd second order Hamiltonian systems, *Boll. Un. Mat. Ital.* (6) **3-B** (1984) 293–304.

[Wil_{11}] WILLEM, M., Perturbations of nondegenerate periodic orbits of Hamiltonian systems, in *Periodic Solutions of Hamiltonian Systems and Related Topics,* Rabinowitz, Ambrosetti, Ekeland, Zehnder, ed., Reidel, Dordrecht, 1987, 261–266.

[Wil_{12}] WILLEM, M., Bifurcation, symmetry and Morse Theory, *Boll. Unione Mat. Ital.,* to appear.

[Wu_1] WU, S.P., The solvability of a class of mountain pass-type operator equations (Chinese), *Chinese Ann. Math.* **2** (1981) 365–376.

[Wu_2] WU, S.P., A resonance case for an asymptotically linear vibrating string equation, *J. Math. Anal. Appl.* **91** (1983) 47–67.

[Wu_3] WU, S.P., An application of variational method to nonlinear eigenvalue problem for a class of wave equation, *J. Nonlinear Anal.* **6** (1982) 649–658.

[WuL_1] WU, S.P. and J. LIU, A note on the resonance case for asymptotically linear wave equations, *Chinese Ann. Math.* **5B** (1984) 653–659.

[Yan_1] YANG, C.T., On the theorems of Borsuk Ulam, Kakutani-Tujabâ and Dysin, *Ann. Math.* **60** (1954) 262–282; **62** (1955) 271–283.

[Yor_1] YORKE, J.A., Periods of periodic solutions and the Lipschitz constant, *Proc. Amer. Math. Soc.* **22** (1969) 509–512.

[You_1] YOUNG, W.H., On classes of summable functions and their Fourier series, *Proc. Roy. Soc.,* (A) **87** (1917) 225–229.

[You_2] YOUNG, L.C., *Lectures on the Calculus of Variations and Optimal Control Theory,* Chelsea, New York, 1980.

[Zeh_1] ZEHNDER, E., Periodic solutions of Hamiltonian equations, *Lect. Notes in Math.* No. 1031, Springer, Berlin, 1987, 172–213.

Index

Applied Mathematical Sciences

cont. from page ii

55. *Yosida:* Operational Calculus: A Theory of Hyperfunctions.
56. *Chang/Howes:* Nonlinear Singular Perturbation Phenomena: Theory and Applications.
57. *Reinhardt:* Analysis of Approximation Methods for Differential and Integral Equations.
58. *Dwoyer/Hussaini/Voigt (eds.):* Theoretical Approaches to Turbulence.
59. *Sanders/Verhulst:* Averaging Methods in Nonlinear Dynamical Systems.
60. *Ghil/Childress:* Topics in Geophysical Dynamics: Atmospheric Dynamics, Dynamo Theory and Climate Dynamics.
61. *Sattinger/Weaver:* Lie Groups and Algebras with Applications to Physics, Geometry, and Mechanics.
62. *LaSalle:* The Stability and Control of Discrete Processes.
63. *Grasman:* Asymptotic Methods of Relaxation Oscillations and Applications.
64. *Hsu:* Cell-to-Cell Mapping: A Method of Global Analysis for Nonlinear Systems.
65. *Rand/Armbruster:* Perturbation Methods, Bifurcation Theory and Computer Algebra.
66. *Hlaváček/Haslinger/Nečas/Lovíšek:* Solution of Variational Inequalities in Mechanics.
67. *Cercignani:* The Boltzmann Equation and Its Applications.
68. *Temam:* Infinite Dimensional Dynamical System in Mechanics and Physics.
69. *Golubitsky/Stewart/Schaeffer:* Singularities and Groups in Bifurcation Theory, Vol. II.
70. *Constantin/Foias/Nicolaenko/Temam:* Integral Manifolds and Inertial Manifolds for Dissipative Partial Differential Equations.
71. *Catlin:* Estimation, Control, and the Discrete Kalman Filter.
72. *Lochak/Meunier:* Multiphase Averaging for Classical Systems.
73. *Wiggins:* Global Bifurcations and Chaos.
74. *Mawhin/Willem:* Critical Point Theory and Hamiltonian Systems.
75. *Abraham/Marsden/Ratiu:* Manifolds, Tensor Analysis, and Applications, 2nd ed.
76. *Lagerstrom:* Matched Asymptotic Expansions: Ideas and Techniques.
77. *Aldous:* Probability Approximations via the Poisson Clumping Heuristic.
78. *Dacorogna:* Direct Methods in the Calculus of Variations.

www.ingramcontent.com/pod-product-compliance
Ingram Content Group UK Ltd.
Pitfield, Milton Keynes, MK11 3LW, UK
UKHW012200240726
13966UKWH00002B/486